PHYSICS OF HOT ELECTRON TRANSPORT IN SEMICONDUCTORS

PHYSICS OF HOT ELECTRON TRANSPORT IN SEMICONDUCTORS

edited by

C. S. Ting

Texas Center for Superconductivity
University of Houston

World Scientific
Singapore • New Jersey • London • Hong Kong

Published by

World Scientific Publishing Co. Pte. Ltd.
5 Toh Tuck Link, Singapore 596224
USA office: 27 Warren Street, Suite 401-402, Hackensack, NJ 07601
UK office: 57 Shelton Street, Covent Garden, London WC2H 9HE

British Library Cataloguing-in-Publication Data
A catalogue record for this book is available from the British Library.

PHYSICS OF HOT ELECTRON TRANSPORT IN SEMICONDUCTORS

ISBN-13 978-981-02-1008-3
ISBN-10 981-02-1008-6

PREFACE

The development of submicrometer semiconductor devices during the past decade has underscored the need for a better theoretical understanding of carrier transport in these small sized systems. Even a few volts applied across the device can lead to a rather high electric field on the order of 10^4 V/cm. In these circumstances, the relationship between current and voltage is no longer linear, and the statistical temperature of the charge carriers becomes higher than that of the heat bath. Thus, the carrier dynamics under a strong electric field are characterized as "hot electron transport" Since the first survey on this subject by E. M. Conwell in her review *High Field Transport in Semiconductors* (New York: Academic Press, 1967), where the experimental data were analyzed with the phenomenological Boltzmann equation, many new experimental techniques and theoretical methods have been developed or proposed to study hot electron transport in various semiconductors. In particular, Monte Carlo simulation methods have been introduced to solve the Boltzmann equation numerically. A useful review of such developments can be found in the book *Hot Electron Transport in Semiconductors* edited by L. Reggiani (New York: Springer-Verlag, 1985). However, in the present volume, we shall review primarily analytical approaches developed since 1984 for treating various kinds of hot electron transport problems.

The balance equation method, originally proposed by X. L. Lei and myself in 1984 for the purpose of investigating steady state hot electron transport, has subsequently been extended to study thermal noise, time dependent transport, and other subjects relating to charge carriers in strong electric fields. The predicted theoretical results are in good agreement with both experimental measurements and Monte Carlo numerical simulations. Much of the simplicity of the mathematical structure of the balance equation approach arises in conjunction with its effectiveness in dealing with many-electron systems having strong inter-carrier collisions, and it is particularly powerful in this respect. Extensive reviews of this method are given by the first two articles, which deal with it in the absence and presence of a magnetic field. When hot electrons are confined to a quantum well, the proper treatment of the electron-phonon-coupling involving phonons which propagate in three dimensions requires special care, because phonon heating produces a bottleneck in the rate energy transfer from electrons to phonons. This subject is carefully reviewed in Article 3. During the past decades, the intra-collisional field effect (ICFE), which has been promoted as a major ingredient of quantum transport theory, has drawn tremendous theoretical efforts in the attempt to understand its influence on hot electron transport (see Section 16, Article 1). Nevertheless, all the numerical calculations to date seem to indicate that this effect yields almost no correction to steady state transport up to a field strength of 10^7 V/cm except possibly minor modifications to the classical overshoot behavior within a time scale $t > 10^{12}$ sec. ($\sim$ electron-phonon scattering time) in the transient region respond to a step electric field. Within the framework of the balance equation formulation, the intra-collisional field term is simply embedded in the structure of the memory function wherein the drift velocity of the charge carriers is taken in its time dependent form — accelerated by the electric field, even with the neglect of contributions from the electron-phonon interaction and other scattering mechanisms. This feature will be discussed in Article 4, where the memory effect is demonstrated to be very important for investigating nonlinear high frequency conductivity. The method of nonequilibrium statistical operators (NSO), developed by Zubarev in the early sixties, has subsequently been employed by Kalashnikov in 1970 to study hot electron transport. In Article 5, it will be shown that had the problem been done correctly, the results obtained from NSO would reduce exactly to those of the balance equation method as described in Article 1. In this connection, I wish to emphasize that Articles 1 to 5 in this book assume charge carriers with strong carrier-carrier collisions. These collisions may originate directly from Coulomb interaction and/or indirectly from phonon mediated electron-electron interaction, with the strength of such interaction mandating a relatively high carrier density. On the other hand, for the case of very low carrier density, the charge carriers are far apart and collisions between carriers are negligible. In this situation, the transport problem needs to be treated in terms of single carrier transport, which will be discussed in Article 6 using the nonequilibrium Green's function techniques and the Feynman path integral theory. At this stage, it is of interest to note that the linear isothermal (I) impurity resistivity derived from the balance equation method outlined in Articles 1 to 5 yields the expression $\rho_i^I = \frac{m}{Ne^2}\left\langle\frac{1}{\tau}\right\rangle$, with $\frac{1}{\tau}$ and $\langle\cdots\rangle$ as defined in Equations (15.38) and (15.39) of Article 1. This form is different from the standard adiabatic (A) result based upon the Kubo formula or the Boltzmann equation in the absence of carrier-carrier scattering, wherein $\rho_i^A = \frac{m}{Ne^2\langle\tau\rangle}$. Although ρ_i^A and ρ_i^I are practically identical with each other at low temperatures, they are quite different at higher temperatures. The question of which formula should be used to compare with experimental measurements has been the subject of debate in the literature for quite some time. In Article 7, the authors introduce an inter-carrier collision time τ_{in} and show that ρ_i^I and ρ_i^A are derived respectively under the conditions $\tau_{in} \ll \tau$ (strong inter-carrier collisions) and $\tau_{in} \ll \tau$ (weak inter-carrier collisions). Finally, in the last paper, the dynamic conductivity and the shot noise suppression of a double-barrier resonant tunneling system, which is of great current interest (but having no links to the previous articles,) is also reviewed.

Finally, I wish to thank all the contributors to this volume for their cooperation and special efforts in writing their review articles to meet the highest standards of quality in research, as well as pedagogical clarity. Acknowledgement is also due to the Texas Center for Superconductivity at the University of Houston for support in every respect.

C. S. Ting

CONTENTS

International Journal of Modern Physics B, Vol. 6, No. 7 (1992) 805–936

Balance-Equation Approach to Hot-Carrier Transport in Semiconductors

X.L. Lei

Department of Solid State Physics, Shanghai Institute of Metallurgy Chinese Academy of Sciences, Shanghai, 200050, China

N.J.M. Horing

Department of Physics and Engineering Physics, Stevens Institute of Technology Hoboken, New Jersey 07030, U.S.A.

Abstract

The balance-equation approach to nonlinear hot-carrier transport theory, formulated by Lei and Ting (1984), is addressed in this comprehensive review. A central feature is the role of strong electron-electron interactions in promoting rapid thermalization about the drifted transport state and the concomitant substantial simplification of the transport theory. This physical feature is embodied in the initial density matrix chosen to represent the unperturbed carrier system.

Force and energy balance equations are formulated for the dc steady state, ac dynamic and transient cases of charge conduction, including memory effects. The scattering mechanisms include impurity and phonon interactions along with dynamic nonlocal screening effects due to carrier-carrier interactions. Both linear and nonlinear resistivities are discussed in the degenerate and nondegenerate statistical regimes. Interesting phenomena such as electron cooling and thermal noise and diffusion are discussed as well.

Semiconductor microstructure transport is described for both linear and nonlinear hot carrier conduction. In this connection, quasi-2D-systems, heterojunctions, and quantum well superlattices are treated with attention to steady state, transient and high frequency transport, including, for example, superlattice plasmon resonance structure. Type-II superlattice transport is reviewed as well as type-I, and electron-hole drag is treated in conjunction with negative minority electron mobility in a quantum well. Multivalley semiconductors are discussed in some detail. Furthermore, attention is also focused on the center-of-mass velocity fluctuation, Langevin-type equation and thermal noise and diffusion for microstructures and multivalley systems. A number of particularly important phenomena are examined from the balance-equation point of view, such as nonequilibrium phonons, higher order scattering effects and weak localization, hydrodynamic equations for weakly nonuniform systems, and the intracollisional field effect.

Alternative formulations and interpretations of the balance-equation approach are reviewed. The distinction between this many-particle, isothermal, balance-equation theory and the non-interacting (single-particle) adiabatic transport theory is discussed to clarify issues subject to controversy in the literature. Finally, we give a brief review of recent developments in the balance-equation approach: investigation of the distribution function in balance-equation theory, improved calculations for GaAs/AlGaAs heterojunctions, extension of the balance equations to an abitrary energy band and recent work on superlattice miniband transport.

1. INTRODUCTION

Innovation in the development of microfabrication techniques has brought modern theoretical physics directly into the arena of semiconductor microstructure device technology. The quantum mechanics of electronic transport first examined theoretically in the landmark paper of Kohn-Luttinger (1957) is no longer merely a subject of important intellectual qualities, but it is vital to the understanding and design of conduction in heterostructures and superlattices whose geometric features lead to quantization phenomena that cannot be ignored in the most practical circumstances.

While focusing attention on the central demand for performance oriented information concerning microstructure transport characteristics, quantum transport theory has had to deal with its own, internal, rather harsh difficulties of rigor and also even of principle. Transport theorists, both classical and quantal, have been obliged to exercise considerable ingenuity in constructing a suitable theoretical framework for analysis. Such vehicles for the description of transport generally have fallen into two classes: (a) transport equations, and (b) identification of transport coefficients in terms of equilibrium correlation functions, as in the Kubo formula. The construction of transport equations is by no means trivial, as there are even difficulties in rigorously deriving the Boltzmann equation. Moreover, the construction of a quantum transport equation is by no means the end of the problem. It must also be solved to yield useful practical information, and the same may be said for the evaluation of equilibrium correlation functions embodying scattering mechanisms.

Over the years, both classical and quantum transport theorists have come to grips with these problems using innovative techniques: superoperators, Feynman diagram expansions of equilibrium correlation functions and more recently nonequilibrium Green's functions, path integrals as well as Wigner distributions and multi-particle density matrices have been brought to bear on quantum transport. With the development of computational capacity, Monte Carlo techniques have been employed in transport calculations. These efforts have had great success, but also have some shortcomings. Some are formidable to deal with accurately even within the linear regime and very hard to extend beyond linear conduction, others involve complicated analytic continuation, etc. Calculations are lengthy and difficult, and few of these techniques seems to offer a simple vehicle for generating useful physical insight into transport dynamics on a systematic basis. Most often, because of technical difficulties, they also ignore electron-electron interaction effects which can be of vital importance. Finally, many theories describe transport in terms of a Liouville equation for the density matrix of a carrier system which is adiabatically sealed after turn-on of the electric field, such that the initial state is an equilibrium one prepared in contact with a bath which is conceptually disconnected after turn-on. This is a problem in regard to the description of transport as measured in the laboratory.

The difficulties of the techniques cited above for three dimensional bulk systems also pertain to the lower dimensional semiconductor microstructures, where interacting quantum well and quantum wire geometries and heterostructures can further complicate the description of transport. The lack of a clear, crisp physical picture of conduction which offers a quasi-analytic sense of the functional dependence of transport on the parameters of the system, has been a serious impediment to practical applications for microstructure device development. Such a quasi-analytic picture of transport could also relieve the need to repeat expensive (computer) time-consuming Monte Carlo calculations to examine the effect of a change of parameters on transport.

On a formal level, the admission to consideration of carrier-carrier (e-e) interaction introduces considerable complication. Physically, however, there is a substantial simplification in the situation when e-e interactions are large enough to induce rapid thermalization of the carriers about the drifted transport state. This opens the possibility of a quasi-analytic description of transport, which is successfully addressed by the Lei-Ting (1984) balance equation approach in developing a perturbation theory about the displaced drifted equilibrium state. This perturbation theory in powers of the scattering interactions yields the leading order resistivity directly in a manner that is devoid of Van Hove's "$\lambda^2 t$"-series divergences due to the very short thermalization time. Moreover, such a perturbative expansion of the resistivity is not subject to the zero-order linear-time divergence of a perturbation expansion of the dc conductivity. As one should expect, the results differ from those analyses which ignore the role of e-e interactions. However, in the classical limit, the quantum mechanical Lei-Ting balance equations exactly reproduce the successful gas plasma transport theory (Spitzer, 1956; Dreicer, 1959; Schmidt, 1979; Lifshitz and Pitaevskii, 1981), which similarly cut through the formidable difficulties of rigorous formal classical transport theory, particularly with the inclusion of e-e interactions which are germane to gas plasma dynamics. The results offer considerable physical insight into the dynamics of both quantum and classical transport, and are informative about the functional dependence of transport on system parameters.

The object of this review article is to provide a comprehensive description of the quantum dynamical formulation of the Lei-Ting balance equation transport theory, clarifying approximations and its relation to other formulations, and to explore in detail its advantages as a quasi-analytic approach to transport in providing physical insight into the role of system parameters in conduction, and in implementing device simulation with high accuracy/CUP cost ratio.

2. FUNDAMENTAL PHYSICAL CONSIDERATIONS

2.1 Center-of-Mass and Relative Variables of Electrons

We consider a model electron-ion system, having N electrons under the influence of a uniform applied electric field $\mathbf{E}$. The electrons, interacting with each other through the Coulomb potential, are coupled with phonons and scattered by randomly distributed impurities. Assuming a single-band effective-mass description for the electrons, we can write the Hamiltonian of the system as

$$H = \sum_i \frac{\mathbf{p}_i^2}{2m} + \sum_{i<j} \frac{e^2}{4\pi\epsilon_0|\mathbf{r}_i - \mathbf{r}_j|} + \sum_{i,a} u(\mathbf{r}_i - \mathbf{R}_a) - \sum_{i,j} \mathbf{u}_l \cdot \nabla v_l(\mathbf{r}_i - \mathbf{R}_l) - e\sum_i \mathbf{r}_i \cdot \mathbf{E}. \quad (2.1)$$

Here, $\mathbf{r}_i$ and $\mathbf{p}_i = -i\nabla_i$ are the coordinate and momentum of the ith electron with effective mass m and charge e. $v_l(\mathbf{r} - \mathbf{R}_l)$ denotes the potential by the lth ion at lattice site $\mathbf{R}_l$, and $\mathbf{u}_l$ is its displacement from the regular (equilibrium) position; $u(\mathbf{r} - \mathbf{R}_a)$ denotes the additional scattering potential due to an impurity sited at $\mathbf{R}_a$, which is randomly located; and the second term on the right-hand side of Eq.(2.1) is the Coulomb interaction between electrons. Balance equation theory proceeds from the separation of the motion of the center of-mass (c.m.), i.e. the motion of the system as an entirety, from the relative motion of the electrons. Denoting the center-of mass momentum and coordinate variables by $\mathbf{P}, \mathbf{R}$ and the relative electron momentum and coordinate variables by $\mathbf{p}_i', \mathbf{r}_i'$, we have (Ting, Ying and Quinn, 1976; Lei and Ting, 1984):

$$\mathbf{P} = \sum_i \mathbf{p}_i, \quad \mathbf{R} = \frac{1}{N}\sum_i \mathbf{r}_i, \quad (2.2)$$

$$\mathbf{p}_i' = \mathbf{p}_i - \frac{1}{N}\mathbf{P}, \quad \mathbf{r}_i' = \mathbf{r}_i - \mathbf{R}. \quad (2.3)$$

$\mathbf{P}, \mathbf{R}$ are canonically conjugate variables:

$$[R_\alpha, P_\beta] = i\delta_{\alpha\beta} \quad (2.4)$$

$(\alpha, \beta = x, y, z)$, and they commute with $\mathbf{r}_i', \mathbf{p}_i'$:

$$[\mathbf{r}_i', \mathbf{P}] = [\mathbf{R}, \mathbf{p}_i'] = 0. \quad (2.5)$$

The N relative electron momenta and coordinates so defined are not completely independent, but subject to constraints:

$$\sum_j \mathbf{p}_j' = 0, \quad \sum_j \mathbf{r}_j' = 0, \quad (2.6)$$

and their commutation relations include a noncanonical $1/N$ term:

$$[r_{i\alpha}', p_{j\beta}'] = i\delta_{\alpha\beta}(\delta_{ij} - \frac{1}{N}). \quad (2.7)$$

The original derivation of the balance equation theory (Lei and Ting, 1984, 1985a,b) involves the reasonable assumption that the $1/N$ term in Eq.(2.7) is negligible for macroscopic large N system, such that one can treat the relative momentum and coordinate as canonical variables:

$$\left[r'_{i\alpha}, p'_{j\beta}\right] \simeq i\delta_{\alpha\beta}\delta_{ij}. \tag{2.8}$$

While Eq.(2.8) seems plausible, it is an approximation. Therefore, care must be taken to avoid using it in a way that can result in a macroscopic error, as, for example, in constructing the translation generator of the system by summing individual momenta over all particles. Nevertheless, it has been shown (Lei, 1990b), that the neglect of noncanonical $1/N$ term in Eq(2.7) induces no error within the framework of balance equation theory if the center-of-mass coordinate $\mathbf{R}(t)$ is reated classically.

In terms of the center-of-mass variables and the relative electron variables we can express the Hamiltonian (2.1) as

$$H = H_c + H_e + H_{ph} + H_{ei} + H_{ep} \tag{2.9}$$

$$H_c = \frac{P^2}{2Nm} - Ne\mathbf{E}\cdot\mathbf{R} \tag{2.10}$$

$$H_e = \sum_{\mathbf{k},\sigma} \varepsilon_{\mathbf{k}} c^{\dagger}_{\mathbf{k}\sigma} c_{\mathbf{k}\sigma} + H_{ee} \tag{2.11}$$

$$H_{ee} = \frac{1}{4\pi\epsilon_0}\sum_{i<j}\frac{e^2}{|\mathbf{r}'_i - \mathbf{r}'_j|} = \frac{1}{2V}\sum_{\mathbf{q}}{}' v_c(q)\left(\rho_{\mathbf{q}}\rho_{-\mathbf{q}} - N\right) \tag{2.12}$$

$$H_{ph} = \sum_{\mathbf{q},\lambda} \Omega_{\mathbf{q}\lambda} b^{\dagger}_{\mathbf{q}\lambda} b_{\mathbf{q}\lambda} \tag{2.13}$$

$$H_{ei} = \frac{1}{V}\sum_{\mathbf{q},a} u(\mathbf{q}) e^{i\mathbf{q}\cdot(\mathbf{R}-\mathbf{R}_a)}\rho_{\mathbf{q}} \tag{2.14}$$

$$H_{ep} = \frac{1}{V^{1/2}}\sum_{\mathbf{q},\lambda} M(\mathbf{q}\lambda)\,\phi_{\mathbf{q}\lambda} e^{i\mathbf{q}\cdot\mathbf{R}}\rho_{\mathbf{q}}, \tag{2.15}$$

where V is the volume of the system, $c^{\dagger}_{\mathbf{k}\sigma}$ and $c_{\mathbf{k}\sigma}$ are creation and annihilation operators for relative electrons with wave vector $\mathbf{k}$, spin σ, and energy $\varepsilon_{\mathbf{k}} = k^2/2m$, $\rho_{\mathbf{q}}$ is the density operator of the relative electrons, given by

$$\rho_{\mathbf{q}} = \sum_{\mathbf{k}} \rho_{\mathbf{kq}}, \quad \rho_{\mathbf{kq}} = \sum_{\sigma} c^{\dagger}_{\mathbf{k}+\mathbf{q}\sigma} c_{\mathbf{k}\sigma} \tag{2.16}$$

$\phi_{\mathbf{q}\lambda} \equiv b_{\mathbf{q}\lambda} + b^{\dagger}_{-\mathbf{q}\lambda}$ is the phonon field operator with $b^{\dagger}_{\mathbf{q}\lambda}$ and $b_{\mathbf{q}\lambda}$ being creation and annihilation operators for phonons of wavevector $\mathbf{q}$ in branch λ with frequency $\Omega_{\mathbf{q}\lambda}$, $v_c(q) = e^2/\epsilon_0\kappa q^2$ is the Coulomb potential (κ is the background dielectric constant), and $u(q)$ and $M(\mathbf{q},\lambda)$ are respectively the electron-impurity potential and the electron-phonon matrix element in the plane wave representation, satisfying $u(\mathbf{q}) = u^*(-\mathbf{q})$ and $M(\mathbf{q},\lambda) = M^*(-\mathbf{q},\lambda)$.

2.2 Center of Mass Treated as a Classical Particle

As described above, our system consists of three parts: center of mass, relative electrons and phonons. The center of mass, considered as a single particle having mass Nm and charge Ne, moves under the influence of the acceleration due to the applied electric field $\mathbf{E}$ and the damping by the relative electron-phonon medium due to electron-impurity and electron-phonon interactions H_{ei} and H_{ep}. Because of its enormous mass, the motion of the center-of-mass is essentially classical, and we can treat the c.m.-related operators classically, regarding $\mathbf{R}$ in H_{ei} and H_{ep} as the real, time-dependent position of the center-of-mass, $\mathbf{R}(t)$. The relative electron system, which is composed of a large number of interacting particles and is treated fully quantum-mechanically, does not directly sense the electric field but is coupled to the center-of-mass motion through the c.m. position $\mathbf{R}(t)$ appearing in both H_{ei} and H_{ep} as a time-dependent parameter. In the following we will denote

$$H_{It} = H_{ei} + H_{ep} \tag{2.17}$$

as a time-dependent interaction with $\mathbf{R}(t)$ being a time-dependent classical parameter. Note that the error arising from treating $\mathbf{R}(t)$ as a classical variable, i.e. from neglecting the noncommutability of the c.m. coordinates $\mathbf{R}(t)$ at different times, is order of $1/N$, and has been shown (Lei, 1990b) to be exactly cancelled by that arising from the neglect of the noncanonical $1/N$ terms in the commutation relations of the relative electron variables, Eq.(2.7). Therefore, the two major approximations made above in formulating the balance equation theory, i.e. treating relative electrons as an unconstrained canonical system and treating the center-of-mass as a classical single particle, induce no error save in regard to the intracollisional field effect (see Section 16). In accordance with this Hamiltonian, it is apparent that the c.m. velocity $\mathbf{V}$, or the time derivative of the c.m. position, is given by

$$\mathbf{V} \equiv \dot{\mathbf{R}} = -i\,[\mathbf{R}, H] = \frac{\partial H}{\partial \mathbf{P}} = \frac{\mathbf{P}}{Nm}. \tag{2.18}$$

Note that the center of mass, which is accelerated by the electric field and damped by the relative-electron-phonon bath medium, is a Brownian particle due to the random force associated with the electron-impurity and electron-phonon interactions. Therefore the c.m. velocity $\mathbf{V}$ is composed of a drift part $\mathbf{v}$ and a fluctuation or random part $\delta\mathbf{V}$:

$$\mathbf{V} = \mathbf{v} + \delta\mathbf{V}. \tag{2.19}$$

We denote $\mathbf{v} = \langle \mathbf{V} \rangle$ or $\langle \delta\mathbf{V} \rangle = 0$, with the $\langle \cdots \rangle$ symbol understood as an average either over an appropriate time period or over a statistical ensemble which eliminates the fluctuations. The forces experienced by the center of mass are obtained by calculating the time derivative of the c.m. momentum $\dot{\mathbf{P}} = -i\,[\mathbf{P}, H] = -\partial H/\partial \mathbf{R}$, yielding

$$\dot{\mathbf{P}} = Ne\mathbf{E} + \mathbf{F} \tag{2.20}$$

with $\mathbf{F} = \mathbf{F}_i + \mathbf{F}_p$, and

$$\mathbf{F}_i = -i \sum_{\mathbf{q},a} u(\mathbf{q}) \mathbf{q}\, e^{i\mathbf{q}\cdot(\mathbf{R}-\mathbf{R}_a)} \rho_{\mathbf{q}} \tag{2.21}$$

$$\mathbf{F}_p = -i \sum_{\mathbf{q},\lambda} M(\mathbf{q},\lambda)\, \mathbf{q}\, \phi_{\mathbf{q}\lambda} e^{i\mathbf{q}\cdot\mathbf{R}} \rho_{\mathbf{q}} \tag{2.22}$$

are identified as forces respectively due to impurity and phonon scatterings. Being quantum-mechanical operators in the relative-electron and phonon space, $\mathbf{F}_i$ and $\mathbf{F}_p$ also comprise an average part and a fluctuation part . The latter contributes a random force exerted on the center of mass. Eq.(2.20) can be regarded as an operator force balance equation.

Similarly, we have the rates of changes of the energies of the phonon and relative electron systems as:

$$\dot{H}_{ph} = -i\,[H_{ph}, H] = W \equiv -\sum_{\mathbf{q},\lambda} M(\mathbf{q},\lambda)\, \dot{\phi}_{\mathbf{q}\lambda} e^{i\mathbf{q}\cdot\mathbf{R}} \rho_{\mathbf{q}} \tag{2.23}$$

$$\begin{aligned} \dot{H}_e = -i\,[H_e, H] = S \equiv & -i \sum_{\mathbf{q},\lambda} M(\mathbf{q},\lambda) \phi_{\mathbf{q}\lambda} e^{i\mathbf{q}\cdot\mathbf{R}} \sum_{\mathbf{k}} (\varepsilon_{\mathbf{k}+\mathbf{q}} - \varepsilon_{\mathbf{k}}) \rho_{\mathbf{k}\mathbf{q}} \\ & -i \sum_{\mathbf{q},a} u(\mathbf{q}) e^{i\mathbf{q}\cdot\mathbf{R}_a} \sum_{\mathbf{k}} (\varepsilon_{\mathbf{k}+\mathbf{q}} - \varepsilon_{\mathbf{k}}) \rho_{\mathbf{k}\mathbf{q}} \end{aligned} \tag{2.24}$$

with

$$\dot{\phi}_{\mathbf{q}\lambda} \equiv -i\,[\phi_{\mathbf{q}\lambda}, H_{ph}] = -i\Omega_{\mathbf{q}\lambda}(b_{\mathbf{q}\lambda} - b^{\dagger}_{-\mathbf{q}\lambda}). \tag{2.25}$$

Energy balance for the electron system requires that the energy supplied per unit time by the electric field, $Ne\mathbf{E}\cdot\mathbf{V}$, matches the sum of the energy increase rate of the center of mass, $d(\frac{1}{2}NmV^2)/dt$, the increase rate of the relative electron internal energy, $\dot{H}_e$, and the energy loss rate of the electron system to the phonon system W:

$$Ne\mathbf{E}\cdot\mathbf{V} = Nm\mathbf{V}\cdot\frac{d\mathbf{V}}{dt} + \dot{H}_e + W. \tag{2.26}$$

Of course Eq.(2.24) itself also states the energy balance for the relative electron system. All these equations are in operator form in the quantum number spaces of relative electrons and phonons.

The balance equations can be derived by taking the ensemble averages of $\mathbf{F}$ and W over the density matrix of the relative electron and phonon systems. The statistical average of a dynamical variable A at time t can be evaluated in Schrödinger picture by

$$\langle A \rangle = Tr\,\{\hat{\rho} A\}, \tag{2.27}$$

as was done in the original paper of Lei and Ting (1984, 1985b). Here A is the operator (possibly time dependent) representing a dynamical variable in the Schrödinger picture and $\hat{\rho}$ is the density matrix satisfying the Liouville equation

$$i\frac{d\hat{\rho}}{dt} = [H_e + H_{ph} + H_{It}, \hat{\rho}] \tag{2.28}$$

subject to the initial condition at initial time t_0

$$\hat{\rho}|_{t_0} = \hat{\rho}_0 \tag{2.29}$$

we will use $t_0 = -\infty$ or $t_0 = 0$ as initial time.

2.3 Initial Density Matrix

Physically, we expect that external conditions (electric field, magnetic field, etc $\cdots$) uniquely determine the final steady state of a system in transport. This means that the final steady state to which the system evolves under the influence of a given electric field, does not depend on the particulars of the initial condition. After a sufficiently long time the system will arrive at a unique steady state no matter what initial state it starts from. However, the time required for the system to go from an initial state to the final state does depend strongly on the initial state. As a parametrized theory, the balance-equation approach allows us to choose an initial state to have its major features resemble, as closely as possible, the final state, such that the system, starting from this virtual initial state, can reach the real final state in a short evolution process. We assume that both the interaction H_{It} and the electric field $\mathbf{E}(t)$ are turned on from the initial time $t = -\infty$. In the balance equation theory the initial state is taken as: a free center-of-mass moving at a constant velocity $\mathbf{v}$, (note that there is no fluctuation associated with the motion of a free particle), an isolated relative electron system in a thermoequilibrium state at a temperature T_e, and a phonon system in equilibrium or quasi-equilibrium with a heat bath at temperature T.

The parameters $\mathbf{v}$ and T_e are chosen in the following way (Lei and Ting, 1985b; Xing and Ting, 1987). Imagine that the electron-phonon and electron-impurity interactions are turned off together with the electric field at time t during the temporal development of the transport process. Then the center of mass, the relative electrons and the phonons are decoupled from each other. As a free particle, the center of mass continues to move with a velocity $\mathbf{v}(t)$ equal to the drift velocity of the electron system at time t. This $\mathbf{v}(t)$ is then chosen as the c.m. velocity for the initial state: $\mathbf{v} = \mathbf{v}(t)$. The relative electrons, as an isolated system, will approach a thermoequilibrium state with a fixed temperature $T_e(t)$, which is dependent on the system state at time t. This value of $T_e\,(t)$ is then used for the initial state: $T_e = T_e\,(t)$. In the same vein, the phonon system, which has been decoupled from the electron system, can also be considered to arrive at an equilibrium state jointly with the bath at the lattice temperature T. This is justified if phonons relax faster than electrons. If this is not the case one may need to consider that during the time period after turning off the interactions when relative electrons have arrived at an equilibrium state, phonons only reach a quasiequilibrium state with a mode(wavevector)- and time-dependent temperature $T_{\mathbf{q}}(t)$. Further discussion of this will be offered in Section 10 below (nonequilibrium phonon phenomena). For the most part in this paper, we assume the former case for phonons and correspondingly the initial density matrix is chosen as:

$$\hat{\rho}_0 = \frac{1}{Z} e^{-H_e/T_e} e^{-H_{ph}/T} \tag{2.30}$$

where Z is a normalization coefficient and H_e is implicitly measured from the Fermi level corresponding to a grand canonical ensemble for the relative electron system. This initial density matrix depends on time t parametrically, $\hat{\rho}_0 \equiv \hat{\rho}_{0t}$.

It should be noted that there are at least two distinct time scales involved. The motion of the center of mass, the time variation of the applied electric field $\mathbf{E}(t)$ and the temporal development of statistically averaged

quantities such as $\mathbf{v}(t)$, $T_e(t)$, occur on a macroscopic time scale, which is assumed to be much larger than the microscopic relaxation time, or the evolution time scale of the Liouville equation. In other words we assume that the evolution from the initial state to the final state in accordance with the Liouville equation is a fast microscopic process, whereas the temporal development is a slow process. The latter is included in the time-variation of the initial density matrix ρ_{0t} and the center-of-mass velocity $\mathbf{v}(t)$.

The initial state described by Eq.(2.30) is one in which the electrons are characterized as being in thermal equilibrium with respect to their center of mass which is moving at a velocity equal to their average drift velocity in the presence of the electric field. Such a selection of the initial state is significant (Lei, 1990a). Since the final state bears the same center-of-mass velocity as the initial state, the time evolution from the initial state to the final state is solely a thermalization process with no need for momentum relaxation. This process involves not only impurity and phonon scatterings, but in most cases, it is determined mainly by intercarrier couplings. In this connection, microscopic-relaxation is very fast, lending validity to the time-dependent description of transport for systems in which there exists a strong carrier-carrier interaction. Even for steady state transport in a constant electric field, a short thermalization time τ_{th} also plays a key role in making it possible to develop a simple perturbative expansion (Lei, 1990a). The underlying phenomenology is that the strong intercarrier scattering promotes a tendency towards rapid thermolization of the drifted electrons, such that the final steady state is only "a perturbation away " from the drifted thermal equilibrium state (2.30). We will return to this matter in Section 15.

3. FORCE- AND ENERGY-BALANCE EQUATIONS TO LOWEST ORDER IN H_I

3.1 Density Matrix and Statistical Averages to Linear Order in H_I

The statistical average of a dynamical variable A at time t (Eq.(2.27)) can alternatively be written as

$$\langle A\rangle = Tr\,\{\hat{\rho}(t)A(t)\} \tag{3.1}$$

where we define

$$A(t) \equiv e^{i(H_e+H_{ph})t}Ae^{-i(H_e+H_{ph})t} \tag{3.2}$$

and

$$\hat{\rho}(t) \equiv e^{i(H_e+H_{ph})t}\hat{\rho}e^{-i(H_e+H_{ph})t}. \tag{3.3}$$

In view of the Liouville equation (2.28) and the initial condition (2.30) $\hat{\rho}(t)$ satisfies the equation

$$i\frac{d}{dt'}\hat{\rho}(t') = \left[H_{It'}(t'), \hat{\rho}(t')\right] \tag{3.4}$$

and subject to the initial condition

$$\hat{\rho}(t')|_{t'=-\infty} = \hat{\rho}_0, \tag{3.5}$$

or, equivalently, obeys the integral equation:

$$\hat{\rho}(t') = \hat{\rho}_0 - i\int_{-\infty}^{t'} dt''\,[H_{It''}(t''), \hat{\rho}(t'')]\,. \tag{3.6}$$

This integral equation facilitates perturbative expansion in the interaction H_I. The first order iteration of Eq.(3.6) yields

$$\hat{\rho}(t') = \hat{\rho}_0 - \int_{-\infty}^{t'} dt''\,[H_{It''}(t''), \hat{\rho}_0]\,. \tag{3.7}$$

The statistical average for the dynamical variable A, Eq.(3.1), can thus be written as

$$\langle A\rangle = \langle A\rangle_0 - i\int_{-\infty}^{t} dt'\langle\left[H_{It'}(t'), A(t)\right]\rangle_0 \tag{3.8}$$

where

$$\langle(\cdots)\rangle_0 \equiv Tr\,\{\hat{\rho}_0(\cdots)\}\,. \tag{3.9}$$

Note that ρ_0, and the average $\langle\cdots\rangle_0$, may be dependent on time t.

3.2 Force Balance Equation

Taking the statistical average of the operator equation (2.20) in the relative electron and phonon space to the lowest order in H_I in accordance with (3.8), and using expressions (2.14) and (2.15) for H_{ei} and H_{ep}, we have (Lei and Ting, 1984, 1985b; Xing and Ting, 1987)

$$nm\frac{d}{dt}\mathbf{v}(t) = ne\mathbf{E}(t) + \mathbf{f}_i + \mathbf{f}_p, \tag{3.10}$$

which is written for unit volume of the system with n being the carrier number density. Here we have identified the average rate of change of c.m. momentum as

$$\langle\dot{\mathbf{P}}\rangle = nm\frac{d}{dt}\mathbf{v}(t) \tag{3.11}$$

$\mathbf{v}(t)$ being the drift part of the c.m. velocity. The frictional forces $\mathbf{f}_i$ and $\mathbf{f}_p$, due to impurity scattering and due to electron-phonon interaction respectively, are given to lowest order by

$$\mathbf{f}_i = -in_i\sum_{\mathbf{q}}\mathbf{q}|u(\mathbf{q})|^2\int_{-\infty}^{\infty}dt' A(\mathbf{q},t,t')\Pi(\mathbf{q},t-t') \tag{3.12}$$

$$\mathbf{f}_p = -i\sum_{\mathbf{q},\lambda}\mathbf{q}|M(\mathbf{q},\lambda)|^2\int_{-\infty}^{\infty}dt' A(\mathbf{q},t,t')\Lambda(\mathbf{q},\lambda;t-t'). \tag{3.13}$$

Here n_i is the impurity density, and the correlation functions are defined by

$$\Pi(\mathbf{q},t-t') = -i\theta(t-t')\langle\left[\rho_{\mathbf{q}}(t),\rho_{-\mathbf{q}}(t')\right]\rangle_0 \tag{3.14}$$

$$\Lambda(\mathbf{q},\lambda;t-t') = -i\theta(t-t')\langle\left[\phi_{\mathbf{q}\lambda}(t)\rho_{\mathbf{q}}(t),\phi_{-\mathbf{q}\lambda}(t')\rho_{-\mathbf{q}}(t')\right]\rangle_0 \tag{3.15}$$

with

$$\rho_{\mathbf{q}}(t) = e^{iH_{et}t}\rho_{\mathbf{q}}e^{-iH_{et}t} = \sum_{\mathbf{k}}e^{i(\varepsilon_{\mathbf{k}+\mathbf{q}}-\varepsilon_{\mathbf{k}})t}\rho_{\mathbf{k},\mathbf{q}} \tag{3.16}$$

and

$$\phi_{\mathbf{q}\lambda}(t) = e^{iH_{ph}t}\phi_{\mathbf{q}\lambda}e^{-iH_{ph}t} = b_{\mathbf{q}\lambda}e^{-i\Omega_{\mathbf{q}\lambda}t} + b^{\dagger}_{-\mathbf{q}\lambda}e^{i\Omega_{\mathbf{q}\lambda}t} \tag{3.17}$$

and

$$A(\mathbf{q},t,t') \equiv \exp\left[i\mathbf{q}\cdot(\mathbf{R}(t)-\mathbf{R}(t'))\right] = \exp\left[i\mathbf{q}\cdot\int_{t'}^{t}\mathbf{v}(s)ds\right]. \tag{3.18}$$

In writing the last equality of Eq.(3.18), we have neglected effects due to velocity fluctuation. Eq. (3.10) is the time-dependent momentum (force) balance equation, or the equation of motion of the center of mass.

3.3 Energy Balance Equation

Taking the statistical average of the right hand side of the operator equation (2.23) in the relative electron and phonon space to the lowest order in H_I , we have (Lei and Ting, 1984, 1985b; Xing and Ting, 1987).

$$w = \langle W\rangle = -\sum_{\mathbf{q}\lambda}|M(\mathbf{q},\lambda)|^2\int_{-\infty}^{\infty}dt' A(\mathbf{q},t,t')\Gamma(\mathbf{q},\lambda;t-t') \tag{3.19}$$

with

$$\Gamma(\mathbf{q},\lambda,t-t') = -i\theta(t-t')\langle\left[\dot{\phi}_{\mathbf{q}\lambda}(t)\rho_{\mathbf{q}}(t),\phi_{-\mathbf{q}\lambda}(t')\rho_{-\mathbf{q}}(t')\right]\rangle_0 \tag{3.20}$$

and

$$\dot{\phi}_{\mathbf{q}\lambda}(t) = e^{iH_{ph}t}\dot{\phi}_{\mathbf{q}\lambda}e^{-iH_{ph}t} = -i\Omega_{\mathbf{q}\lambda}\left(b_{\mathbf{q}\lambda}e^{-i\Omega_{\mathbf{q}\lambda}t} - b^{\dagger}_{-\mathbf{q}\lambda}e^{i\Omega_{\mathbf{q}\lambda}t}\right). \tag{3.21}$$

w is the energy increase rate of the phonon system due to electron-phonon interaction, i.e. the energy transfer rate from the electron system to the phonon system. The energy balance equation results from the statistical average of the operator equation (2.26) if we retain only the drift part of c.m. velocity and identify $\langle\dot{H}_e\rangle$ as the rate of increase of relative electron internal energy, $dE_e/dt \equiv d\langle H_e\rangle/dt$:

$$ne\mathbf{E}\cdot\mathbf{v} = nm\mathbf{v}\cdot\frac{d\mathbf{v}}{dt} + \frac{dE_e}{dt} + w. \tag{3.22}$$

Using the first order perturbation expansion, Eq.(3.7), for the density matrix to calculate $E_e \equiv \langle H_e\rangle$, we find that only the $\hat{\rho}_0$ term in (3.7) contributes, leading to

$$E_e = \sum_{\mathbf{k},\sigma}\varepsilon_{\mathbf{k}}f(\varepsilon_{\mathbf{k}},T_e) \tag{3.23}$$

where

$$f(\varepsilon_{\mathbf{k}},T_e) = 1/\left[\exp(\varepsilon_{\mathbf{k}}-\mu)/T_e+1\right] \tag{3.24}$$

is the Fermi distribution function at the electron temperature, and μ is the chemical potential. Thus we have

$$\delta\frac{dE_e}{dt} = C_e\frac{dT_e}{dt}, \tag{3.25}$$

with $C_e = \partial E_e/\partial T_e$ as the specific heat of the relative electron system. Eliminating $nm\mathbf{v}\cdot d\mathbf{v}/dt$ by means of the force balance equation (3.10) we can write the energy balance equation(3.22) as

$$-C_e\frac{dT_e}{dt} = \mathbf{v}\cdot\mathbf{f} + w. \tag{3.26}$$

Eq. (3.22) or Eq. (3.26) is the time-dependent energy balance equation of the system.

It should be noted that the energy balance equation can also be obtained by taking the average of operator equation (2.24) and identifying $\langle\dot{H}_e\rangle = dE_e/dt$, such that

$$\frac{dE_e}{dt} = \langle S\rangle \tag{3.27}$$

It has been shown (Lei and Ting, 1985b) that the balance equation thus obtained is equivalent to Eq.(3.22) for steady state transport and for time-dependent drift motion involving negligible memory effects (see Section 6.1). However, there may be some differences when discussing nonlinear processes in which memory effects are significant, and in processes involving effects of the velocity fluctuation in higher order (see Section 8.3).

4. STEADY-STATE DC TRANSPORT IN A 3D SYSTEM

As an immediate test of the method, Lei and Ting (1984, 1985a,b) first applied the force- and energy-balance equations developed in the preceding sections to the examination of steady state dc transport in a 3-dimensional isotropic system.

4.1 Balance Equations for Steady State

When the uniform applied electric field is constant in time, the system approaches a steady state with a constant drift velocity $\mathbf{v}_d$ and a constant electron temperature T_e after transients die out. For steady state transport we can take $\mathbf{v}(s) = \mathbf{v}_d$ in Eq.(3.18) such that (Lei and Ting, 1984, 1985b)

$$\exp\left[i\mathbf{q}\cdot\int_{t'}^{t}\mathbf{v}(s)ds\right] = \exp\left[i\mathbf{q}\cdot\mathbf{v}_d(t-t')\right]. \tag{4.1}$$

Substituting (4.1) into (3.12),(3.13) and (3.19), we obtain the steady state force and energy balance equations as (Lei and Ting, 1984, 1985b):

$$ne\mathbf{E} + \mathbf{f} = 0 \tag{4.2}$$

$$\mathbf{v}_d \cdot \mathbf{f} + w = 0 \tag{4.3}$$

where $\mathbf{f} = \mathbf{f}_i + \mathbf{f}_p$, with

$$\mathbf{f}_i = n_i \sum_{\mathbf{q}} \mathbf{q}|u(\mathbf{q})|^2 \Pi_2(\mathbf{q}, \omega_0) \tag{4.4}$$

$$\mathbf{f}_p = 2\sum_{\mathbf{q},\lambda} \mathbf{q}|M(\mathbf{q},\lambda)|^2 \Pi_2(\mathbf{q}, \Omega_{\mathbf{q}\lambda} + \omega_0)\left[n(\frac{\Omega_{\mathbf{q}\lambda}}{T}) - n(\frac{\Omega_{\mathbf{q}\lambda}+\omega_0}{T_e})\right] \tag{4.5}$$

and

$$w = 2\sum_{\mathbf{q},\lambda} \Omega_{\mathbf{q}\lambda}|M(\mathbf{q},\lambda)|^2 \Pi_2(\mathbf{q}, \Omega_{\mathbf{q}\lambda} + \omega_0)\left[n(\frac{\Omega_{\mathbf{q}\lambda}}{T}) - n(\frac{\Omega_{\mathbf{q}\lambda}+\omega_0}{T_e})\right]. \tag{4.6}$$

In these equations, $\omega_0 \equiv \mathbf{q}\cdot\mathbf{v}_d$, $n(x) \equiv 1/(e^x - 1)$ is the Bose distribution function, and $\Pi_2(\mathbf{q},\omega)$ is the imaginary part of the Fourier spectrum of the density-density correlation function:

$$\Pi(\mathbf{q},\omega) = \int_{-\infty}^{\infty} e^{i\omega t}\Pi(\mathbf{q},t)dt. \tag{4.7}$$

The function $\Pi(\mathbf{q},t)$, defined by Eq.(3.14), is the density-density correlation function in a thermoequilibrium state with temperature T_e for the relative electron system described by the Hamiltonian

$$H_e = \sum_{\mathbf{k}\sigma} \varepsilon_{\mathbf{k}} c_{\mathbf{k}\sigma}^{\dagger} c_{\mathbf{k}\sigma} + H_{ee}. \tag{4.8}$$

i.e. the correlation function of the system without impurity and phonon interactions and without electric field. This function has been widely investigated for different types of systems (Keldysh, Kirzhnitz and Maradudin,1989). For a 3-dimensional single parabolic band system in the absence of the Coulomb interaction between carriers, $\Pi(\mathbf{q},\omega)$ function is well known as

$$\Pi_0(\mathbf{q},\omega)=2\sum_{\mathbf{k}}\frac{f(\varepsilon_{\mathbf{k+q}},T_e)-f(\varepsilon_{\mathbf{k}},T_e)}{\varepsilon_{\mathbf{k+q}}-\varepsilon_{\mathbf{k}}+\omega+i\delta}. \tag{4.9}$$

The real part $\Pi_{01}(\mathbf{q},\omega)$ and the imaginary part $\Pi_{02}(\mathbf{q},\omega)$ of this function are, respectively, given by

$$\Pi_{01}(\mathbf{q},\omega) = -2\sum_{\mathbf{q}} f(\varepsilon_{\mathbf{k}},T_e)\left[\frac{1}{\varepsilon_{\mathbf{k+q}}-\varepsilon_{\mathbf{k}}+\omega}-\frac{1}{\varepsilon_{\mathbf{k+q}}-\varepsilon_{\mathbf{k}}-\omega}\right] \tag{4.10}$$

$$\Pi_{02}(\mathbf{q},\omega) = 2\pi\sum_{\mathbf{k}}\left[f(\varepsilon_{\mathbf{k}},T_e)-f(\varepsilon_{\mathbf{k+q}},T_e)\right]\delta(\varepsilon_{\mathbf{k+q}}-\varepsilon_{\mathbf{k}}+\omega). \tag{4.11}$$

For the case of a parabolic band, $\varepsilon_{\mathbf{k}}=k^2/2m$, the summation over $\mathbf{k}$ can easily be carried out, yielding

$$\Pi_{02}(\mathbf{q},\omega)=-\frac{m^2T_e}{2\pi q}P_2(\mathbf{q},\omega) \tag{4.12}$$

with

$$P_2(\mathbf{q},\omega)=\ln\left\{\frac{1+\exp\left[-\frac{1}{2mT_e}(\frac{q}{2}-\frac{m\omega}{q})^2+\frac{\mu}{T_e}\right]}{1+\exp\left[-\frac{1}{2mT_e}(\frac{q}{2}+\frac{m\omega}{q})^2+\frac{\mu}{T_e}\right]}\right\}. \tag{4.13}$$

For the case of a degenerate system at low temperature, we have the explicit expressions ($T=0$ K)

$$\begin{aligned}\Pi_{01}(\mathbf{q},\omega) &= -\frac{mk_F^2}{4\pi^2 q}\left\{4\mu+\left[1-(\mu+\frac{\nu}{\mu})^2\right]\ln\left|\frac{1+(\mu+\nu/\mu)}{1-(\mu+\nu/\mu)}\right|\right.\\ &\left.+\left[1-(\mu-\frac{\nu}{\mu})^2\right]\ln\left|\frac{1+(\mu-\nu/\mu)}{1-(\mu-\nu/\mu)}\right|\right\}\end{aligned} \tag{4.14}$$

$$\begin{aligned}\Pi_{02}(\mathbf{q},\omega) &= -\frac{mk_F^2}{4\pi q}\left\{\left[1-(\mu-\frac{\nu}{\mu})^2\right]\theta(1-(\mu-\frac{\nu}{\mu})^2)\right.\\ &\left.-\left[1-(\mu+\frac{\nu}{\mu})^2\right]\theta(1-(\mu+\frac{\nu}{\mu})^2)\right\},\end{aligned} \tag{4.15}$$

where $\theta(x)$ is the Heaviside unit step function, k_F is the Fermi wavevector, $v_F=k_F/m$ is the Fermi velocity, $\varepsilon_F=k_F^2/2m$ is the Fermi level from band bottom at $T=0$ K, and $\mu\equiv q/2k_F$, $\nu\equiv\omega/4\varepsilon_F$. The introduction of carrier-carrier Coulomb interactions H_{ee} may substantially modify the correlation function. Within the random phase approximation the density-density correlation function including Coulomb interactions is given by

$$\Pi(\mathbf{q},\omega)=\frac{\Pi_0(\mathbf{q},\omega)}{|\epsilon(\mathbf{q},\omega)|^2} \tag{4.16}$$

where

$$\epsilon(\mathbf{q},\omega)=1-v_c(q)\Pi_0(\mathbf{q},\omega) \tag{4.17}$$

is the RPA dielectric function. We may consider the factor $|\epsilon(\mathbf{q},\omega)|^2$ as dynamic screening by carriers of the electron-impurity potential $|u(\mathbf{q})|^2$ and of the electron-phonon interaction matrix element $|M(\mathbf{q},\lambda)|^2$. The frictional forces $\mathbf{f}_i$ and $\mathbf{f}_p$, and the energy transfer rate w may be written in terms of screened potentials

$$u(\mathbf{q})/\epsilon(\mathbf{q},\omega_0) \tag{4.18}$$

and

$$M(\mathbf{q},\lambda)/\epsilon(\mathbf{q},\Omega_{\mathbf{q}\lambda}+\omega_0) \tag{4.19}$$

together with the Π_{02} function. This screening is dynamic in that the dielectric functions appearing in these equations are at the frequency $\omega_0 \equiv \mathbf{q}\cdot\mathbf{v}_d$ or at frequency $\Omega_{\mathbf{q}\lambda}+\omega_0$, which are dependent on the drift velocity $\mathbf{v}_d$. For an isotropic system the total electric resistivity ρ is defined as

$$\rho = \mathbf{E}\cdot\mathbf{v}_d/(\mathbf{J}\cdot\mathbf{v}_d) \tag{4.20}$$

in which

$$\mathbf{J} = ne\mathbf{v}_d \tag{4.21}$$

is the current density. The force balance equation can thus be written as

$$\rho = \rho_i + \rho_p \tag{4.22}$$

where

$$\rho_i = -\mathbf{v}_d\cdot\mathbf{f}_i/n^2e^2v_d^2 \tag{4.23}$$

and

$$\rho_p = -\mathbf{v}_d\cdot\mathbf{f}_p/n^2e^2v_d^2 \tag{4.24}$$

are, respectively, the resistivities due to impurity and phonon scatterings. The energy balance equation can be written in the form

$$\rho - \rho_E = 0 \tag{4.25}$$

in terms of an equivalent resistivity related to the electron energy loss rate:

$$\rho_E = w/n^2e^2v_d^2. \tag{4.26}$$

4.2 Weak Current Ohmic Resistivity

We first consider the solution of the balance equations (4.2)-(4.3) in the limit of weak current at finite lattice temperature. In this connection, the energy balance equation (4.3) has the solution

$$T_e/T = 1 + O(v_d^2) \tag{4.27}$$

Therefore the impurity- and phonon-induced resistivities in the limit $v_d \to 0$ are obtained from Eqs.(4.23) and (4.24) with $T_e = T$, whence

$$\rho_i = -\frac{n_i}{n^2e^2}\sum_{\mathbf{q}} q_x^2|\bar{u}(\mathbf{q})|^2\frac{\partial}{\partial w}\Pi_{02}(\mathbf{q},\omega)|_{\omega=0} \tag{4.28}$$

and

$$\rho_p = -\frac{2}{n^2e^2}\sum_{\mathbf{q},\lambda} q_x^2|\bar{M}(\mathbf{q},\lambda)|^2\Pi_{02}(\mathbf{q},\Omega_{\mathbf{q}\lambda})\left[-\frac{1}{T}n'(\frac{\Omega_{\mathbf{q}\lambda}}{T})\right] \tag{4.29}$$

with

$$|\bar{u}(\mathbf{q})|^2 \equiv |u(\mathbf{q})/\epsilon(\mathbf{q},0)|^2 \tag{4.30}$$

and

$$|\bar{M}(\mathbf{q},\lambda)|^2 \equiv |M(\mathbf{q},\lambda)/\epsilon(\mathbf{q},\Omega_{\mathbf{q}\lambda})|^2. \tag{4.31}$$

In degenerate case, for temperature $T \ll \varepsilon_F$ (ε_F is the zero-temperature Fermi level), we can use zero temperature results (4.14) and (4.15) for Π_{01} and Π_{02}, leading to the well-known result for impurity-induced Ohmic resistivity:

$$\rho_i(v_d \to 0) = \rho_{i0} = \frac{m^2n_i}{12\pi^3n^2e^2}\int_0^{2k_F}|\bar{u}(\mathbf{q})|^2q^3dq \tag{4.32}$$

For the evaluation of phonon-induced resistivity at low temperature, we consider long wavelength longitudinal acoustic phonons with a Debye-type spectrum

$$\Omega_{\mathbf{q}} = v_sq, \tag{4.33}$$

v_s being the longitudinal sound velocity. Transverse acoustic phonons may also contribute via piezoelectric interaction. This may be accounted for by including an additional phonon-induced resistivity term. In the case $v_s/v_F \ll 1$, it is appropriate to use the approximate expression for Π_{20},

$$\Pi_{20}(\mathbf{q},\omega) \simeq -\frac{m^2}{2\pi q}\theta(1-(q/2k_F)^2)\omega, \tag{4.34}$$

whence we obtain

$$\rho_p(v_d \to 0) = \rho_{p0} = \frac{m^2v_s}{6\pi^3n^2e^2}\int_0^{2k_F}dqq^4|\bar{M}(\mathbf{q},\lambda)|^2\left[-\frac{1}{T}n'(\frac{\Omega_{\mathbf{q}\lambda}}{T})\right]. \tag{4.35}$$

This is the Bloch-Grüneissen formula (Ziman 1960, 1972). Assuming that

$$|\bar{M}(\mathbf{q},\lambda)|^2 = \text{const.} \times q \tag{4.36}$$

(e.g. deformation potential) we can write

$$\rho_{p0} = \rho^* g(t_F) \tag{4.37}$$

where ρ^* is a resistivity-dimension coefficient, $t_F \equiv T/\Theta_F$, $\Theta_F \equiv 2k_F v_s$, and

$$g(t) = \frac{1}{t}\int_0^1 dy \frac{y^5}{(e^{y/t}-1)(1-e^{-y/t})}. \tag{4.38}$$

All these formulas for impurity- and phonon-induced dc steady-state resistivities are known as isothermal force-balance equation results, in contrast to those obtained from conventional adiabatic transport theories (see Section 15.3 for further discussion).

4.3 Nonlinear Zero-Lattice-Temperature Limit

It is of particular interest to examine nonlinear conduction in the zero lattice temperature limit. When $T \to 0$, $n(\Omega_{\mathbf{q}\lambda}/T)$ vanishes and only the electron temperature T_e is involved in the balance equations. We find that the electron temperature T_e approaches a finite value T^* which depends on the electric field and impurity resistivity. As a simple illustration, we assume $M(\mathbf{q},\lambda)$ to be of the deformation potential type, (4.36). For $v_s \ll v_F$, the energy-balance equation and the phonon-induced resistivity at $T \to 0$ can be written, respectively, as

$$\frac{\rho_{i0}}{\rho^*} = \frac{3}{2}\left(\frac{v_d}{v_s}\right)^2 \int_0^1 dy \int_{-1}^1 dx y^4 \left(1+\frac{v_d}{v_s}x\right)^2 n\left[\frac{y}{t_F}\left(1+\frac{v_d}{v_s}x\right)\right] \tag{4.39}$$

$$\frac{\rho_p}{\rho^*} = -\frac{3}{2}\frac{v_s}{v_d}\int_0^1 dy \int_{-1}^1 dx y^4 x \left(1+\frac{v_d}{v_s}x\right) n\left[\frac{y}{t_F}\left(1+\frac{v_d}{vs}x\right)\right]. \tag{4.40}$$

If $v_d/v_s < 1$, the integration in Eqs.(4.39) and (4.40) are easily carried out, yielding

$$\left[\frac{T^*}{\Theta_F}\right]^5 = 0.0134\frac{\rho_{i0}}{\rho^*}\frac{(v_d/v_s)^2}{[1-(v_d/v_s)^2]^2} \tag{4.41}$$

and

$$\frac{\rho_p}{\rho^*} = \frac{4}{3}\frac{\rho_{i0}}{\rho^*}\frac{(v_d/v_s)^2}{1-(v_d/v_s)^2}. \tag{4.42}$$

Furthermore the impurity resistivity ρ_i is almost equal to its weak field value ρ_{i0}. The electric field $E = (\rho_p + \rho_i)J$ can be expressed as

$$\frac{E}{E^*} = \frac{\rho_{i0}}{\rho^*}\frac{v_d}{v_s}\left[1 + \frac{4}{3}\frac{(v_d/v_s)^2}{1-(v_d/v_s)^2}\right], \tag{4.43}$$

with $E^* = \rho^* n e v_s$. It is easily seen from Eqs.(4.41) and (4.43) that for $v_d/v_s \ll 1$, or $E \ll E^*$,

$$\frac{T^*}{\Theta_F} = 0.422\left(\frac{\rho^*}{\rho_{i0}}\right)^{1/5}\left(\frac{E}{E^*}\right)^{2/5} \tag{4.44}$$

and

$$\frac{\rho_p}{\rho^*} = \frac{4}{3}\frac{\rho^*}{\rho_{i0}}\left(\frac{E}{E^*}\right)^2. \tag{4.45}$$

Equation (4.44) is in agreement with the result obtained by Arai(1983). It should be noted that Eqs.(4.44) and (4.45) are valid only in a rather narrow electric field range, depending on the impurity resistivity ρ_{i0}/ρ^*, but they are completely invalid when $\rho_{i0} = 0$. However, Eqs.(4.41)-(4.43) can describe a system devoid of impurities with $v_d/v_s < 1$, i.e., when $\rho_{i0} = 0$ we have $T^* = 0$, $\rho_p = 0$, and $E = 0$ for $v_d < v_s$. This means that, in the absence of impurity scattering, the current can flow without resistance up to the drift velocity $v_d = v_s$ at $T = 0K$. But, once $v_d > v_s$, resistance appears. The calculated results of T^*/Θ_F and the phonon-induced resistivity ρ_p/ρ^* were given by Lei and Ting(1984) as functions of electric field for different values of impurity scattering, as shown in Fig.4.1. It is worth noting that in the presence of a finite electric field the phonon contribution to resistivity does not vanish at $T = 0K$.

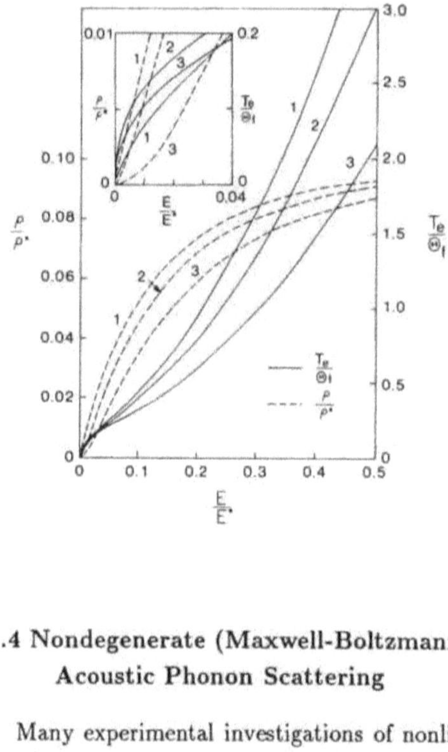

Fig.4.1 The electron temperature T_e and phonon resistivity ρ_p in the limit of $T \to 0$ are shown as functions of dimensionless electric field E/E^* for several impurity resistivities in a degenerate electron system with acoustic phonon and impurity scatterings. $\rho_{i0}/\rho^* = 0, 0.03$, and 0.1 for curves 1,2, and 3, respectively. In the figure ρ stands for ρ_p. (From Lei and Ting, 1984)

4.4 Nondegenerate (Maxwell-Boltzmann) Case Acoustic Phonon Scattering

Many experimental investigations of nonlinear transport have been carried out with nondegenerate semiconductors, in which carriers in thermal equilibrium obey a Maxwell-Boltzmann distribution even at low temperatures. In this case the chemical potential μ is negative and $|\mu|/T_e \gg 1$, such that the function $P_2(\mathbf{q},\omega)$ (Eq.(4.13)) reduces to

$$P_2(\mathbf{q},\omega) = e^{-|\mu|/T_e}\left\{\exp\left[-\frac{1}{2mT_e}(\frac{q}{2} - \frac{m\omega}{q})^2\right] - \exp\left[-\frac{1}{2mT_e}(\frac{q}{2} + \frac{m\omega}{q})^2\right]\right\}. \quad (4.46)$$

We assume that acoustic phonons are the only scatterers and $M(\mathbf{q})$ is of the deformation potential type (4.36). For convenience, we introduce a temperature-dimension parameter

$$\Theta_s = mv_s^2, \tag{4.47}$$

which is about $1K$ for $v_s = 5\times 10^5 cm/s$ and $m = 0.6m_e$ (m_e is the free-election mass), and denote $\alpha \equiv T_e/T$ and $t_s \equiv T/\Theta_s$. We consider solutions of the energy-balance equation under the condition $\alpha t_s \gg 1$: to the lowest order of the power expansion in $(\alpha t_s)^{-1}$, the energy-balance equation (4.25) reduces to

$$\left(\frac{v_d}{v_s}\right)^2 = \frac{6\alpha^3 t_s^3}{S(\alpha, t_s)} - 3, \tag{4.48}$$

where

$$S(\alpha, t_s) = \int_0^\infty e^{-y^2/2\alpha t_s} y^4 \coth\left(\frac{y}{t_s}\right) dy. \tag{4.49}$$

If it is also true that $\alpha/t_s \gg 1$, we have

$$S(\alpha, t_s) \simeq 3\frac{\pi}{\sqrt{2}}(\alpha t_s)^{5/2} \tag{4.50}$$

and Eq.(4.47) yields

$$\alpha = \frac{\pi}{8t_s}\left(\frac{v_d}{v_s}\right)^4. \tag{4.51}$$

The dimensionless resistivity ρ_p/ρ_{p0} for $\alpha t_s \gg 1$ can be thus expressed as

$$\frac{\rho_p}{\rho_{p0}} = \frac{S(\alpha, t_s)}{2\alpha^{3/2} g_s(t_s)} \tag{4.52}$$

with

$$g_s(t_s) = \int_0^\infty \exp\left[-\frac{(y-1)^2}{2t_s}\right] [-n'(2y/t_s)]\, dy, \tag{4.53}$$

which, for $t_s \gg 1$, reduces to

$$g_s(t_s) \simeq t_s^3. \tag{4.54}$$

Therefore, when $\alpha/t_s \gg 1$ and $t_s \gg 1$, we have

$$\frac{\rho_p}{\rho_{p0}} \simeq 1.88\frac{\alpha}{t_s^{1/2}}. \tag{4.55}$$

Combining Eqs.(4.51) and (4.55) we can easily obtain the relation between $\alpha = T_e/T$ and the electric field $E = \rho_p n e v_d$:

$$\frac{T_e}{T} = 0.5\left(\frac{T}{\Theta_s}\right)^{1/5}\left(\frac{E}{E_0}\right)^{4/5}, \tag{4.56}$$

which is valid for $t/\Theta_s \gg 1$ and $E/E_0 \gg 1$. Here $E_0 = \rho_{p0} n e v_s = v_s/\mu_0$, μ_0 being the carrier mobility in the zero-field limit. Equation (4.56) differs from the result of the carrier temperature model analysis (Seeger, 1982). The latter predicts a linear dependence of

T_e/T on the electric field with a temperature-independent coefficient. The results for pure acoustic phonon scattering in the general case have been carried out numerically by Lei and Ting (1985b) using Eq.(4.13) for $P_2(\mathbf{q},\omega)$; one of their original figure is shown in Fig.4.2.

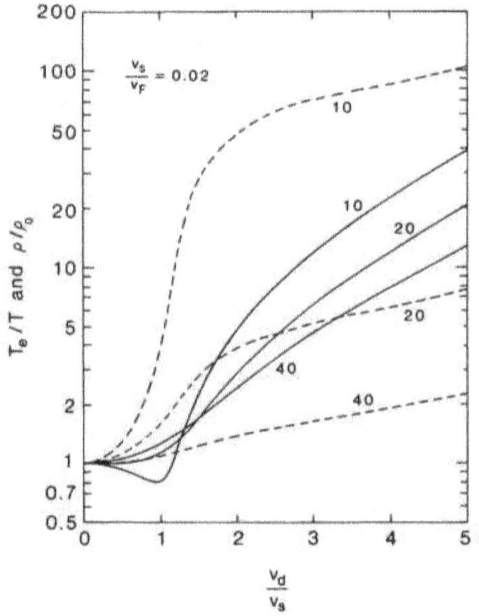

Fig.4.2 Temperature ratio T_e/T (solid lines) and the dimensionless phonon-induced resistivity (dashed lines) are shown as functions of v_d/v_s at several lattice temperatures for a system with $v_s/v_F = 0.02$. The scattering is entirely due to acoustic phonons, $\rho = \rho_p$. The numbers near the curves are the values of T/Θ_s. (From Lei and Ting, 1985b)

4.5 Optic Phonon Scattering

The case in which the optic phonon scattering dominates is also interesting. As is usually done in discussing optic-phonon scattering, we employ the Einstein model for the optic phonon spectrum: $\Omega_{\mathbf{q}\lambda} = \Omega_o$, and assume $|\bar{M}(\mathbf{q},\lambda)|^2$ to be a constant (nonpolar optic deformation potential).

It is convenient to introduce a velocity scale v_o defined by

$$mv_o^2 \equiv \Omega_o, \tag{4.57}$$

and denote

$$t_o \equiv T/\Omega_o. \tag{4.58}$$

The salient feature of optic-phonon scattering is saturation of the current density at high electric field. That is, the solution of the energy-balance equation exists only when $v_d < v_m$ with a saturation maximum v_m. To determine the saturation value of the drift velocity, we examine the asymptotic behavior of the energy-balance equation at large αt_o. To the leading term of its expansion in $(\alpha t_o)^{-1}$, the energy-balance equation yields

$$\left(\frac{v_d}{v_o}\right)^2 = 3\left[2 - \frac{K_1'(1/2\alpha t_o)}{\alpha t_o K_1(1/2\alpha t_o)}\right]^{-1} \tanh\left(\frac{1}{2t_o}\right), \tag{4.59}$$

where $K_1(x)$ and $K_1'(x)$ are the modified Bessel function and its derivative. The saturation value v_m can be obtained by letting $\alpha t_o \to \infty$ in Eq. (4.57), whence

$$\left(\frac{v_m}{v_o}\right)^2 = \frac{3}{4}\tanh\left(\frac{1}{2t_o}\right), \tag{4.60}$$

which is temperature dependent, having a maximum $v_m = 0.866 v_o$. In the limit $E \to \infty$ the above result is the same as that obtained from the Boltzmann equation (Seeger, 1982). It is also consistent with the result of Thornber and Feynman (1970) in regard to saturation at high field. Fig.4.3 shows the dimensionless resistivity and electron temperature calculated by Lei and Ting (1985b) as functions of drift velocity for the case of pure nonpolar optic phonon scattering.

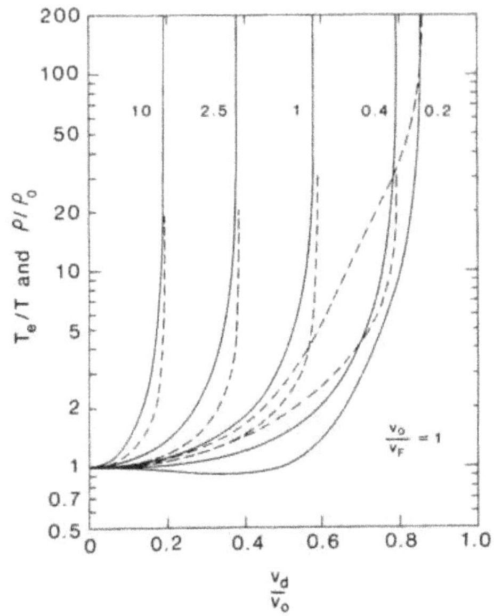

Fig.4.3 Temperature ratio T_e/T (solid lines) and the dimensionless phonon-induced resistivity (dashed lines) are shown as functions of v_d/v_s at several lattice temperatures for a system with $v_o/v_F = 1$. The scattering is entirely due to nonpolar-optical phonons, $\rho = \rho_p$. The numbers near the curves are the values of T/Θ_o. In the figure ρ_0 stands for ρ_{p0}. (From Lei and Ting, 1985b)

4.6 Electron Cooling

An interesting consequence of the force and energy balance equations is the possible cooling of the relative electrons, in that the electron temperature T_e may be lower than the lattice temperature T in the presence of a finite drift v_d (Lei and Ting, 1985b,c). Such a lowering of the electron temperature below the lattice temperature occurs in low-impurity samples. At low lattice temperature, when acoustic phonons dominate the scattering, the maximum cooling occurs at a current density for which the electron drift velocity is near the sound velocity. Fig.4.4 provides an example of electron cooling in the case of dominant acoustic phonon scattering (Lei and Ting, 1985b). If optic phonons dominate, cooling is most likely to occur at much higher drift velocities in the range $v_d = 0.3 - 0.5v_o$. The problem of the electron cooling was further examined in GaAs/AlGaAs systems by Lei and Horing (1987b). It was found that in these selectively doped high-mobility systems, in which polar optic phonon scattering dominates, that not only the electron temperature T_e itself, but also the longitudinal and transverse noise temperatures T_{nl} and T_{nt}, can be lower than the lattice temperature $T = 77K$ for wide range of dc electric fields (section 8.2).

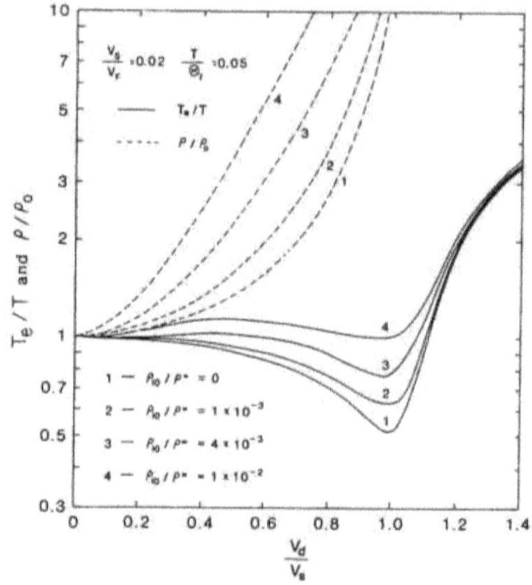

Fig.4.4 Temperature ratio T_e/T and the dimensionless phonon-induced resistivity (dashed lines) are shown as functions of v_d/v_s at several lattice temperatures for a system with $v_s/v_F = 0.02$ at $T/\Theta_F = 0.05$. Zero-field phonon resistivity $\rho_{p0} = 1.94 \times 10^{-3}\rho^*$. In the figure ρ_0 stands for ρ_{p0}. (From Lei and Ting, 1985b)

4.7 Role of Dynamic Screening

In most earlier investigations of transport, the carrier-carrier Coulomb interactions were taken into account only in the form of static screening. On the contrary, the balance equation theory incorporates the dynamic effects of carrier-carrier interactions within the framework of the density-density correlation function $\Pi_2(\mathbf{q}, \omega)$, which is expressed by Eq.(4.16) in the random phase approximation. The role of this feature in impurity-induced dc resistivity was discussed in detail by Lei and Ting (1985a). The full effect of the RPA density-density correlation function on the behavior of resistivity and the electron temperature as functions of drift velocity was investigated numerically by the same authors (Lei and Ting, 1985b). Although the inclusion of the dynamic Coulomb interaction generally does not qualitatively change transport behavior at relatively small drift velocity, as is obvious from the analytical expression, it has significant quantitative influence at high current density in the following two respects.

Since $\epsilon(\mathbf{q},\omega) \to 1$ (or $\Pi_{10}(\mathbf{q},\omega) \sim 0$ and $\Pi_{20}(\mathbf{q},\omega) \sim 0$) for very large ω, in the case of large drift velocity v_d, $\epsilon(\mathbf{q},\omega_0)$ and $\epsilon(\mathbf{q},\omega_0 + \Omega_{\mathbf{q}\lambda})$ can be replaced by unity for the dominant $\mathbf{q}$-integration region. Therefore the contributions from very large ω should be equivalent to that without screening. This is a high-field descreening effect.

The other dynamic screening effect comes from plasmon contributions. It is seen from the RPA expression for $\Pi_{02}(\mathbf{q},\omega)$, Eq.(4.16), that significant contributions to the frictional forces and energy transfer rate may occur in the region where both $[1 - \nu_c(q)\Pi_{01}(\mathbf{q},\omega)]$ and $\Pi_{02}(\mathbf{q},\omega)$ nearly vanish, the region of plasmon excitations. This effect on dc steady state transport is significant in the impurity-induced resistivity at low temperature for a degenerate 3D system (Lei and Ting, 1985a). The plasmon contribution to impurity-induced resistivity shows up as a broad bump in the range of $v_d/v_F \sim 1.5 - 2.0$, as is shown in Fig.4.5.

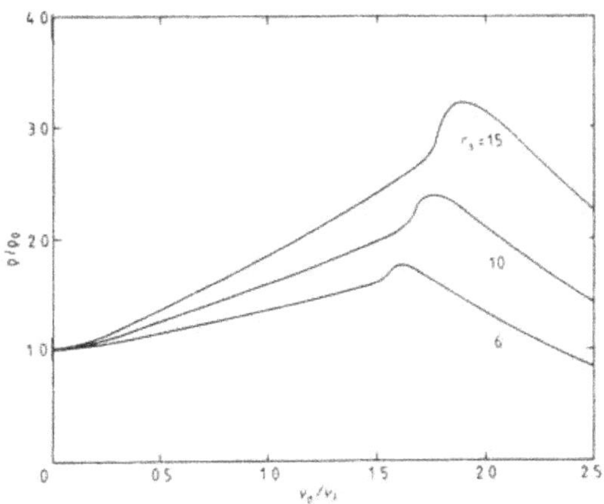

Fig.4.5 Impurity-induced resistivity ρ_i/ρ_{i0} is plotted as a function of v_d/v_F in the degenerate case at several values of r_s (the radius of the electron sphere in atomic units) for charged-impurity scattering. For large value of v_d/v_F the plasmon pole of the electron gas contributes to the resistivity. In the figure ρ stands for ρ_i, and ρ_0 stands for ρ_{i0}. (From Lei and Ting, 1985a)

5. DC STEADY STATE TRANSPORT IN QUASI-TWO-DIMENSIONAL SYSTEMS AND SUPERLATTICES

The balance equation approach expounded in the preceding sections for isotropic three-dimensional systems can be easily extended to quasi-low-dimensional structures, such as inversion layers, heterojunctions, quantum wells, superlattices and quantum wires (Lei, Birman and Ting, 1985; Lei and Horing, 1986; Lei, Horing and Zhang, 1986; Cai, 1989; Yang, Lei and Tsai, 1990), and it thus provides a convenient basis to analyse hot carrier transport in these systems of intense current interest.

5.1 Balance Equations for Quasi-Two-Dimensional Systems

Following the discussion of Lei, Birman and Ting (1985), we consider mobile carriers in, for example, a GaAs/AlGaAs heterojunction. These carriers form a quasi-two-dimensional electron system. In the effective-mass approximation the electron system can be described generally by the following Hamiltonian (Ando, Fowler and Stern, 1982)

$$H = \sum_i \left[\frac{\mathbf{p}_i^2}{2m} + \frac{p_{zi}^2}{2m_z} + U(z_i)\right] + \sum_{i<j} V(\mathbf{r}_i - \mathbf{r}_j, z_i, z_j), \tag{5.1}$$

where $\mathbf{p}_i = (p_{xi}, p_{yi})$ and $\mathbf{r}_i = (x_i, y_i)$ are the two-dimensional momentum and coordinate of the ith electron along the layer plane and p_{zi} and z_i are those perpendicular to the interface; m and m_z are, respectively, the electron band-effective-masses parallel and perpendicular to the plane; $U(z)$ is the confining potential, and the last term is the electron-electron interaction. Under the influence of a uniform electric field in the plane electronic transport takes place only within the plane, and it is convenient to introduce two-dimensional center-of-mass variables $\mathbf{P} = (P_x, P_y)$ and $\mathbf{R} = (X, Y)$:

$$\mathbf{P} = \sum_i \mathbf{p}_i, \quad \mathbf{R} = \frac{1}{N}\sum_i \mathbf{r}_i, \tag{5.2}$$

and two-dimensional relative electron variables

$$\mathbf{p}_i' = \mathbf{p}_i - \frac{1}{N}\mathbf{P}, \quad \mathbf{r}_i' = \mathbf{r}_i - \mathbf{R}, \tag{5.3}$$

where N is the total number of electrons (normalized to unit area). In terms of these variables the Hamiltonian of the electron system in the presence of a uniform electric field $\mathbf{E}$ parallel to the layer plane can be written as the sum of a 2D center-of-mass part H_c and a relative electron part H_e, with

$$H_c = \frac{\mathbf{P}^2}{2Nm} - Ne\mathbf{E}\cdot\mathbf{R} \tag{5.4}$$

and H_e is also expressed by (5.1) but in terms of $\mathbf{p}_i'$ and $\mathbf{r}_i'$. Hereafter we shall omit the primes and always understand reference to the relative motion of the electrons. In a

GaAs/AlGaAs structure the mobile carriers are electrons released from donors located in the heavily doped AlGaAs region and are essentially confined in the GaAs region, formed by the band discontinuity of GaAs/AlGaAs interfaces.

In the absence of carrier-carrier interaction the electron state can be characterized by a subband index n and a two dimensional wavevector $\mathbf{k}_\parallel = (k_x, k_y)$ with the eigen energy

$$\varepsilon_{n\mathbf{k}_\parallel} = \varepsilon_n + k_\parallel^2/2m \tag{5.5}$$

and the wave function(unit junction area)

$$\psi_{n\mathbf{k}_\parallel}(\mathbf{r}, z) = e^{i\mathbf{k}_\parallel\cdot\mathbf{r}}\zeta_n(z), \tag{5.6}$$

$\zeta_n(z)$ being the envelope function (Stern and Howard, 1964).

Using this set of single electron states as basis, we write the Hamiltonian of the relative electrons as

$$\begin{aligned} H_e = &\sum_{n,\mathbf{k}_\parallel,\sigma} \varepsilon_{n\mathbf{k}_\parallel} c^\dagger_{n\mathbf{k}_\parallel\sigma} c_{n\mathbf{k}_\parallel\sigma} \\ &+ \frac{1}{2} \sum_{\substack{m',m,n',n \\ \mathbf{k}_\parallel,\mathbf{k}'_\parallel,\mathbf{q}_\parallel \\ \sigma,\sigma'}} V_{m'm,n'n}(q_\parallel) c^\dagger_{m'\mathbf{k}_\parallel+\mathbf{q}_\parallel\sigma} c^\dagger_{n'\mathbf{k}'_\parallel-\mathbf{q}_\parallel\sigma'} c_{n\mathbf{k}'_\parallel\sigma'} c_{m\mathbf{k}_\parallel\sigma'} \end{aligned} \tag{5.7}$$

with

$$V_{m'm,n'n}(q_\parallel) = \frac{e^2}{2\epsilon_0\kappa q_\parallel} H_{m'm,n'n}(q_\parallel) \tag{5.8}$$

$$H_{m'm,n'n}(q_\parallel) = \int\!\!\int dz_1 dz_2 \zeta^*_{m'}(z_1)\zeta_m(z_1)\zeta^*_{n'}(z_2)\zeta_n(z_2)(e^{-q_\parallel|z_1-z_2|} + \text{images}), \tag{5.9}$$

where κ is the background dielectric constant. The image terms depend on the difference of the dielectric constants on the two sides of the interface in the usual way. In the case of GaAs/AlGaAs system the dielectric constants on both sides are almost equal, such that the image terms can be neglected, and we will use this approximation for simplicity.

We assume that the elastic scatterings in the system are mainly due to ionized impurities, and the impurity potential takes the form (neglecting images)

$$U(\mathbf{r}-\mathbf{r}_a, z-z_a) = \frac{Ze^2}{4\pi\epsilon_0\kappa}\left[(\mathbf{r}-\mathbf{r}_a)^2 + (z-z_a)^2\right]^{-1/2} \tag{5.10}$$

where $(\mathbf{r}_a, z_a)$ is the location of the impurity and Z is its charge number. These potentials yield the electron-impurity interaction

$$H_{ei} = \sum_{\substack{n',n,\sigma \\ \mathbf{k}_\parallel,\mathbf{q}_\parallel,a}} U_{n'n}(q_\parallel, z_a) e^{-\mathbf{q}\cdot(\mathbf{R}-\mathbf{r}_a)} c^\dagger_{n'\mathbf{k}_\parallel+\mathbf{q}_\parallel\sigma} c_{n\mathbf{k}_\parallel\sigma}, \tag{5.11}$$

in which

$$U_{n'n}(q_\parallel, z_a) = \frac{Ze^2}{2\epsilon_0 \kappa q_\parallel} F_{n'n}(q_\parallel, z_a) \tag{5.12}$$

with

$$F_{n'n}(q_\parallel, z_a) = \int_0^\infty e^{-q_\parallel |z - z_a|} \zeta^*_{n'}(z) \zeta_n(z) dz. \tag{5.13}$$

For impurities on the AlGaAs side, $z_a < 0$, $F_{n'n}(q_\parallel, z_a) = \exp(q_\parallel z_a) I_{n'n}(q_\parallel)$, with

$$I_{n'n}(q_\parallel) = \int_0^\infty e^{-q_\parallel z} \zeta^*_{n'}(z) \zeta_n(z) dz. \tag{5.14}$$

For GaAs/AlGaAs heterojunctions, we take the phonons to be the same as in bulk GaAs:

$$H_{ph} = \sum_{\mathbf{q}\lambda} \Omega_{\mathbf{q}\lambda} b^\dagger_{\mathbf{q}\lambda} b_{\mathbf{q}\lambda}. \tag{5.15}$$

In the quasi-two-dimensional representation, the electron-phonon interaction takes the form:

$$H_{ep} = \sum_{\substack{n',n \\ \mathbf{q}_\parallel, q_z, \lambda}} M_{n'n}(\mathbf{q}, \lambda) e^{i\mathbf{q}_\parallel \cdot \mathbf{R}} (b_{\mathbf{q}\lambda} + b^\dagger_{-\mathbf{q}\lambda}) \sum_{\mathbf{k}_\parallel, \sigma} c^\dagger_{n'\mathbf{k}_\parallel + \mathbf{q}_\parallel \sigma} c_{n\mathbf{k}_\parallel \sigma}, \tag{5.16}$$

with $M_{n'n}(\mathbf{q}, \lambda) = M(\mathbf{q}, \lambda) I^*_{n'n}(iq_z)$. Following the procedure described above in Section 3 we obtain force- and energy-balance equations similar to Eqs.(3.10) and (3.26). For steady state dc transport these equations become

$$N_s e\mathbf{E} + \mathbf{f} = 0 \tag{5.17}$$

$$\mathbf{v} \cdot \mathbf{f} + w = 0, \tag{5.18}$$

where N_s is the carrier sheet density, $\mathbf{f}$ and w are, respectively, the frictional force and the energy-loss rate for the electron system of unit area:

$$\mathbf{f} = \mathbf{f}_i + \mathbf{f}_p, \tag{5.19}$$

$$\mathbf{f}_i = \sum_{n', n, \mathbf{q}_\parallel} |U_{n'n}(q_\parallel)\Pi_2(n', n, \mathbf{q}_\parallel, \omega_0) \tag{5.20}$$

$$\mathbf{f}_p = 2 \sum_{n', n, \mathbf{q}, \lambda} |M_{n'n}(\mathbf{q}, \lambda)|^2 \mathbf{q}_\parallel \Pi_2(n', n, \mathbf{q}_\parallel, \Omega_{\mathbf{q}\lambda} + \omega_0) \left[n(\frac{\Omega_{\mathbf{q}\lambda}}{T}) - n(\frac{\Omega_{\mathbf{q}\lambda} + \omega_0}{T_e}) \right] \tag{5.21}$$

$$w = 2 \sum_{n', n, \mathbf{q}, \lambda} |M_{n'n}(\mathbf{q}, \lambda)|^2 \Omega_{\mathbf{q}\lambda} \Pi_2(n', n, \mathbf{q}_\parallel, \Omega_{\mathbf{q}\lambda} + \omega_0) \left[n(\frac{\Omega_{\mathbf{q}\lambda}}{T}) - n(\frac{\Omega_{\mathbf{q}\lambda} + \omega_0}{T_e}) \right] \tag{5.22}$$

in which

$$|U_{n'n}(q_\parallel)|^2 = \int dz n_i(z) |U_{n'n}(q_\parallel, z)|^2 \tag{5.23}$$

is the effective impurity potential and $n_i(z)$ is the impurity volume density at z. In Eqs (5.20)-(5.22) $\Pi_2(n', n, \mathbf{q}_\parallel, \omega)$ is the Fourier spectrum of the electron correlation function

$$\Pi(n', n, \mathbf{q}_\parallel, t) = -i\theta(t) \sum_{\mathbf{k}_\parallel \sigma} \langle \left[c^\dagger_{n'\mathbf{k}_\parallel + \mathbf{q}_\parallel \sigma}(t) c_{n\mathbf{k}_\parallel \sigma}(t) \, , \, c^\dagger_{n\mathbf{k}_\parallel \sigma} c_{n'\mathbf{k}_\parallel + \mathbf{q}_\parallel \sigma} \right] \rangle, \tag{5.24}$$

where $c(t) \equiv \exp(iH_e t) c \exp(-iH_e t)$, and the statistical average is taken over the relative electron equilibrium density matrix $\hat{\rho}_0 \sim e^{-H_e/T_e}$ at the electron temperature T_e.

5.2 Hot Electron Transport in GaAs/AlGaAs Heterojunctions

The above balance equations have been applied to the discussion of hot-electron transport in a GaAs/AlGaAs heterosystem by Lei et al. (1986). They made use of random-phase approximation to deal with the electron density correlation function $\Pi(n', n, \mathbf{q}_{\parallel}, \omega)$ in the presence of electron-electron Coulomb interactions:

$$\Pi(n', n, \mathbf{q}_{\parallel}, \omega) = \Pi_0(n', n, \mathbf{q}_{\parallel}, \omega)\left[1 + \tilde{V}_{n'n,n'n}(q_{\parallel})\Pi(n', n, \mathbf{q}_{\parallel}, \omega)\right] \quad (5.25)$$

and the renormalized potentials $\tilde{V}_{n'n,n'n}(q_{\parallel})$ satisfy the following equations

$$\tilde{V}_{n'n,m'm}(q_{\parallel}, \omega) = V_{n'n,m'm}(q_{\parallel}) + \sum_{l',l} V_{n'n,l'l}(q_{\parallel})\Pi_0(l', l, \mathbf{q}_{\parallel}, \omega)\tilde{V}_{l'l,m'm}(q_{\parallel}, \omega), \quad (5.26)$$

in which

$$\Pi_0(n', n, \mathbf{q}_{\parallel}, \omega) = 2\sum_{k_{\parallel}} \frac{f(\varepsilon_{n'\mathbf{k}_{\parallel}+\mathbf{q}_{\parallel}}, T_e) - f(\varepsilon_{n\mathbf{k}_{\parallel}}, T_e)}{\varepsilon_{n'\mathbf{k}_{\parallel}+\mathbf{q}_{\parallel}} - \varepsilon_{n\mathbf{k}_{\parallel}} + \omega + i\delta} \quad (5.27)$$

is the density-density correlation function for electrons in the n and n' subbands in the absence of Coulomb interactions.

In numerical calculations Lei et al (1986) considered only the lowest and the next lowest subbands, and thus truncated the RPA equation (5.26) by taking $n', n, m', m = 0, 1$ only. The Fang-Howard-Stern variational functions (Ando, Fowler and Stern, 1982)

$$\zeta_0(z) = (\frac{b_0^3}{2})^{1/2} z \exp(-\frac{b_0 z}{2}) \quad (5.28)$$

$$\zeta_1(z) = \frac{3}{2}\left(\frac{b_1^5}{b_0^2 - b_0 b_1 + b_1^2}\right)^{1/2} z\left[1 - \left(\frac{b_0 + b_1}{6}\right)z\right]\exp\left(-\frac{b_1 z}{2}\right) \quad (5.29)$$

were used as the envelope functions for the $n = 0$ and 1 subbands respectively. The parameters b_0 and b_1 were determined by minimizing the energies ε_0 and ε_1 as usual. With these envelope functions the integrations in Eq.(5.9) were performed to give a closed expression for $H_{n'n,m'm}(q_{\parallel})$. The density-density correlation function was then obtained from Eq.(5.25) and the truncated RPA equation (5.26).

Remote impurities were assumed to be the dominant elastic scatterers, located within a narrow-space charge layer with a total sheet density N_I at a distance s from the interface in the barrier of the AlGaAs side. The effective potential appearing in the impurity-induced frictional force Eq.(5.20) was thus written as

$$|U_{n'n}(q_{\parallel})|^2 = (\frac{Ze^2}{2\epsilon_0\kappa})^2 N_I e^{-2q_{\parallel}s} q_{\parallel}^{-2} I_{n'n}(q_{\parallel})^2. \quad (5.30)$$

Both the electron-acoustic-phonon interaction (via piezoelectric coupling and deformation potential) and electron-polar-optic-phonon interaction (via Fröhlich coupling) were taken into account for GaAs. For a given acoustic phonon branch, the matrix elements of the piezoelectric interaction and that of the deformation potential are $\pi/2$ out of phase so that their contributions to $|M(\mathbf{q},\lambda)|^2$ are additive (Mahan, 1981). Only the longitudinal mode gives rise to deformation potential coupling with matrix element (Vogl, 1980)

$$|M(\mathbf{q},l)|^2 = \frac{\Xi^2 q}{2dv_{sl}}, \tag{5.31}$$

where d is the mass density of the crystal; v_{sl} is the velocity of the longitudinal sound wave and Ξ is the shift of the band edge per unit dilation. Both longitudinal and transverse phonons contribute to the piezoelectric interaction. In GaAs (zinc-blend structure) there is only one nonzero independent piezoelectric constant $e_{14} = e_{25} = e_{36}$ and the dielectric constant and sound speed are isotropic. The piezoelectric matrix element is (Mahan, 1972)

$$M(\mathbf{q},\lambda) = \frac{8\pi e e_{14}}{\kappa q^2}\left(\frac{1}{2dv_{s\lambda}q}\right)^{1/2}\left[q_x q_y e_z(\mathbf{q},\lambda) + q_z q_x e_y(\mathbf{q},\lambda) + q_y q_z e_x(\mathbf{q},\lambda)\right], \tag{5.32}$$

in which $v_{s\lambda}(\lambda = l, t)$ is the velocity of the λ-th branch phonon and $\mathbf{e}(\mathbf{q},\lambda)$ is the unit polarization vector of the phonon with wavevector $\mathbf{q} = (q_x, q_y, q_z)$ in branch λ. For the longitudinal branch we have

$$|M(\mathbf{q},l)|^2 = \frac{32\pi^2 e^2 e_{14}^2}{\kappa^2 d\, v_{sl}}\frac{(3q_x q_y q_z)^2}{q^7}. \tag{5.33}$$

There are two independent transverse branches. Their total contribution is (Lei, Birman and Ting, 1985)

$$\sum_{j=1,2}|M(\mathbf{q},t_j)|^2 = \frac{32\pi^2 e^2 e_{14}^2}{\kappa^2 d\, v_{st} q^5}\left(q_x^2 q_y^2 + q_y^2 q_z^2 + q_z^2 q_x^2 - \frac{(3q_x q_y q_z)^2}{q^7}\right). \tag{5.34}$$

In GaAs, the transverse sound speed, $v_{st} = 2.48 \times 10^5$cm/s, is less than half of the longitudinal sound speed $v_{sl} = 5.29 \times 10^5$cm/s. Therefore, it is expected that the transverse acoustic phonon, thus the piezoelectric interaction, plays a major role in determining the phonon-limited mobility and energy transfer rate at very low temperatures. When the temperature rises above 40 K, polar optical scattering begins to contribute significantly. Polar interaction between longitudinal optical phonons and electrons has the Fröhlich matrix element:

$$|M(\mathbf{q},LO)|^2 = \frac{e^2}{2\epsilon_0 q^2}\left(\frac{1}{\kappa_\infty} - \frac{1}{\kappa}\right)\Omega_o, \tag{5.35}$$

in which κ_∞ is the optical dielectric constant and Ω_o is the longitudinal optical phonon frequency.

With these considerations in view, Lei et al. (1986) calculated the nonlinear mobility and the electron temperature from the balance equations (5.17) and (5.18). In Fig.5.1 the calculated normalized mobility $\mu/\mu_0(\tau)$ (solid curves) and electron temperature T_e (dashed curves) are shown as functions of the drift velocity v for a GaAs/AlGaAs system with carrier sheet density $N_s = 4.0 \times 10^{11}\text{cm}^{-2}$ and zero temperature Ohmic mobility $\mu_0(0) = 2.2 \times 10^5\text{cm}^2/\text{Vs}$, at lattice temperatures $T = 10, 77$ and 100 K, respectively. Here $\mu_0(T)$ is the Ohmic mobility at temperature T.

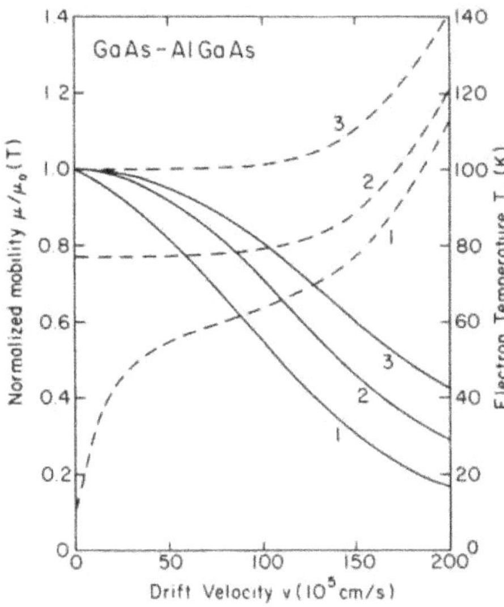

Fig.5.1 Calculated normalized mobility $\mu/\mu_0(T)$ (solid curves) and electron temperature T_e (dashed curves) are shown as functions of drift velocity v for a GaAS-AlGaAs system with $N = 4.0 \times 10^{11}$ cm^{-2} and $\mu_0(0) = 2.2 \times 10^5$ cm^2/Vs. The impurities are located at a distance $s = 125\text{Å}$ from the interface. The lattice temperatures are 1, 10 K; 2, 77 K; 3, 100 K. (From Lei et al.,1986)

In Fig.5.2 the calculated normalized mobilities μ/μ_0 (solid curves) and electron temperature T_e (dashed curves) are shown as functions of applied electric field E at lattice temperature $T = 4.2$ K for two GaAs-AlGaAs systems with $N_s = 2.5 \times 10^{11}$ cm^{-2} and $\mu_0 = 1.0 \times 10^6$ cm^2/Vs, and with $N = 3.9 \times 10^{11}$ cm^{-2} and $\mu_0 = 7.9 \times 10^4\text{cm}^2/\text{Vs}$, respectively, together with some experimental data on μ/mu_0. The crosses are taken from the dark data at 4.2 K of sample 2 of Höpfel and Weimann (1985). The dots are experimental results of Shah et al.(1984) for a multi-quantum-well sample of width 260Å and layer carrier density $N_s = 3.9 \times 10^{11}\text{cm}^{-2}$ at $T = 2$ K. The GaAs/AlGaAs material parameters used in the calculations are as follows: density $d = 5.31$ g/cm^3, effective mass $m = 0.07m_e$, transverse sound velocity $v_{st} = 5.29 \times 10^3$ m/s, longitudinal-optic-phonon energy $\Omega_0 = 35.4$ meV, low-frequency dielectric constant $\kappa = 12.9$, optical dielectric constant $\kappa_\infty = 10.8$, acoustic deformation potential $\Xi = 8.5$ eV, piezoelectric constant $e_{14} = 1.41 \times 10^9$ V/m. Fig.5.3 is taken from Hirakawa and Sakaki (1988), who compared their experimental data with the electron temperature model, and with the balance equation approach, i.e. the theoretical curve shown in Fig.5.2. They found that the measured mobilities are in excellent agreement with the theoretical results of Lei et al. (1986).

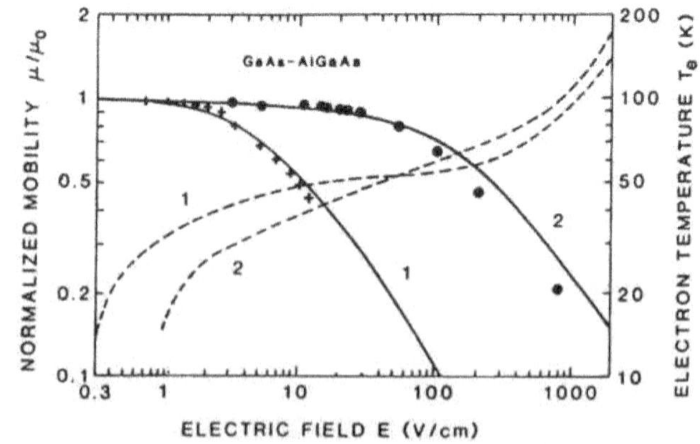

Fig.5.2 Calculated normalized mobility μ/μ_0 (solid curves) and electron temperature T_e (dashed curves) vs electric field E at lattice temperature $T = 4.2$ K for two systems: (1) $N = 2.5 \times 10^{11}$ cm^{-2}, $\mu_0 = 1 \times 10^6$ cm^2/Vs, $s = 250Å$; (2) $N = 3.9 \times 10^{11}$ cm^{-2}, $\mu_0 = 7.9 \times 10^4$ cm^2/Vs, $s = 125Å$. The experimental data shown in crosses are taken from Höpfel and Weimann (1985) and those shown in dots from Shah et al.(1984). (From Lei et al.,1986)

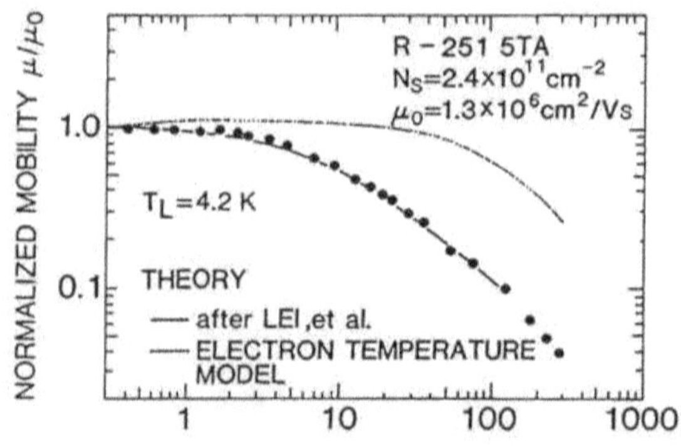

Fig.5.3 Measured normalized mobility μ/μ_0 is plotted as a function of the electric field. The normalized mobilities calculated by the electron temperature model (dotted line) and by the balance-equation approach (solid line; Lei et al., 1986) are also shown. (This figure is taken from Hirakawa and Sakaki, 1988).

5.3 Balance Equations for Quantum-Well Superlattices

For quantum well superlattices, balance equation analyses of hot carrier transport were carried out by Lei, Horing and Zhang (1986b). The model employed for the superlattice consists of an infinite number of periodically arranged quantum wells of width a, and d is the spatial period, or distance between two adjacent layer centers. In the effective-mass approximation the electron system can be described also by the Hamiltonian (5.1), with the confining potential $U(z)$ adjusted to reflect the superlattice structure of the quantum wells. Considering the applied electric field parallel to the layer plane, we again introduce two-dimensional center-of-mass variables and relative electron variables, and separate H_c and H_e as in Section 5.1 for quasi-two dimensional systems. We assume that (i) the potential-well depth is deep enough and tunneling is small enough, so that electrons are confined to just one well, and (ii) the width of the well is narrow and the electron density is not too high, such that electrons occupy only the lowest subband. The lowest subband wave function of the electron in lth well is taken as (for unit transverse area of the superlattice)

$$\psi_{l\mathbf{k}_\parallel}(\mathbf{r},z) = e^{i\mathbf{k}_\parallel\cdot\mathbf{r}}\zeta(z-ld) \qquad (l=0,\pm1,\ldots) \tag{5.36}$$

with

$$\zeta(z) = \begin{cases} (2/a)^{1/2}\sin(\pi z/a) & \text{for} \quad 0<z<a, \\ 0 & \text{elsewhere}, \end{cases} \tag{5.37}$$

and the corresponding energy $\varepsilon_{\mathbf{k}_\parallel} = k_\parallel^2/2m$ is degenerate with respect to the layer indices. The Hamiltonian H_e for relative electrons can be written in second quantized notation as

$$H_e = \sum_{l,\mathbf{k}_\parallel,\sigma} \varepsilon_{\mathbf{k}_\parallel} c^\dagger_{l\mathbf{k}_\parallel\sigma} c_{l\mathbf{k}_\parallel\sigma} + \frac{1}{2}\sum_{\substack{l,l',\sigma,\sigma' \\ \mathbf{k}_\parallel,\mathbf{k}'_\parallel,\mathbf{q}_\parallel}} V_{ll'}(q_\parallel) c^\dagger_{l\mathbf{k}_\parallel+\mathbf{q}_\parallel\sigma} c^\dagger_{l'\mathbf{k}_\parallel-\mathbf{q}_\parallel\sigma'} c_{l'\mathbf{k}'_\parallel\sigma'} c_{l\mathbf{k}_\parallel\sigma}, \tag{5.38}$$

where $c^\dagger_{l\mathbf{k}_\parallel\sigma}(c_{l\mathbf{k}_\parallel\sigma})$ are creation (annihilation) operators for relative electrons with lowest subband wave function $\psi_{l\mathbf{k}_\parallel}$, and $V_{lm}(q_\parallel)$ is the corresponding matrix element of the Coulomb potential, which can be written as

$$V_{lm}(q_\parallel) = \frac{e^2}{2\epsilon_0\kappa q_\parallel} F_{lm}(q_\parallel), \tag{5.39}$$

In this, κ is the background dielectric constant and the form factor $F_{lm}(q_\parallel)$ has a simple expression if the image charge contribution is negligible:

$$\begin{aligned} F_{lm}(q_\parallel) &= F_{l-m}(q_\parallel) \\ &= \int dz \int dz' e^{-q_\parallel|z-z'|}\zeta(z-ld)^2\zeta(z'-md)^2 \\ &= e^{-q_\parallel|l-m|d}\left[\exp(q_\parallel a)I(q_\parallel)^2(1-\delta_{lm}) + \delta_{lm}H(q_\parallel)\right]. \end{aligned} \tag{5.40}$$

Here, δ_{lm} is the Kronecker delta, and the form factors $I(q_\|)$ and $H(q_\|)$ are determined by the electron wave function within the well ($u \equiv q_\| a$)

$$I(q_\|) = \int e^{-q_\| z}\zeta(z)^2 dz$$
$$= 4\pi^2 \left[1 - \exp(-u)\right] / \left[u(u^2 + 4\pi^2)\right], \tag{5.41}$$
$$H(q_\|) = \int\int e^{-q_\| |z-z'|}\zeta(z)^2\zeta(z')^2 dz dz'$$
$$= 3\frac{1-\exp(-u)}{u^2+4\pi^2} + \frac{u}{u^2+4\pi^2}$$
$$- \frac{1-\exp(-u)}{(u^2+4\pi^2)^2}(u^2 - 4\pi^2) + \frac{2}{u}\left[1 - \frac{1-\exp(-u)}{u}\right]. \tag{5.42}$$

The charged-impurity–electron interaction takes the form

$$H_{ei} = \sum_{\substack{l,\sigma,a \\ \mathbf{k}_\|,\mathbf{q}_\|}} \frac{Z_a e^2}{2\epsilon_0 \kappa q_\|} F_l(q_\|, z_a) e^{i\mathbf{q}_\| \cdot (\mathbf{R} - \mathbf{r}_a)} c^\dagger_{l\mathbf{k}_\| + \mathbf{q}_\| \sigma} c_{l\mathbf{k}_\| \sigma}, \tag{5.43}$$

where $(\mathbf{r}_a, \mathbf{z}_a)$ and $\mathbf{Z}_a$ are location and the charge number of the ath impurity and

$$F_l(q_\|, z_a) = \int e^{-q_\| |z - z_a|}\zeta(z - ld)^2 dz. \tag{5.44}$$

Image terms have been neglected again. We consider two kinds of impurities: remote impurities and background impurities. The former are the ionized dopants in AlGaAs regions, which are located in planar sheets at a distance s from the center of each layer with area density N_r per sheet; the latter are distributed uniformly within the GaAs well regions with area density N_b per layer. Here, we assume that there is the same impurity number density in each remote sheet (or in each layer). For the expressions of $F_l(q_\|, z_a)$ for remote and background impurities we refer the reader to the original paper (Lei, Horing and Zhang, 1986b).

The electron-phonon interaction can be written as

$$H_{ep} = \sum_{\mathbf{q},l,\lambda} M(\mathbf{q},\lambda) I(iq_z) e^{i\mathbf{q}_\| \cdot \mathbf{R}} (b_{\mathbf{q}\lambda} + b^\dagger_{-\mathbf{q}\lambda}) e^{-iq_z l d} \sum_{\mathbf{k}_\|,\sigma} c^\dagger_{l\mathbf{k}_\| + \mathbf{q}_\| \sigma} c_{l\mathbf{k}_\| \sigma}, \tag{5.45}$$

in which form factor $I(iq_z)$ is (from Eq.(5.41)):

$$|I(iq_z)|^2 = \pi^4 \sin^2 y / \left[y^2 (y^2 - \pi^2)^2\right], \tag{5.46}$$

with $y \equiv q_z a/2$. For steady state, when the center of mass moves at a constant velocity v, the frictional force due to phonons can be shown to be

$$2\sum_{\mathbf{q},\lambda} |M(\mathbf{q},\lambda)|^2 |I(iq_z)|^2 \mathbf{q}_\| \sum_{l,l'} e^{-iq_z(l-l')d}\ \Pi_2(l,l',\mathbf{q}_\|,\Omega_{\mathbf{q}\lambda} + \omega_0)$$
$$\left[n\left(\frac{\Omega_{\mathbf{q}\lambda}}{T}\right) - n\left(\frac{\Omega_{\mathbf{q}\lambda} + \omega_0}{T_e}\right)\right]. \tag{5.47}$$

Here, $\Pi_2(l, l', \mathbf{q}, \Omega)$ is the imaginary part of the Fourier transform of the electron density-density correlation function of the superlattice

$$\Pi(l, l', \mathbf{q}_\parallel, t) = -i\theta(t) \sum_{\mathbf{k}_\parallel, \mathbf{k}'_\parallel, \sigma} \langle \left[c^\dagger_{l\mathbf{k}_\parallel+\mathbf{q}_\parallel\sigma}(t) c_{l\mathbf{k}_\parallel\sigma}(t), c^\dagger_{l'\mathbf{k}'_\parallel-\mathbf{q}_\parallel\sigma} c_{l'\mathbf{k}'_\parallel\sigma} \right] \rangle. \tag{5.48}$$

In the absence of Coulomb interaction between carriers, the density-density correlation function has nonzero value only when $l = l'$:

$$\Pi(l, l', \mathbf{q}_\parallel, \omega) = \delta_{ll'} \Pi_0(\mathbf{q}_\parallel, \omega), \tag{5.49}$$

and

$$\Pi_0(\mathbf{q}_\parallel, \omega) = 2 \sum_{\mathbf{k}} \frac{f(\varepsilon_{\mathbf{k}_\parallel+\mathbf{q}_\parallel}, T_e) - f(\varepsilon_{\mathbf{k}_\parallel}, T_e)}{\omega + \varepsilon_{\mathbf{k}_\parallel+\mathbf{q}_\parallel} - \varepsilon_{\mathbf{k}_\parallel} + i\delta} \tag{5.50}$$

is the density-density correlation function for a single sheet of 2D electrons without Coulomb interaction. The inclusion of both intralayer and interlayer carrier Coulomb interactions for the superlattice can be achieved using the random-phase approximation, which leads to the following equation:

$$\Pi(l, l', \mathbf{q}_\parallel, \omega) = \delta_{ll'} \Pi(\mathbf{q}_\parallel, \omega) + \Pi(\mathbf{q}_\parallel, \omega) \sum_m V_{lm}(q_\parallel) \Pi(m, l', \mathbf{q}_\parallel, \omega).\delta \tag{5.51}$$

Assuming periodic boundary conditions and neglecting image contributions, we have $V_{lm}(q) = V_{l-m}(q_\parallel)$ and $\Pi(l, l', \mathbf{q}_\parallel, \omega) = \Pi(l - l', \mathbf{q}_\parallel, \omega)$, whence Eq.(5.51) reduces to

$$\mathbf{\Pi}(l, \mathbf{q}_\parallel, \omega) = \mathbf{\Pi}_0(\mathbf{q}, \omega) \left[\delta_{l0} + \sum_m V_{l-m}(q_\parallel) \Pi(m, \mathbf{q}_\parallel, \omega) \right]. \tag{5.52}$$

By introducing

$$\Pi(q_z, \mathbf{q}_\parallel, \omega) = \sum_l e^{-iq_z l d} \Pi(l, \mathbf{q}_\parallel, \omega), \tag{5.53}$$

Eq.(5.52) is easily solved to give

$$\Pi(q_z, \mathbf{q}_\parallel, \omega) = \frac{\Pi_0(\mathbf{q}_\parallel, \omega)}{1 - V(q_\parallel, q_z)\Pi_0(\mathbf{q}_\parallel, \omega)} \tag{5.54}$$

with

$$V(q_\parallel, q_z) = \sum_l V_l(q) e^{-iq_z l d} = \frac{e^2}{2\epsilon_0 \kappa q_\parallel} \left[H(q_\parallel) + S(q_\parallel, q_z) \right], \tag{5.55}$$

where $S(q_\parallel, q_z)$ comes from interlayer carrier interactions:

$$S(q_\parallel, q_z) = \frac{\cos(q_z d) - \exp(-q_\parallel d)}{\cosh(q_\parallel d) - \cos(q_z d)} \exp(qa) I(q_\parallel)^2. \tag{5.56}$$

In terms of the correlation function $\Pi(q_z, \mathbf{q}_\parallel, \omega)$, we can write the phonon-induced frictional force $\mathbf{f}_p$ (per-layer) as

$$\mathbf{f}_p = 2 \sum_{\mathbf{q}_\parallel, q_z, \lambda} |M(q, \lambda)|^2 |I(iq_z)|^2 \mathbf{q}_\parallel \; \Pi_2(q_z, \mathbf{q}_\parallel, \Omega_{\mathbf{q},\lambda} + \omega_0)$$
$$\left[n\left(\frac{\Omega_{\mathbf{q}\lambda}}{T}\right) - n\left(\frac{\Omega_{\mathbf{q}\lambda} + \omega_0}{T_e}\right) \right]. \tag{5.57}$$

Note that although the correlation function $\Pi_2(q_z, \mathbf{q}_\parallel, \omega)$ is a periodic function of q_z, the factor $|M|^2|I|^2$ is not. The sum over q_z ranges from $-\infty$ to ∞ as usual:

$$\sum_{q_z} \rightarrow \frac{d}{2\pi}\int_{-\infty}^{\infty} dq_z. \tag{5.58}$$

The energy loss of the relative electron system is due to inelastic scattering associated with the electron-phonon coupling. Consideration along these lines yields the energy-transfer rate (per layer) from the electron system to the phonon system:

$$w = 2\sum_{\mathbf{q}_\parallel, q_z, \lambda} |M(\mathbf{q},\lambda)|^2 |I|(iq_z)|^2 \Omega_{\mathbf{q}\lambda} \\ \Pi_2(q_z, \mathbf{q}_\parallel, \Omega_{\mathbf{q}\lambda}+\omega_0)\left[n\left(\frac{\Omega_{\mathbf{q}\lambda}}{T}\right) - n\left(\frac{\Omega_{\mathbf{q}\lambda}+\omega_0}{T_e}\right)\right]. \tag{5.59}$$

In deriving the impurity-induced frictional force, we deal with impurity site averaging:

$$A = \langle \sum_{\alpha,\beta} e^{i\mathbf{q}_\parallel\cdot\mathbf{r}_{m\alpha} + i\mathbf{q}'_\parallel\cdot\mathbf{r}_{n\beta}} \rangle. \tag{5.60}$$

If the impurity distribution is random within each sheet (layer) and there is no correlation between different sheets (layers), the quantity in Eq.(5.60)) is a two-dimensional extension of the averaging discussed by many authors in the 3D case (Kohn and Luttinger, 1957). Thus, to lowest order in N_I ($N_I = N_r$ or N_b is the impurity sheet density),

$$A = N_I \delta_{\mathbf{q}'_\parallel, -\mathbf{q}_\parallel}, \tag{5.61}$$

independent of the layer indices m and n. The assumption of noncorrelation of the impurity distributions in different sheets generally seems reasonable. Nevertheless, the possibility of correlation between the impurity distributions in different layers could be included in the following way (Lei, Horing and Zhang, 1986a). The periodicity of the superlattice system enables us to assume that the quantity in Eq. (5.60) is a function of $m-n$ after configuration averaging:

$$\langle \sum_{\alpha,\beta} e^{i\mathbf{q}_\parallel\cdot\mathbf{r}_{m\alpha} + i\mathbf{q}'_\parallel\cdot\mathbf{r}_{n\beta}} \rangle = N_I \delta_{\mathbf{q}'_\parallel, -\mathbf{q}_\parallel} g(m-n), \tag{5.62}$$

where the function $g(m)$ ($m = 0, \pm 1, \cdots$) may be expressed as a Fourier coefficient:

$$g(m) = \frac{d}{2\pi}\int_{-\pi/d}^{\pi/d} dq_z g(q_z) e^{iq_z md} \tag{5.63}$$

with the corresponding inverse series

$$g(q_z) = \sum_m g(m) e^{-iq_z md}. \tag{5.64}$$

The final expression for the frictional force per-layer due to remote- and background-impurity scatterings jointly includes $g(q_z)$:

$$\mathbf{f}_i = \left(\frac{e^2}{2\epsilon_0\kappa}\right)^2 \left(\frac{d}{2\pi}\right) \int_{-\pi/d}^{\pi/d} dq_z \sum_{\mathbf{q}_\parallel} \frac{\mathbf{q}_\parallel}{q_\parallel^2} \tilde{N}(q_\parallel, q_z) g(q_z) \Pi_2(q_z, \mathbf{q}_\parallel, \mathbf{q}_\parallel \cdot \mathbf{v}), \tag{5.65}$$

in which $\tilde{N}(q_\parallel, q_z)$is an effective combined impurity density:

$$\begin{aligned} \tilde{N}(q_\parallel, q_z) &= N_r Z_r^2 \left| \frac{\sinh\left[q_\parallel(d-s)\right] + \exp(iq_z d)\sinh(q_\parallel s)}{\cosh(q_\parallel d) - \cos(q_z d)} \right|^2 \exp(q_\parallel a) I(q_\parallel)^2 \\ &+ N_b Z_b^2 \left[\frac{\cos(q_z d) - \exp(-q_z d)}{\cosh(q_\parallel d) - \cos(q_z d)} \frac{\exp(q_\parallel a) - 1}{q_\parallel a} I(q_\parallel) + K(q_\parallel) \right]^2, \end{aligned} \tag{5.66}$$

Z_r and Z_b are equivalent charge numbers of the remote and background impurities, and $(u \equiv q_\parallel a)$

$$K(q_\parallel) = \frac{8\pi^2}{(4\pi^2 + u^2)u} \left[1 + \frac{u^2}{4\pi^2} - \frac{1 - \exp(-u)}{u} \right]. \tag{5.67}$$

$I(q_\parallel)$ has been given by Eq(5.46). Now the force and the energy balance equations for a superlattice can be written for each layer in exactly the same form as for a quasi-2D system, Eqs.(5.17) and (5.18).

The force and energy balance equations were applied to the calculations of Ohmic mobility as a function of temperature, and of nonlinear mobility and electron temperature as functions of drift velocity, by Lei, Horing and Zhang (1986b) for GaAs/AlGaAs quantum well superlattice, including remote and background impurity scatterings and acoustic and polar-optic phonon scatterings, as well as encompassing the full effects of intralayer and interlayer carrier-carrier interaction within the framework of random-phase-approximation. A figure taken from Lei, Horing and Zhang (1986b) is shown in Fig.5.4.

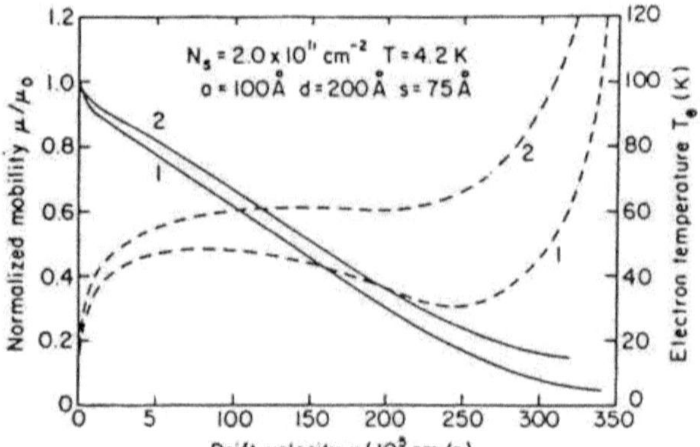

Fig.5.4 Nonlinear mobilities normalized to the Ohmic limit μ/μ_0 (solid curves) and electron temperatures T_e (dashed curves) are shown as functions of drift velocity v at lattice temperature T = 4.2 K for GaAs-$Al_xGa_{1-x}As$ quantum well superlattices with carrier density $N_s = 2.0 \times 10^{11}$ cm^{-2} per sheet, but different values of impurity-limited Ohmic mobilities: 1 - 2.5×^{5} cm^2/Vs; 2 - 8.0 × 10^4 cm^2/Vs.

6. TRANSIENT TRANSPORT

The time-dependent force-balance equation (3.10) and energy-balance equation (3.22) or (3.26) also form the basic starting point to investigate transient transport in the balance-equation approach. Notwithstanding the fact that the derivation of these equations treats the center of mass classically due to its large mass, memory effects associated with c.m. motion are retained in the expression (3.18) for the factor $A(\mathbf{q},t,t')$. This form of $A(\mathbf{q},t,t')$, however, renders the time-dependent force- and energy-balance equations (3.10) and (3.26) as complicated coupled integro-differential equations, and significant computational labor is required to solve them numerically.

6.1 Transient Balance Equations without Memory

While memory effects may be essential in the analysis of very rapid processes or very high frequency problems (next section), they play no role in steady state dc transport (preceding section). It is to be expected that in ordinary conditions the transient response of a realistic semiconductor to a suddenly impressed constant electric field will remain a relatively slow process involving negligible memory effects. With this, one may simplify the analytical and numerical calculations by making the following approximation (Xing and Ting, 1987; Lei, Cui and Horing, 1987; Lei at el, 1987)

$$A(\mathbf{q},t,t') = \exp\left[i\mathbf{q}\cdot\int_{t'}^{t}\mathbf{v}(s)ds\right] \simeq \exp\left[i\mathbf{q}\cdot\mathbf{v}(t)(t-t')\right]. \tag{6.1}$$

With this approximation for $A(\mathbf{q},t,t')$, the expressions for the frictional forces $\mathbf{f}_i$, $\mathbf{f}_p$ and the energy transfer rate w, which appear in the time-dependent force and energy balance equations,

$$nm\frac{d}{dt}\mathbf{v}(t) = ne\mathbf{E}(t) + \mathbf{f}_i + \mathbf{f}_p \tag{6.2}$$

$$C_e\frac{dT_e}{dt} = -\mathbf{v}\cdot(\mathbf{f}_i + \mathbf{f}_p) - w, \tag{6.3}$$

can be written in exactly the same form as in the steady state case:

$$\mathbf{f}_i = n_i\sum_{\mathbf{q}}\mathbf{q}|u(\mathbf{q})|^2\Pi_2(\mathbf{q},\omega_0) \tag{6.4}$$

$$\mathbf{f}_p = 2\sum_{\mathbf{q},\lambda}\mathbf{q}|M(\mathbf{q},\lambda)|^2\Pi_2(\mathbf{q},\Omega_{\mathbf{q}\lambda}+\omega_0)\left[n(\frac{\Omega_{\mathbf{q}\lambda}}{T}) - n(\frac{\Omega_{\mathbf{q}\lambda}+\omega_0}{T_e})\right] \tag{6.5}$$

$$w = 2\sum_{\mathbf{q},\lambda}\Omega_{\mathbf{q}\lambda}|M(\mathbf{q},\lambda)|^2\Pi_2(\mathbf{q},\Omega_{\mathbf{q}\lambda}+\omega_0)\left[n(\frac{\Omega_{\mathbf{q}\lambda}}{T}) - n(\frac{\Omega_{\mathbf{q}\lambda}+\omega_0}{T_e})\right] \tag{6.6}$$

with the replacement $\omega_0 \equiv \mathbf{q}\cdot\mathbf{v}(t)$. They are dependent on time t through the time dependence of the variables $v(t)$ and $T_e(t)$. Eqs.(6.2) and (6.3) are ordinary coupled differential equations. They are easily integrated for a specified $\mathbf{E}(t)$ by starting from given initial values of $\mathbf{v}(t)$ and $T_e(t)$.

Xing and Ting (1987) calculated the transient response of the drift velocity $v(t)$ and the electron temperture $T_e(t)$ of p-type Ge at lattice temperature $T = 300K$ to suddenly-impressed constant electric fields of various strengths. The results are shown in Fig.6.1. Pronounced overshoot velocities appear near $t = 0.5ps$ after turn-on of the electric field in the curves for $E = 20$ and $10kV/cm$. The transport is assumed to be primarily due to heavy holes with effective mass $m = 0.34\, m_e$ (m_e being the free electron mass). Scatterings due to longitudinal acoustic phonons and due to nonpolar optic phonons are considered with $\Xi = 4.6eV$ and $D = 9 \times 10^8 eV/cm$ for acoustic and optic deformation potential parameters, $v_s = 5.4 \times 10^5 cm/s$ as the longitudinal sound velocity, and $\Omega_o = 430K$ as the optic phonon frequency. Fig.6.2 shows the drift velocity and electron temperature response to an impulse electric field in p-type Ge at $T = 173K$, calculated by Xing and Ting (1987).

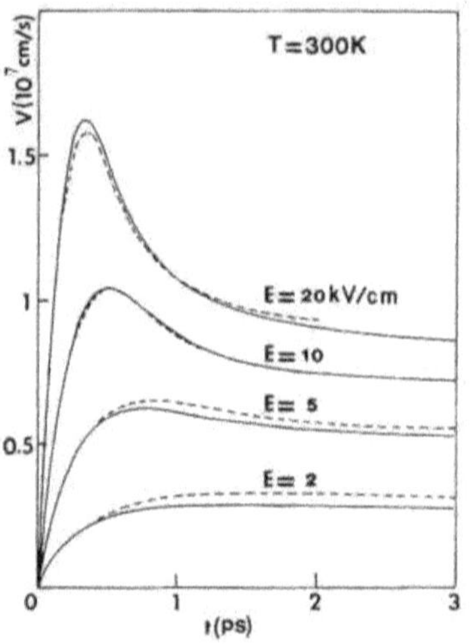

Fig. 6.1 Transient drift velocity $v(t)$ (solid curves) plotted against time when a time-step electric field is applied to p-type Ge at $T = 300$ K, obtained from Eqs.(6.2) and (6.3). The dashed curves correspond to the iterative calculation of the Boltzmann equation (Nougier and Rolland, 1973). (From Xing and Ting, 1987)

Lei, Cui and Horing (1987) analyzed the hot carrier transient transport in GaAs-based heterojunctions and superlattices using the time-dependent force- and energy-balance equations (6.2) and (6.3) with the force- and energy transfer rate expressions appropriate to such quasi-two-dimensional systems. Fig.6.3 shows the calculated transient response of the drift velocity v and the electron temperature T_e to a step electric field of $1.0kV/cm$ turned on at time $t = 0$ in a single quantum well heterosystem of well width $a = 150\mathring{A}$, carrier sheet density $N_s = 2.0 \times 10^{11}\ cm^{-2}$ and zero-temperature mobility $\mu_0 = 5.0 \times 10^5$ cm^2/Vs.

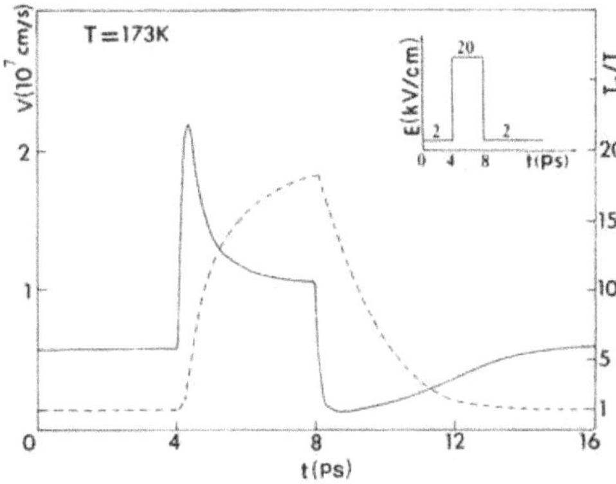

Fig.6.2 Transient drift velocity $v(t)$ (solid curve) and temperature ratio $T_e(t)/T$ (dashed curve) versus time when a time pulse and a dc electric field are applied simultaneously to a p-type Ge at $T = 173$ K. (From Xing and Ting, 1987)

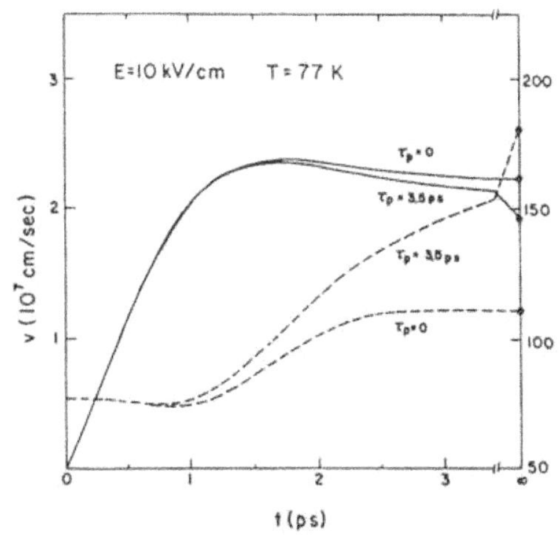

Fig.6.3 The drift velocity v (full curves) and electron temperature T_e (broken curves) are shown as functions of time t after an electric field of $E = 1.0$ kV/cm is suddenly turned on at $t = 0$ in a quantum well GaAs heterosystem at lattice temperature $T = 4.2$ K for the phonon relaxation time $\tau_p = 0$ and $\tau_p = 3.5$ ps. (From Lei, Cui and Horing, 1987)

6.2 Memory Effects in Transient Response for the Weak Current Case

Ting and Nee (1986) and Xing and Ting (1987) carried out detailed examinations of memory effects involved in transient response to a weak electric field. For small c.m. velocity one can linearize both the force and energy-balance equations by expanding the factor $A(\mathbf{q},t,t')$ to the linear order in $\mathbf{v}$:

$$A(\mathbf{q},t,t') = \exp\left[i\mathbf{q}\cdot\int_{t'}^{t}\mathbf{v}(s)ds\right] \simeq 1 + i\mathbf{q}\cdot\int_{t'}^{t}\mathbf{v}(s)ds. \tag{6.7}$$

Substituting this into Eqs. (3.12) and (3.13), one readily finds that the contributions to the frictional forces from the first term of the right hand side of Eq.(6.7) vanish. By exchanging the order of time integrations, the force-balance equation (along the field direction) may be written as (Ting and Nee, 1986)

$$\frac{dv(t)}{dt} = \frac{e}{m}E(t) + \int_{-\infty}^{t} K(t-s)v(s)ds, \tag{6.8}$$

with the kernel given by

$$K(t-s) = \frac{1}{nm}\sum_{\mathbf{q}} q_x^2 \int_{-\infty}^{s} dt' \left[n_I|u(\mathbf{q})|^2\Pi(\mathbf{q},t-t') + \sum_{\lambda}|M(\mathbf{q},\lambda)|^2\Lambda(\mathbf{q},\lambda,t-t')\right]. \tag{6.9}$$

Note that in addition to the explicit t-dependence shown in Eq.(6.9), $K(t-s)$ depends implicitly on t through the t-dependence of the electron temperature $T_e(t)$ in the initial density matrix $\hat{\rho}_0$. However, this dependence of $T_e(t)$ is of higher order in $\mathbf{v}$, such that to lowest order in $\mathbf{v}$ we can set $T_e = T$ and neglect this implicit t-dependence of $K(t-s)$. Assuming that the applied electric field is suddenly turned on at $t = 0:\ E(t) = E\theta(t)$, and the c.m. velocity $v(t) = 0$ at $t \le 0$, we can take Laplace transformation of Eq.(6.8), yielding

$$v(\varepsilon) = \frac{eE}{m\varepsilon\left[\varepsilon - K(\varepsilon)\right]} \tag{6.10}$$

The time-dependent c.m. velocity may then be obtained by inverting the Laplace transform of Eq.(6.10):

$$v(t) = -\frac{eE}{2\pi m}\int_{-\infty+ic}^{\infty+ic}\frac{e^{izt}dz}{z+M(z)} \tag{6.11}$$

where $K(\varepsilon)$ is the Laplace transform of the kernel $K(t)$ defined by Eq.(69) and $M(z) = -iK(-iz)$ is the memory function (Götze and Wölfle 1972, Ting, Ying and Quinn 1976b, 1977; Ting and Nee, 1986). Dividing $M(\omega+i\delta)$ (for real ω) into a real and an imaginary part

$$M(\omega+i\delta) = M_1(\omega) + iM_2(\omega) \tag{6.12}$$

and noting the relations $M_1(\omega) = -M_1(-\omega)$ and $M_2(\omega) = M_2(-\omega)$, we can rewrite Eq.(6.11) as (Ting and Nee, 1986)

$$v(t) = \frac{2eE}{\pi m}\int_0^{\infty}\frac{M_2(\omega)\sin(\omega t)d\omega}{\omega\left[(\omega+M_1(\omega))^2 + M_2(\omega)^2\right]}. \tag{6.13}$$

The memory function $M(\omega)$ consists of contributions from impurities and from phonons:

$$M(\omega) = M^i(\omega) + M^p(\omega) \tag{6.14}$$

with (Ting and Nee, 1986; Lei, Horing and Zhang, 1987)

$$M^{i}(\omega) = \frac{n_i}{nm\omega}\sum_{\mathbf{q}}|u(\mathbf{q})|^2 q_x^2\,[\Pi(\mathbf{q},0) - \Pi(\mathbf{q},\omega)] \tag{6.15}$$

$$M^{p}(\omega) = \frac{1}{nm\omega}\sum_{\mathbf{q},\lambda}|M(\mathbf{q},\lambda)|^2 q_x^2\,[\Lambda(\mathbf{q},\lambda,0) - \Lambda(\mathbf{q},\lambda,\omega)]\,, \tag{6.16}$$

where $\Pi(\mathbf{q},\omega)$ and $\Lambda(\mathbf{q},\lambda,\omega)$ are, respectively, the electron density-density correlation function and the electron-phonon correlation function. The imaginary part $\Lambda_2(\mathbf{q},\lambda,\omega)$ and the real part $\Lambda_1(\mathbf{q},\lambda,\omega)$ of the correlation function $\Lambda(\mathbf{q},\lambda,\omega)$ can be expressed as

$$\begin{aligned}\delta\Lambda_2(\mathbf{q},\lambda,\omega) &= \Pi_2(\mathbf{q},\omega-\Omega_{\mathbf{q}\lambda})\left[n(\frac{\Omega_{\mathbf{q}\lambda}}{T}) - n(\frac{\Omega_{\mathbf{q}\lambda}-\omega}{T})\right]\\ &\quad+ \Pi_2(\mathbf{q},\omega+\Omega_{\mathbf{q}\lambda})\left[n(\frac{\Omega_{\mathbf{q}\lambda}}{T}) - n(\frac{\Omega_{\mathbf{q}\lambda}+\omega}{T})\right],\end{aligned} \tag{6.17}$$

$$\begin{aligned}\Lambda_1(\mathbf{q},\lambda;\omega) &= [\Pi_1(\mathbf{q},\omega-\Omega_{\mathbf{q}\lambda}) + \Pi_1(\mathbf{q},\omega+\Omega_{\mathbf{q}\lambda})]\,n(\frac{\Omega_{\mathbf{q}\lambda}}{T})\\ &\quad- \frac{1}{\pi}\int_{-\infty}^{\infty} d\omega_1 \Pi_2(\mathbf{q},\omega_1) n(\frac{\omega_1}{T})(\frac{1}{\omega_1+\omega-\Omega_{\mathbf{q}\lambda}} + \frac{1}{\omega_1-\omega-\Omega_{\mathbf{q}\lambda}}).\end{aligned} \tag{6.18}$$

If we use low frequency values

$$M_1(0) = 0 \quad\text{and}\quad M_2(0) = \frac{1}{\tau_i} + \frac{1}{\tau_p} \equiv \frac{1}{\tau} \tag{6.19}$$

to replace $M_1(\omega)$ and $M_2(\omega)$ in Eq.(6.13), the classical Drude result emerges

$$v(t) = \frac{eE}{m}\tau(1 - e^{-t/\tau}) \tag{6.20}$$

as a simple relaxation time approximation.

Ting and Nee (1986) calculated the drift velocity response from (6.13) in two cases: (a) low-temperature degenerate case, impurity scattering only; (b) high-temperature Maxwell-Boltzmann case, nonpolar optic phonon scattering only. They found that the behavior of $v(t)$ appreciably differs from the results without memory only in case (a), provided that $\varepsilon_F\tau \lesssim 1$.

The approximation discussed in the preceding subsection is a nonlinear generalization of the approximation in which the low frequency evaluation of the memory function is used. It may be expected to be valid in discussing the transient response of a realistic semiconductor in ordinary conditions. Nevertheless, the validity of this approximation in highly nonlinear cases is still a question to be investigated.

7. HIGH-FREQUENCY STEADY STATE TRANSPORT

The time-dependent force-balance equation (3.10) and energy-balance equation (3.22) or (3.26) also provide a very convenient basis to investigate high frequency steady-state transport. From these equations Lei and Zhang (1986, 1987), Lei, Horing and Zhang (1986), and Horing, Lei and Cui (1986) derived the small signal linear high-frequency ac conductivity expressed in terms of the memory function $M(\omega)$ (Götze and Wölfle, 1972), along with explicit expressions for the memory functions due to impurity and phonon scatterings. In the presence of a strong dc bias the linear ac signal conductivity is more complicated because of possible oscillation of the electron temperature with ac signal (Das and Ferry, 1976). Its distinction from the conductivity related to noise was discussed by Hu and Ting (1987) and by Lei and Horing (1987b). A balance equation formulation to determine the high-frequency small signal conductivity with a strong dc bias, and an analysis of nonlinear high-frequency transport has recently be carried out by Cai et al. (1989).

7.1 High-Frequency Conductivity for Small Signal

We consider a small amplitude high-frequency signal electric field $2\mathbf{E}_1 \cos\omega t$ applied to the system together with a constant dc bias field $\mathbf{E}_0$:

$$\mathbf{E}(t) = \mathbf{E}_0 + 2\mathbf{E}_1 \cos\omega t = \mathbf{E}_0 + \mathbf{E}_1 e^{-i\omega t} + \mathbf{E}_1 e^{i\omega t}. \tag{7.1}$$

Assuming the effects of the associated high-frequency magnetic field to be negligible we can still describe the system by the Hamiltonian (2.1) with $\mathbf{E} = \mathbf{E}(t)$ given by Eq.(7.1). It is obvious that after a transient process the system will reach an oscillatory steady state in which the center of mass moves at a constant drift velocity $\mathbf{v}_d$ and oscillates with a small amplitude at the single driving frequency ω, if the high-frequency signal field is sufficiently small:

$$\mathbf{v}(t) = \mathbf{v}_d + \mathbf{v}_1 e^{-i\omega t} + \mathbf{v}_1^* e^{i\omega t}. \tag{7.2}$$

Substituting it into Eq.(3.18), and expanding to linear order in $\mathbf{v}_1$, we obtain

$$\begin{aligned} A(\mathbf{q}, t, t') &= \exp\left[i\mathbf{q}\cdot\int_{t'}^{t} \mathbf{v}(s)ds\right] \\ &= \exp\left\{i\mathbf{q}\cdot\mathbf{v}_d(t-t') - \left[\mathbf{q}\cdot\mathbf{v}_1 e^{-i\omega t}(1-e^{i\omega(t-t')})/\omega - c.c.\right]\right\} \\ &\simeq \exp\left[i\mathbf{q}\cdot\mathbf{v}_d(t-t')\right]\left\{1 - \left[\mathbf{q}\cdot\mathbf{v}_1 e^{-i\omega t}(1-e^{i\omega(t-t')})/\omega - c.c.\right]\right\}. \end{aligned} \tag{7.3}$$

In the ac steady state the electron temperature $T_e(t)$ will also perform a small amplitude harmonic oscillation of frequency ω about a constant value T_e:

$$T_e(t) = T_e + T_1 e^{-i\omega t} + T_1^* e^{i\omega t}. \tag{7.4}$$

Taken jointly, Eq.(7.3) for $A(\mathbf{q},t,t')$ and Eq.(7.4) for $T_e(t)$ provide the frictional force $\mathbf{f} = \mathbf{f}_i + \mathbf{f}_p$, and the energy transfer rate w, (Eqs. (3.12), (3.13) and (3.19)), in the following forms:

$$\mathbf{f} = \mathbf{f}_0 + \mathbf{f}_0^{(1)}(T_1 e^{-i\omega t} + T_1^* e^{i\omega t}) + (\mathbf{f}_1 e^{-i\omega t} + \mathbf{f}_1^* e^{i\omega t}), \tag{7.5}$$

$$w = w_0 + w_0^{(1)}(T_1 e^{-i\omega t} + T_1^* e^{i\omega t}) + (w_1 e^{-i\omega t} + w_1^* e^{i\omega t}). \tag{7.6}$$

Here $\mathbf{f}_0 = \mathbf{f}_0(v_d, T_e)$ and $w_0 = w_0(v_d, T_e)$ are given by the force- and energy-transfer rate expressions for the dc steady state, Eqs.(4.4)-(4.6), with drift velocity v_d and electron temperature T_e; $\mathbf{f}_0^{(1)} \equiv \partial \mathbf{f}_0(v_d, T_e)/\partial T_e$ and $w_0^{(1)} \equiv \partial w_0(v_d, T_e)/\partial T_e$; and

$$\mathbf{f}_1 = \mathbf{f}_1(v_d, T_e) = \{in_i \sum_{\mathbf{q}} (\mathbf{v}_1 \cdot \mathbf{q})\mathbf{q}|u(\mathbf{q})|^2 \left[\Pi(\mathbf{q},\omega_0) - \Pi(\mathbf{q},\omega_0+\omega)\right]/\omega$$

$$+ i \sum_{\mathbf{q},\lambda} (\mathbf{v}_1 \cdot \mathbf{q})\mathbf{q}|M(\mathbf{q},\lambda)|^2 \left[\Lambda(\mathbf{q},\lambda;\omega_0) - \Lambda(\mathbf{q},\lambda;\omega_0+\omega)\right]\}/\omega, \tag{7.7}$$

$$w_1 = w_1(v_d, T_e) = \sum_{\mathbf{q},\lambda} (\mathbf{v}_1 \cdot \mathbf{q})|M(\mathbf{q},\lambda)/|^2 \left[\Gamma(\mathbf{q},\lambda;\omega_0) - \Gamma(\mathbf{q},\lambda;\omega_0+\omega)\right]/\omega, \tag{7.8}$$

with $\omega_0 \equiv \mathbf{q} \cdot \mathbf{v}_d$.

The force- and the energy-balance equations (3.10) and (3.22) can then be written to linear order in the small amplitude ac quantities. For zero-th order we have exactly the same balance equations as those in the dc steady state:

$$ne\mathbf{E}_0 + \mathbf{f}_0 = 0 \tag{7.9}$$

$$\mathbf{v}_d \cdot \mathbf{f}_0 + w_0 = 0, \tag{7.10}$$

which determine the dc drift velocity v_d and the electron temperature T_e for a given bias field $\mathbf{E}_0$. The linear order ac balance equations are

$$-i\omega nm\mathbf{v}_1 = ne\mathbf{E}_1 + \mathbf{f}_0^{(1)}T_1 + \mathbf{f}_1 \tag{7.11}$$

and

$$i\omega C_e T_1 = \mathbf{v}_1 \cdot \mathbf{f}_0 + \mathbf{v}_d \cdot \mathbf{f}_1 + w_0^{(1)}T_1 + w_1, \tag{7.12}$$

from which $\mathbf{v}_1$ and T_1 can be determined for given ac signal field $\mathbf{E}_1$(Cai et al. 1989).

In the absence of a dc bias, $\mathbf{E}_0 = 0$, the zero order equations (7.9) and (7.10) require that $\mathbf{f}_0 = 0, w_0 = 0, \mathbf{v}_d = 0$ and $T_e = T$. Furthermore it can be easily seen from Eq.(7.8) that $w_1 = 0$ because the system is isotropic when $\omega_0 = 0$, and thus Eq.(7.12) yields $T_1 = 0$. Therefore when only a small-amplitude signal is applied, the electron temperature remains constant to linear order in the signal. Any change of the electron temperature must be of higher order. This reduces the linear order force-balance equation (7.11) to

$$-i\omega nm\mathbf{v}_1 = ne\mathbf{E}_1 + \mathbf{f}_1 \tag{7.13}$$

Moreover, since $\omega_0 = 0$, $\mathbf{f}_1$ is parallel to $\mathbf{v}_1$:

$$\mathbf{f}_1 = inm\mathbf{v}_1 M(\omega). \tag{7.14}$$

Eq.(7.13) then yields the complex dynamic resistivity in the form

$$\rho(\omega) = \frac{E_1}{nev_1} = -i\frac{m}{ne^2}\left[\omega + M(\omega)\right] \tag{7.15}$$

corresponding to the conductivity

$$\sigma(\omega) = i\frac{ne^2}{m}\frac{1}{\omega + M(\omega)}. \tag{7.16}$$

Here the memory function $M(\omega)$ consists of contributions from impurities $M^i(\omega)$ and from phonons $M^p(\omega)$. Their expressions have been given in Eqs.(6.15)-(6.18). The form of Eq.(7.16) for dynamic conductivity was first given in the memory function theory (Götze and Wölfle, 1972). Similar results for impurity scatterings were also obtained in the energy-loss method for transport (see Gerlach, 1986).

7.2 Plasmon Contribution to High-Frequency Conductivity in Quasi-2D Systems and Superlattices

The formula (7.15) for linear high-frequency resistivity $\rho(\omega)$ is also valid for a quasi-two-dimensional system, as well as for a superlattice (on a per-layer basis), if $\rho(\omega)$ is replaced by the complex high-frequency sheet resistivity $R(\omega)$ (per layer) and n by the carrier sheet density N_s (per layer):

$$R(\omega) = -i\frac{m}{N_s e^2}\left[\omega + M(\omega)\right] \tag{7.17}$$

The memory function $M(\omega) = M^i(\omega) + M^p(\omega)$, which includes all the information relating to small-signal high-frequency response at zero dc bias, depends on the system structure and scattering mechanisms. For remote and background charged impurity scatterings we have (Lei, Horing and Zhang, 1986)

$$M^i(\omega) = \frac{1}{mN_s\omega}\left(\frac{e^2}{2\epsilon_0\kappa}\right)^2\frac{d}{2\pi}\int_{-\pi/d}^{\pi/d} dq_z \sum_{\mathbf{q}_\parallel}\frac{q_x^2}{q_\parallel^2}\tilde{N}(\mathbf{q}_\parallel, q_z)g(q_z)$$
$$\left[\Pi(q_z, \mathbf{q}_\parallel, 0) - \Pi(q_z, \mathbf{q}_\parallel, \omega)\right], \tag{7.18}$$
$$M^p(\omega) = \frac{1}{mN_s\omega}\sum_{\mathbf{q}_\parallel, q_z, \lambda} q_x^2 |M(\mathbf{q}, \lambda)|^2 |I(iq_z)|^2 \left[\Lambda(q_z, \mathbf{q}_\parallel, \lambda; 0) - \Lambda(q_z, \mathbf{q}_\parallel, \lambda; \omega)\right], \tag{7.19}$$

where $\Lambda(q_z, q_\parallel, \lambda; \omega)$ can be expressed in terms of $\Pi(q_z, \mathbf{q}_\parallel, \omega)$ by exactly the same expressions as those given in Eqs.(6.17) and (6.18). The density-density correlation function $\Pi(q_z, \mathbf{q}_\parallel, \omega)$ includes the intralayer and interlayer electron-electron interactions of the superlattice structure. Within the framework of the random-phase-approximation $\Pi(q_z, \mathbf{q}_{\parallel}, \omega)$ is given by Eq. (5.54). The effective impurity density $\tilde{N}(q_{\parallel}, q_z)$, correlation factor $g(q_z)$ and form factor $I(iq_z)$ are also as given in Section 5.

Calculations of linear dynamical conductivity for a two-dimensional electron gas were carried out by several authors (Tzoar, Platzman and Simons, 1976; Ando, 1976; Ting, Ying and Quinn, 1976b; Ganguly and Ting, 1977). Dynamic conductivity effects due to Coulomb interactions among carriers, especially those of the plasmon modes in a 2D system, have long been recognized by some authors (Kennedy et al., 1975, Ting, Ganguly, and Lai, 1981). Previous estimates relating to this, however, have been very crude. The balance-equation formulation facilitates accurate evaluation of such dynamic effects of intercarrier Coulomb interactions. Based on Eqs.(7.17) and (7.18), Lei and Zhang (1986) carried out a full RPA calculation of the real and imaginary parts of the impurity-limited high-frequency conductivity in a quasi-two-dimensional electron gas at zero temperature. The calculated functions of the real part, $M_1^i(\omega)$, and the imaginary part, $M_2^i(\omega)$, are shown as solid curves in Fig.7.1 for a pure 2D electron gas of sheet density $N_s = 1.8 \times 10^{11} cm^{-2}$, assuming scattering only by remote impurities, which are located at a distance $s = 5\mathring{A}$ from the mobile electrons. The discrete points in the figure are the calculated values of $M_2(\omega)/M_2(0)$ obtained by neglecting the plasma contribution, which should correspond to single layer result of Tzoar and Zhang (1985). The plasma pole enhancement in $M_2(\omega)$ is significant.

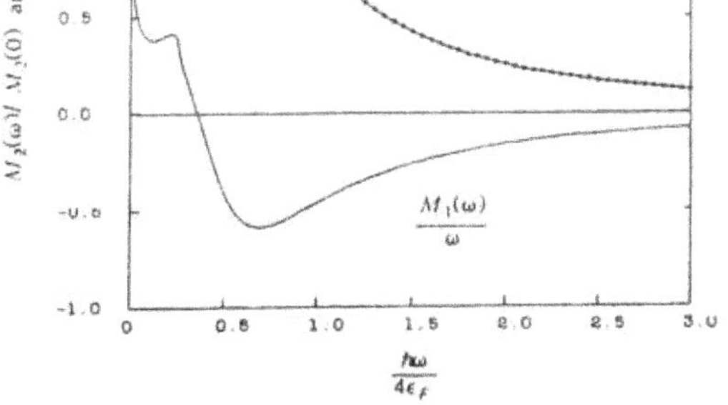

Fig.7.1 $M_2(\omega)/M_2(0)$ and $M_2(\omega)/\omega$ (in units of $mM_2(0)/(2\pi\hbar N_s)$) are shown as functions of $\hbar\omega/4\varepsilon_F$ for a GaAs quantum well with well width $a = 100\mathring{A}$ and electron density $N = 1.8 \times 10^{11}$ cm^{-1}. The impurities are located at $s = 25\mathring{A}$ from the interfaces. The discrete points are the $M_2(\omega)/M_2(0)$ calculated by neglecting the plasma-pole contribution. (From Lei and Zhang, 1986)

For a superlattice, the contribution to high-frequency conductivity by electron collective modes is even more distinctive. Lei, Horing and Zhang (1986a) calculated $M_1(\omega)$ and $M_2(\omega)$ due to remote and background impurity scatterings for several different geometrical parameters of the GaAs/AlGaAs superlattice at zero temperature. An example of their calculated results is shown in Fig.7.2.

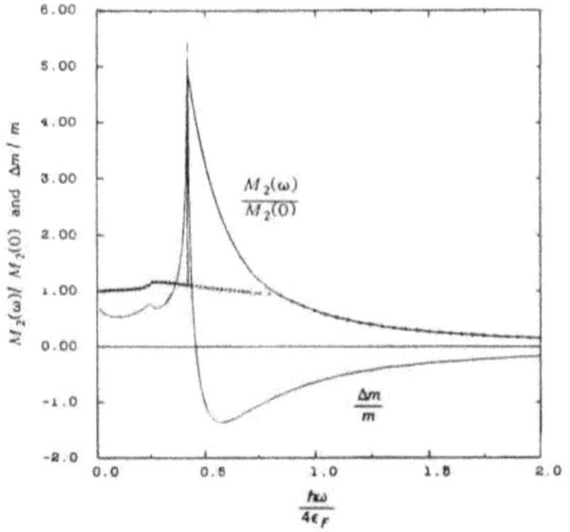

Fig.7.2 Calculated $M_2(\omega)/M_2(0)$ and $\Delta m/m$ (in units of $mM_2(0)/(2\pi\hbar N_s)$) are shown as functions of $\hbar\omega/4\varepsilon_F$ for a quantum well superlattice with $a = 100\text{Å}$, $d = 200\text{Å}$ and $N_s = 2.3 \times 10^{11}$ cm^{-2}. The scatterings are due to remote impurities and background impurities. The former are located at a distance $s = 75\text{Å}$ from the center of each quantum well. The discrete points are $M_2(\omega)/M_2(0)$ calculated by neglecting the plasma-pole contribution. (From Lei, Horing and Zhang, 1986a)

For comparison, the values of $M_2(\omega)/M_2(0)$. calculated by neglecting the plasma pole contribution are shown as discrete points in this figure. The enhancement due to collective modes is very pronounced, resulting in the most striking feature of the frequency dependence of the conductivity: the sharp rise of $M_2(\omega)$ at the bulk superlattice plasmon frequency $\omega_p = (e^2N_s/\epsilon_0\kappa md)^{1/2}$. This sharp rise is characteristic of a closely packed infinitely-repeated-layer system, in which the electrons in one layer are scattered by charged impurities located in all different layers, giving rise to a $(qd)^{-1}$ divergence of $\tilde{N}(\mathbf{q}_{\|}, 0)$ at small q. This jump arises from the long-wavelength ($q \sim 0$ and $q_z = 0$) plasmon contribution at $\omega \cong \omega_p$ and is easily obtained from the behavior of $\tilde{N}(\mathbf{q}_{\|}, 0)$ and $\Pi_2(0, \mathbf{q}_{\|}, \omega)$ at small q as

$$\Delta M_2(\omega_p) = \frac{1}{12\pi}\left(\frac{e^2}{\epsilon_0\kappa}\right)^{3/2}\frac{m^{1/2}}{(N_sd)^{3/2}}(N_rZ_r^2 + N_bZ_b^2). \tag{7.20}$$

It is worth noting that the assumption that the impurity distribution is random within each layer, and has no correlation from one layer to another is relevant to the sharp edge in the plasma resonance of the conductivity for a superlattice. While some such correlation may round off the sharp edge of resonance, the plasmon contribution will remain distinctive (Lei, Horing and Zhang, 1986a)

7.3 Nonlinear High-Frequency Conductivity

Employing time-dependent force- and energy-balance equations equivalent to Eqs. (3.10), (3.22) or (3.26), Cai et al. (1989) carried out a systematic study of steady state high-frequency nonlinear transport. They assume that, in the high-frequency steady state, the applied electric field $\mathbf{E}(t)$, the c.m. velocity $\mathbf{v}(t)$ and the electron temperature $T_e(t)$ can all be written as a sum of a dc part and several harmonic components:

$$\mathbf{E}(t) = \mathbf{E}_0 + \sum_n \mathbf{E}_n e^{-in\omega t} \tag{7.21}$$

$$\mathbf{v}(t) = \mathbf{v}_d + \sum_n \mathbf{v}_n e^{-in\omega t} \tag{7.22}$$

$$T_e(t) = T_e + \sum_n T_n e^{-in\omega t} \equiv T_e + \delta T \tag{7.23}$$

where $n = \pm 1, \pm 2, \cdots$. This form of $\mathbf{v}(t)$ yields

$$\exp\left[i\mathbf{q}\cdot\int_{t'}^{t}\mathbf{v}(s)ds\right] = \exp\left[i\mathbf{q}\cdot\mathbf{v}_d(t-t')\right]\exp\left\{\sum_n \frac{\mathbf{q}\cdot\mathbf{v}_n}{n\omega}e^{-in\omega t}\left[e^{in\omega(t-t')}-1\right]\right\}. \tag{7.24}$$

Expanding the second exponential factor of the right hand side of (7.24) as well as the function of $T_e(t)$ to a given order,

$$g(T_e(t)) = g(T_e) + (\frac{\partial g(T_e)}{\partial T_e})\delta T + \cdots\cdots, \tag{7.25}$$

they are able to express the frictional force $\mathbf{f}$ and the energy transfer rate w as the sum of harmonic components:

$$\mathbf{f} = \mathbf{f}_0 + \sum_n f_n e^{-in\omega t} \tag{7.26}$$

$$w = w_0 + \sum_n w_n e^{-in\omega t}. \tag{7.27}$$

With this, Cai et al. establish force- and the energy-balance equations for each harmonic component. Retaining terms up to the second harmonic component (n=2), they were able to develop a nonlinear high-frequency transport theory for the nondegenerate limit in particular. In principle one can retain harmonic components to any order and such a procedure is valid if the partial sums of the Taylor series of (7.24) and (7.25) can be made sufficiently accurate. However, to keep the approach tractable one has to truncate the expansions to the first few terms (Cai et al. cut off them at third order). This of course limits the maximum amplitude of the harmonic drift velocity v_m for validity of the procedure. Nevertheless, interesting nonlinear results have been obtained by Cai et al., and we refer the reader to their paper (Cai et al. 1989) for detail.

8. C.M. VELOCITY FLUCTUATIONS AND RELATED MATTERS

In Section 3, force and energy balance equations were derived by taking averages of Eqs.(2.20) and (2.26), and employing the drift velocity $v(t)$ in the exponential factor $A(\mathbf{q},t,t')$ of Eq.(3.18) (Lei and Ting, 1985b; Hu and O'Connell, 1987). This involved the implicit neglect of fluctuations of the force and other physical quantities, and thus only information relating to the drift motion of the center-of-mass is involved. To take account of the Brownian motion of the center of mass, we must retain the fluctuating quantities at least to their lowest order and establish Langevin-type equations. This can be achieved using the Heisenberg equation of motion (Hu and Ting, 1986; Xing and Ting, 1987; Lei and Horing, 1987b; Hu and O'Connell, 1987).

8.1 Langevin-Type Equation

Before we proceed, let us first return to the statistical average of a dynamical variable. Instead of using Eq.(2.26) of the Schrödinger picture, or Eq.(3.1) of the interaction picture, we can formulate the statistical average of a dynamic variable A in the Heisenberg picture:

$$\langle A\rangle = Tr\left\{\hat{\rho}_0\tilde{A}(t)\right\} \tag{8.1}$$

where $\tilde{A}(t)$ is the Heisenberg operator of the dynamic variable A, and $\hat{\rho}_0$ is the initial ($t=-\infty$) density matrix of the system. The Heisenberg operator $\tilde{A}(t)$ is given by

$$\tilde{A}(t) = U^{\dagger}(t,-\infty)A(t)U(t,-\infty), \tag{8.2}$$

where $A(t)$ is the interaction picture of the operator A, defined by Eq.(3.2), and

$$U(t,-\infty) \equiv T\exp\left[i\int_{-\infty}^{t} H_{It'}(t')dt'\right] \tag{8.3}$$

is the evolution operator in which T stands for the time-ordering operation.

Eqs.(2.20) in Section 2 can be regarded as the Heisenberg equation of motion in the relative electron space, if all the operators in the equation are considered to be Heisenberg operators. Thus we can write (per unit volume)

$$nm\frac{d}{dt}\mathbf{V}(t) = \dot{\mathbf{P}} = ne\mathbf{E}(t) + \tilde{\mathbf{F}}(t). \tag{8.4}$$

Expanding the evolution operator to lowest order in H_I (Ting and Nee, 1986; Hu and Ting, 1986; Lei and Horing, 1987b) we have

$$\tilde{\mathbf{F}}(t) = \mathbf{F}(t) - i\int_{-\infty}^{t}[\mathbf{F}(t), H_{It'}(t')]\,dt'. \tag{8.5}$$

Here $\mathbf{F}(t)$, the interaction picture representation of the force operator $\mathbf{F}$,

$$\mathbf{F}(t) = e^{i(H_e+H_{ph})t}\mathbf{F}e^{-i(H_e+H_{ph})t} \tag{8.6}$$

is a fluctuating(random) force, being of linear order in H_I, which has null contribution to the statistical average

$$\langle F(t) \rangle = 0. \tag{8.7}$$

The second term of the right hand side of Eq.(8.5), which is of order of H_I^2, includes a nonzero average and a fluctuating part. Being of higher order than H_I, this fluctuating part can be neglected in comparison with $\mathbf{F}(t)$. Thus, keeping both the average part and the fluctuating part to their leading orders, we can take the statistical average for the relative electron and phonon variables over the second term of the right-hand-side of Eq.(8.5), yielding

$$nm\frac{d}{dt}\mathbf{V}(t) = ne\mathbf{E} + \mathbf{F}(t) + \mathbf{f}. \tag{8.8}$$

Here $\mathbf{f} = \mathbf{f}_i + \mathbf{f}_p$ has the same expressions as Eqs.(3.12) and (3.13), but with the factor

$$A(\mathbf{q}, t, t') \equiv \exp\left[i\mathbf{q} \cdot (\mathbf{R}(t) - \mathbf{R}(t'))\right] = \exp\left[i\mathbf{q} \cdot \int_{t'}^{t} \mathbf{V}(s)ds\right] \tag{8.9}$$

in its original form Eq.(3.18) without elimination of the fluctuations by averaging. $\mathbf{V}(t)$ now consists of both the drift velocity $\mathbf{v}(t)$ and a small fluctuation $\delta\mathbf{V}(t)$.

$$\mathbf{V}(t) = \mathbf{v}(t) + \delta\mathbf{V}(t) \tag{8.10}$$

with

$$\langle \delta\mathbf{V}(t) \rangle = 0. \tag{8.11}$$

Expanding the exponential factor of the right-hand-side of Eq.(8.9) to linear order in δV(Ting and Nee, 1986), we have

$$\exp\left[i\mathbf{q} \cdot \int_{t'}^{t} \mathbf{V}(s)ds\right] \simeq \exp\left[i\mathbf{q} \cdot \int_{t'}^{t} \mathbf{v}(s)ds\right]\left[1 + i\mathbf{q} \cdot \int_{t'}^{t} \delta\mathbf{V}(s)ds\right]. \tag{8.12}$$

Substituting this into Eq.(8.8) we can write (Hu and Ting, 1986)

$$nm\frac{d}{dt}\left[\mathbf{v}(t) + \delta\mathbf{V}(t)\right] = ne\mathbf{E}(t) + \mathbf{f} + \mathbf{F}(t) + nm\int_{-\infty}^{t} \mathcal{K}(t-s) \cdot \delta\mathbf{V}(s)ds \tag{8.13}$$

$$\mathcal{K}(t-s) = \frac{1}{nm}\sum_{\mathbf{q}} \mathbf{q}\mathbf{q}\int_{-\infty}^{s} dt' \exp\left[i\mathbf{q} \cdot \int_{t'}^{t} \mathbf{v}(s)ds\right] \times$$
$$\left[n_i|u(\mathbf{q})|^2\Pi(\mathbf{q}, t-t') + \sum_{\lambda}|M(\mathbf{q},\lambda)|^2\Lambda(\mathbf{q},\lambda; t-t')\right]. \tag{8.14}$$

Taking the ensemble average over Eq.(8.13) we obtain exactly the same momentum balance equation as given in Eq.(3.10) of Section 3:

$$nm\frac{d}{dt}\mathbf{v}(t) = ne\mathbf{E}(t) + \mathbf{f}. \tag{8.15}$$

This equation, together with the energy balance equation obtained following the same procedure,

$$-C_e\frac{dT_e}{dt} = \mathbf{v}(t) \cdot \mathbf{f} + w, \tag{8.16}$$

determines $\mathbf{v}(t)$ anf $T_e(t)$.

It should be noted that noise is different in nature from a small alternating signal. An ac signal generally induces an additional change of the electron temperature in linear order in the case of a nonzero dc bias. However, since noise always consists of various components of the frequency spectrum with random phases, it does not influence the electron temperature in linear order (Lei and Horing 1987b). With this in view, the remaining part of Eq.(8.13) becomes a Langevin-type equation relating the random velocity $\delta\mathbf{V}(t)$ and the fluctuation force (Ting and Nee, 1986; Hu and Ting,1986; Lei and Horing, 1987b):

$$\frac{d}{dt}\delta\mathbf{V}(t) = \frac{1}{nm}\mathbf{F}(t) + \int_{-\infty}^{t} \mathcal{K}(t-s)\cdot\delta\mathbf{V}(s)ds. \tag{8.17}$$

8.2 Thermal Noise and Diffusion

Based on the Langevin equation (8.17) we can determine the thermal noise temperatures and diffusion coefficients of the system in the presence of a strong dc bias. In the following analysis we assume that the fluctuating force $\mathbf{F}(t)$ is either parallel to (longitudinal) or perpendicular to (transverse) the dc electric field $\mathbf{E}$ along the x-direction. For $\mathbf{F}(t)$ in an arbitrary direction we can resolve it into longitudinal and transverse components and calculate the longitudinal and transverse fluctuating velocities respectively. In either the longitudinal or transverse configuration the random velocity $\delta\mathbf{V}$ is along the same direction as $\mathbf{F}(t)$, and can be written as

$$\frac{d}{dt}\delta V = \frac{1}{nm}F(t) + \int_{-\infty}^{t} K(t-s)\delta V(s)ds, \tag{8.18}$$

with $K = \mathcal{K}_{xx}$ for longitudinal and $K = \mathcal{K}_{yy}$ for transverse configurations respectively. Eq.(8.18) is readily solved with Fourier transformation, leading to

$$\delta V(\omega) = \frac{1}{n^2}\sigma_n(\omega)F(\omega). \tag{8.19}$$

Here, $\delta V(\omega)$ and $F(\omega)$ are the Fourier transforms of the random velocity $\delta V(t)$ and the fluctuation force $F(t)$, and $\sigma_n(\omega)$ is a dynamic conductivity defined by

$$\sigma_n(\omega) = i\frac{ne^2}{m}\frac{1}{\omega + M(\omega)}, \tag{8.20}$$

which is different from the high-frequency small-signal conductivity discussed in Section 7.1 in the case of a nonzero dc bias. It may be identified as a noise conductivity (Lei and Horing, 1987b). The memory function $M(\omega) = M^i(\omega) + M^p(\omega)$ in Eq.(8.20) consists of an impurity part $M^i(\omega)$ and a phonon part $M^p(\omega)$, with (Hu and Ting, 1986; Lei and Horing, 1987b)

$$M^i(\omega) = \frac{n_i}{nm\omega}\sum_{\mathbf{q}} q_\alpha^2 |u(\mathbf{q})|^2 \left[\Pi(\mathbf{q},\omega_0) - \Pi(\mathbf{q},\omega_0+\omega)\right], \tag{8.21}$$

and

$$M^{p}(\omega)=\frac{1}{nm\omega}\sum_{\mathbf{q},\lambda}q_\alpha^2|M(\mathbf{q},\lambda)|^2\left[\Lambda(\mathbf{q},\lambda;\omega_0)-\Lambda(\mathbf{q},\lambda;\omega_0+\omega)\right] \tag{8.22}$$

in which $\alpha=x$ or y for the longitudinal or transverse configuration. The electron density correlation function $\Pi(\mathbf{q},\omega)$ and the electron-phonon correlation function $\Lambda(\mathbf{q},\lambda;\omega)$ are as given by Eqs.(4.16), (6.17) and (6.18).

Defining a generalized diffusion coefficient

$$D(\omega)=\frac{N}{4}\int_{-\infty}^{\infty}\langle\delta V(t)\delta V(0)+\delta V(0)\delta V(t)\rangle e^{i\omega t}dt \tag{8.23}$$

($N=nV$, is the total number of carriers in the real volume V of the system), and noting that $\delta V(t)=N^{-1}\sum_i\delta v_i(t)$, $\delta v_i(t)$ being the random velocity of the ith particle, we have

$$D(\omega)=\frac{1}{4N}\sum_{i,j}\int_{-\infty}^{\infty}\langle\delta v_i(t)\delta v_j(0)+\delta v_i(0)\delta v_j(t)\rangle e^{i\omega t}dt. \tag{8.24}$$

If velocities of different particles are uncorrelated, $\langle\delta v_i(t)\delta v_j(0)\rangle=\delta_{ij}\langle\delta v_i(t)\delta v_i(0)\rangle$, then $D(\omega)$ reduces to the noise diffusion coefficient as usually defined for independent particles (Nougier, 1980). However, the definition (8.23) is more generally valid for the interacting particle system under consideration here. Employing Eq.(8.19) one can verify that (Hu and Ting, 1986; Lei and Horing, 1987b)

$$D(\omega)=\frac{1}{2n^2e^4}|\sigma_n(\omega)|^2S(\omega) \tag{8.25}$$

where $S(\omega)$ is the Fourier transform of the symmetrized fluctuating-force correlation function:

$$S(\omega)=\frac{V}{2n}\int_{-\infty}^{\infty}\langle F(t)F(0)+F(0)F(t)\rangle e^{i\omega t}dt. \tag{8.26}$$

On the other hand, this generalized diffusion coefficient is proportional to the power spectrum of the current-density fluctuation, and thus we can define the thermal noise temperature $T_n(\omega)$ by a generalized Einstein relation connecting $D(\omega)$ and the real part of the noise conductivity $\sigma_n(\omega)$:

$$D(\omega)=\frac{T_n(\omega)}{ne^2}\mathrm{Re}\left[\sigma_n(\omega)\right]. \tag{8.27}$$

This noise temperature has the physical meaning that the maximum available noise power in frequency range $\omega\rightarrow\omega+d\omega$ for a passive two-terminal network, made of material with noise temperature $T_n(\omega)$, is $T_n(\omega)d\omega$. (Hu and Ting, 1987a; Lei and Horing, 1987b). Combining Eqs.(8.25), (8.27) and using $\sigma_n(\omega)$ given by Eq.(8.20) we obtain a simple expression for $T_n(\omega)$:

$$T_n(\omega)=\frac{S(\omega)}{2mM_2(\omega)}. \tag{8.28}$$

The fluctuating-force correlation function $S(\omega)$, defined in Eq. (8.26), is also composed of an impurity contribution $S^i(\omega)$ and a phonon contribution $S^p(\omega)$:

$$S(\omega) = S^i(\omega) + S^p(\omega), \tag{8.29}$$

which may be determined as (Hu and Ting, 1986; Lei and Horing, 1987b)

$$S^i(\omega) = -\frac{n_i}{n}\sum_{\mathbf{q}} q_\alpha^2|u(\mathbf{q})|^2\Pi_2(\mathbf{q},\omega_0+\omega)\left[2n\left(\frac{\omega_0+\omega}{T_e}\right)+1\right], \tag{8.30}$$

$$\begin{aligned}S^p(\omega) = &-\frac{\omega}{2n}\sum_{\mathbf{q},\lambda} q_\alpha^2|M(\mathbf{q},\lambda)|^2 \\ &\Big\{\Pi_2(\mathbf{q},\Omega_{\mathbf{q}\lambda}+\omega_0+\omega)\left[\left[2n\left(\frac{\Omega_{\mathbf{q}\lambda}+\omega_0+\omega}{T_e}\right)+1\right]\left[2n\left(\frac{\Omega_{\mathbf{q}\lambda}}{T}\right)+1\right]-1\right] \\ &+\Pi_2(\mathbf{q},\Omega_{\mathbf{q}\lambda}+\omega_0-\omega)\left[\left[2n\left(\frac{\Omega_{\mathbf{q}\lambda}+\omega_0-\omega}{T_e}\right)+1\right]\left[2n\left(\frac{\Omega_{\mathbf{q}\lambda}}{T}\right)+1\right]-1\right]\Big\},\end{aligned} \tag{8.31}$$

with $\omega_0 \equiv q_x v_d$, and $\alpha = x$ or y for longitudinal or transverse configuration.

In the case of zero dc bias, $\omega_0 \to 0$ and $T_e \to T$, it is easily seen that

$$T_{nl} = T_{nt} = \frac{\omega}{2}\coth(\frac{\omega}{2T}) \tag{8.32}$$

which is the Nyquist relation (Seeger, 1982). Moreover, for low frequency $\mathrm{Re}[\sigma(\omega)]/ne$ is just the linear mobility μ_0, such that Eq.(8.26) reduces to the conventional Einstein relation

$$D = T\frac{\mu_0}{e}. \tag{8.33}$$

In the case of nonzero dc bias, we find that Nyquist relation still holds for high frequency $\omega \gg qv_d$, but at the electron temperature:

$$T_{nl} = T_{nt} = \frac{\omega}{2}\coth(\frac{\omega}{2T_e}). \tag{8.34}$$

The results cited above assume that the phonons are in an equilibrium state with a unique temperature T. The effect of hot phonons (nonequilibrium phonons, see Section 10) can be incorporated by replacing the Bose functions $n(\Omega_{\mathbf{q}\lambda}/T)$ in the expressions for $\Lambda(\mathbf{q},\lambda;\omega)$ and $S^p(\omega)$ by $n_{\mathbf{q}\lambda}$ as given by

$$n_{\mathbf{q}\lambda} = n(\frac{\Omega_{\mathbf{q}\lambda}-\omega_0}{T_e}) + \left[n(\frac{\Omega_{\mathbf{q}\lambda}}{T}) - n(\frac{\Omega_{\mathbf{q}\lambda}-\omega_0}{T_e})\right]\left(1+\frac{\tau_p}{\tau}\right)^{-1} \tag{8.35}$$

with

$$1/\tau = -2|M(\mathbf{q},\lambda)|^2\Pi_2(\mathbf{q},\Omega_{\mathbf{q}\lambda}-\omega_0), \tag{8.36}$$

where τ_p is the phonon relaxation time (Lei and Horing, 1987b). All the results can easily be extended to quasi-2D systems and superlattices, and we refer the reader to the original paper of Lei and Horing (1987b) for further detail.

Hu and Ting (1986) provide a detailed discussion of the low frequency ($\omega \sim 0$) noise temperatures and diffusion coefficients in degenerate electron systems for the zero dc field limit, the warm electron region, and the zero lattice temperature limit assuming impurity and acoustic phonon scatterings. They also treat nondegenerate systems, assuming nonpolar optic phonon scattering.

More careful calculations were carried out for GaAs/AlGaAs heterosystems by Lei and Horing (1987b), incorporating the Coulomb interaction between carriers by using the full temperature-, wavevector-, and frequency-dependent density-density correlation function in random-phase-approximation. In this, scatterings due to impurities, acoustic phonons (assuming equilibrium occupations) and polar optic phonons (for both equilibrium and nonequilibrium occupations) are taken into account.

In Fig.8.1 we present the longitudinal and transverse noise temperatures $T_{nl}(\omega)$ and $T_{nt}(\omega)$ for an AlGaAs/GaAs/AlGaAs quantum well of width $a = 250\mathring{A}$, zero temperature Ohmic mobility $\mu_0 = 1.0 \times 10^6\ cm^2/Vs$, and carrier sheet density $N_s = 2.0 \times 10^{11}\ cm^{-2}$, which were calculated as functions of ω/Ω_o, at lattice temperature $T = 77K$. The curve labeled 1 represents the Nyquist relation Eq.(8.32) at $T = 77K$. The behavior of the noise temperature is essentially determined by the electron temperature. Generally, at lower dc bias, $T_{nl}(\omega)$ is lower than $T_{nt}(\omega)$, whereas at higher dc bias $T_{nl}(\omega)$ is higher than $T_{nt}(\omega)$. Since the optic-phonon frequency is large ($\Omega_o \sim 5.4 \times 10^{13}\ s^{-1}$), ω/Ω_o is always small even for the millimeter microwave range.

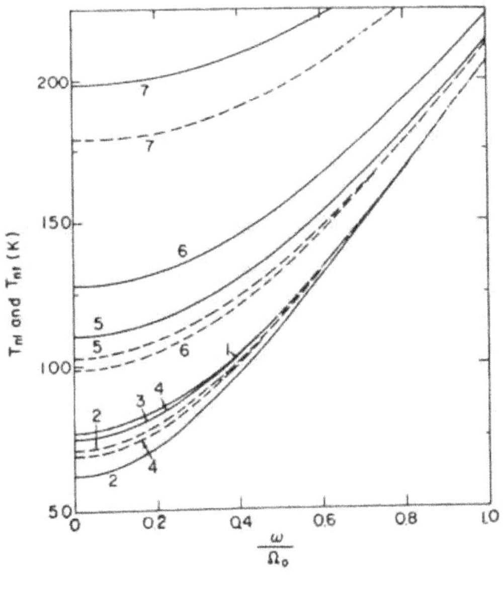

Fig.8.1 Longitudinal (solid curves) and transverse (dashed curves) noise temperatures, $T_{nl}(\omega)$ and $T_{nt}(\omega)$, calculated using Eq.(8.28), are shown as functions of the normalized frequency ω/Ω_0 in a GaAs quantum well at lattice temperature $T = 77$ K for different dc bias points (τ_p is the optic-phonon relaxation time). 1: $E \sim 0$, i.e., Nyquist relation Eq.(8.32). 2: $E = 100$ V/cm, $\tau_p = 0$. 3: $E = 100$ V/cm, $\tau_p = 3.5$ ps; T_{nl} and T_{nt} are indistinguishable in this curve. 4: $E = 400$ V/cm, $\tau_p = 0$. 5: $E = 400$ V/cm, $\tau_p = 3.5$ ps. 6: $E = 1$ kV/cm, $\tau_p = 0$. 7: $E = 100$ V/cm, $\tau_p = 3.5$ ps.

In Fig.8.2 we plot the longitudinal and transverse noise temperatures T_{nl} and T_{nt} at $\omega \sim 0$ as functions of the bias dc field E for the same system as described above at $T = 77K$ for $\tau_p = 0$ and 3.5 ps, together with the curve of the electron temperature T_e. Although (depending on the bias field) both T_{nl} and T_{nt} may be higher or lower than T_e, their main trends are consonant with that of the electron temperature T_e.

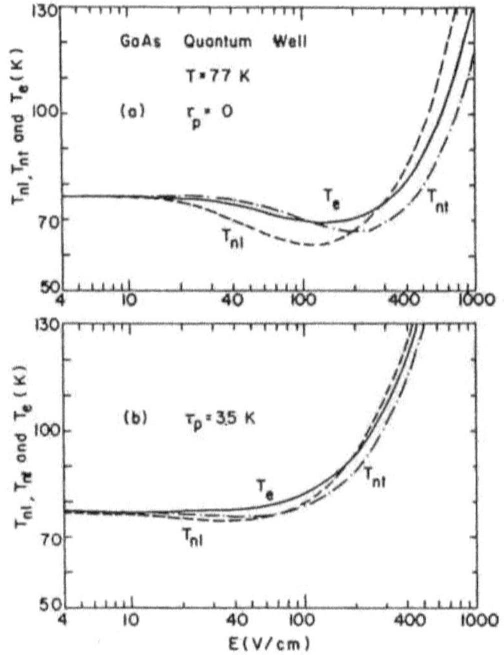

Fig.8.2 The electron temperature T_e and the longitudinal and transverse noise temperatures T_{nl} and T_{nt} at $\omega \sim 0$ are shown as functions of bias dc field for the same GaAs single-layer quantum well at lattice temperature $T = 77$ K for optic-phonon relaxation times (a) $\tau_p = 0$ and $\tau_p = 3.5$ ps. (From Lei and Horing, 1987b)

An interesting feature is that for the case of $\tau_p = 0$, both $T_{nl}(\omega)$ and $T_{nt}(\omega)$ can be lower than the lattice temperature $T = 77K$ over wide ranges of dc bias and frequency. This is the electron cooling phenomenon (Lei and Ting, 1985b,c). Such electron cooling here in GaAs heterosystems is due to polar optic-phonon scattering. Note that the electron temperature T_e in balance equation transport theory is not a directly measurable quantity, but the noise temperatures T_{nl} and T_{nt} are directly measurable. Therefore, measurement of the noise temperature not only has its own merit but is also of interest in the manifestation of electron cooling and hot phonon effects.

Thermal noise of electrons in semiconductors in the presence of crossed magnetic and electric fields has also been discussed by Hu and Ting (1987b) on the basis of the balance equation approach. We refer the reader to their original paper for details.

8.3 Role of Velocity Fluctuations in Drift Motion

The expansion of the exponential factor to linear order of the velocity fluctuation (in the preceding section) gives the appearance that δV does not affect the drift motion. This may not be true, as can be seen by continuing the expansion to higher order of δV. In fact, in deriving the force- and energy-balance equations, we may take the average over the whole exponential factor without expanding it, to obtain the following result from the cumulant approximation:

$$\langle \exp\left[i\mathbf{q}\cdot\int_{t'}^{t}(\mathbf{v}(s)+\delta\mathbf{V}(s))\,ds\right]\rangle \simeq \exp\left[i\mathbf{q}\cdot\int_{t'}^{t}\mathbf{v}(s)ds-\frac{q^2}{6}\int_{t'}^{t}du\int_{t'}^{t}ds\langle\delta\mathbf{V}(u)\cdot\delta\mathbf{V}(s)\rangle\right] \tag{8.37}$$

Thus, the correlation of velocity fluctuations enters the frictional force $\mathbf{f}$ and the energy transfer rate w. Furthermore, in view of $\langle(\mathbf{v}+\delta\mathbf{V})^2\rangle = v^2+\langle\delta V^2\rangle$ there may appear an extra term

$$nm\frac{d}{dt}\langle\delta V(t)^2\rangle \tag{8.38}$$

in the energy-balance equation.

In this way velocity fluctuations affect the drift motion, through the structure of the velocity fluctuation autocorrelation function:

$$\gamma(u-s)\equiv\langle\delta\mathbf{V}(u)\cdot\delta\mathbf{V}(s)\rangle. \tag{8.39}$$

Assuming τ_v to be the characteristic correlation time of this correlation function, such that

$$\gamma(u-s)\simeq\begin{cases}\gamma(0) & \text{if } \ |u-s|\ll\tau_v\\[2ex] 0 & \text{if } \ |u-s|\gg\tau_v,\end{cases} \tag{8.40}$$

we have for $|t-t'|\ll\tau_v$

$$\begin{aligned}B &\equiv \frac{q^2}{6}\int_{t'}^{t}du\int_{t'}^{t}ds\langle\delta\mathbf{V}(u)\cdot\delta\mathbf{V}(s)\rangle\\ &\simeq\begin{cases}\frac{1}{6}q^2\gamma(0)(t-t')^2 & \text{for } \ |t-t'|\ll\tau_v \qquad (8.41)\\[2ex] \frac{1}{3N}q^2D_n(0)|t-t'| & \text{for } \ |t-t'|\gg\tau_v, \qquad (8.42)\end{cases}\end{aligned}$$

where

$$D_n(0)=\frac{N}{2}\int_{-\infty}^{\infty}\gamma(\tau)d\tau \tag{8.43}$$

is the generalized diffusion coefficient defined by the autocorrelation of c.m. velocity fluctuations (Lei and Horing, 1987b).

Two different approximations have been used in examining this. Xing and Ting (1987) estimated the autocorrelation function of velocity fluctuations using the high-temperature result obtained from the Langevin equation with zero dc bias (Ting and Nee, 1986)

$$\langle \delta \mathbf{V}(u) \cdot \delta \mathbf{V}(s) \rangle = \frac{3T_e}{Nm} \exp(-\frac{u-s}{\tau}), \tag{8.44}$$

which involves a characteristic correlation time $\tau_v = \tau = -1/M_2(0)$. They assumed that τ is much larger than the correlation time τ_c of the density-density correlation functions $\Pi(\mathbf{q}, t-t')$ and $\Lambda(\mathbf{q}, \lambda, t-t')$ and used the approximation (8.41) for B:

$$B = \frac{q^2 T_e}{2Nm}(t-t')^2. \tag{8.45}$$

Here N is the total number of particles. Since the effective $t-t'$ range of integration is limited by τ_c, we can use $t-t' = \tau_c = 10^{-13}$s in estimating B. Assuming $T_e = 3\times 10^3 K, q = 1\times 10^{10} m^{-1}$ and $m = 9\times 10^{-31} kg$, we find that only when the total number of particles N is as low as 10^5 can the velocity fluctuation autocorrelations be appreciable.

Hu and O'Connell (1988a), on the other hand, used the approximation (8.42). This carries the implication that $\tau_c \gg \tau_v$, such that the major contributions to the frictional force and the energy transfer rate come from the range $t-t' \gg \tau_v$. Replacing B by $(D_n/2N)q^2(t-t')$ we obtain the same force- and energy-balance equations as those without velocity fluctuation, save for the replacement

$$\omega_0 \to \omega_0 + i\frac{D_n}{3N}q^2, \tag{8.46}$$

leading to an energy broadening. Its effects have been discussed extensively by Hu and O'Connell (1988a,e,f; 1989a,b) and we refer the reader to their original papers for further detail. For ordinary macroscopic systems in which the total number of particles is large, this velocity fluctuation autocorrelation effect is always negligible. For small systems, e.g. in microstructures, where N may be as low as 10^5, velocity fluctuation autocorrelations become appreciable.

It is worth noting that, although choosing Eq.(3.22) or Eq.(3.27) as the energy balance equation does not make any difference up to the linear order in δV, it may have some difference in the terms of order of δV^2 This problem merits further investigation.

9. SEVERAL SPECIES OF CARRIERS

Although the balance equation theory discussed thus far is for one species of carrier having a parabolic band, it is easy to extend it to the case of several species of carriers, such as electron-hole plasmas and type-*II* superlattices (Lei, 1987a; Lei, Cui and Horing, 1988; Zheng, Cai and Lax, 1988), multivalley semiconductors (Lei et al., 1987; Liu et al., 1988; Xing, Liu and Ting, 1988; Xing et al. 1988; Liu, Xing and Ting, 1989), and other systems (Cai and Lei, 1990).

9.1 Type-*II* Superlattices, Electron-Hole Plasma

9.1.1 Balance equations for a type-*II* superlattice

A type-*II* superlattice is a material composed of alternating layers of differing semiconductor constituents, like a InAs-GaSb system in which the band discontinuities at the heterojunction interfaces are so large that the bottom of the conduction band of InAs lies below the top of the valence band of GaSb. There is a transfer of electrons from GaSb layers to the conduction band of InAs, producing a periodic structure with spatially separated electrons and holes in adjacent layers. Such a structure forms one of the simplest systems having two species of carriers: electrons and holes. The total numbers of electrons and of holes are fixed. The Coulomb interaction between them plays an important role in determining the transport properties of the system.

We consider a model system consisting of two sets of periodically arranged quantum wells of width a_1 and a_2 respectively, but with the same spatial period d. The central positions of the quantum wells in the first set are at $z = ld$ ($l = 0, \pm 1 \ldots$), and the carriers in them are electrons with charge $-e$, effective mass m_e and sheet density N_e per layer. The central positions of the quantum wells of the second set are at $z = ld + b$ ($l = 0, \pm 1 \ldots$), and the carriers in them are holes with charge e, effective mass m_h and per-layer density N_h. For simplicity we assume that the electrons and holes populate only their respective first subbands and that the wavefunctions of different wells do not overlap, irrespective of type. The lowest subband wave function for electrons in lth well is given by (5.36) and that for holes is

$$\psi^{(2)}_{l\mathbf{k}_\parallel}(\mathbf{r}, z) = e^{i\mathbf{k}_\parallel \cdot \mathbf{r}} \zeta(z - ld - b), \qquad (l = 0, \pm \cdots). \tag{9.1}$$

The electrons and holes are subject to scatterings by impurities and by phonons just as in their respective type-*I* superlattices. In order to investigate transport parallel to the layers we introduce 2D center-of -mass coordinates $\mathbf{R}_1 = (X_1, Y_1)$ and $\mathbf{R}_2 = (X_2, Y_2)$ for electrons and holes respectively. Under the influence of a uniform electric field $\mathbf{E}$ applied parallel to the plane, the Hamiltonian can be written as

$$H = H_{ph} + H_e + H_h + H_{12}. \tag{9.2}$$

H_{ph}, H_1 and H_2 are the Hamiltonians of the phonon system, and of the electron- and hole-sublattices, having the form as those given in Section 5, and H_{12} represents the Coulomb interaction between electrons and holes

$$H_{12} = \sum_{\mathbf{q}_\parallel} e^{i\mathbf{q}_\parallel\cdot(\mathbf{R}_1-\mathbf{R}_2)} \sum_{l,m} V_{lm}^{12}(\mathbf{q}_\parallel)\rho^{(1)}_{l\mathbf{q}_\parallel}\rho^{(2)}_{m,-\mathbf{q}_\parallel}. \tag{9.3}$$

In this, $\rho^{(1)}_{l\mathbf{q}}$ and $\rho^{(2)}_{m,-\mathbf{q}_\parallel}$ are density operators for electrons in lth well and holes in mth well, and

$$V_{lm}(q_\parallel) = -\frac{e^2}{2\epsilon_0\kappa q}\int\int dz dz' e^{-q_\parallel|z-z'|}\zeta_1(z-ld)^2\zeta_2(z-md-b)^2. \tag{9.4}$$

Taking the initial density matrix to be

$$\hat{\rho}|_{t=-\infty} = \hat{\rho}_0 = \frac{1}{Z}e^{-H_{ph}/T-H_e/T_e-H_h/T_h} \tag{9.5}$$

where T, T_e and T_h are, respectively, the lattice, electron and hole temperatures, and treating H_{12} as a perturbation along with the carrier-impurity and carrier-phonon interactions, we obtain the force and energy balance equations for nonlinear transport as follows(Lei, 1987a; Lei, Cui and Horing, 1988)

$$N_e m_e\frac{dv_1}{dt} = -N_e eE + f_1(v_1) + f_{12}(v_1-v_2) \tag{9.6}$$

$$N_h m_h\frac{dv_2}{dt} = N_h eE + f_2(v_2) + f_{21}(v_2-v_1) \tag{9.7}$$

$$-\frac{dE_e}{dt} = v_1 f_1(v_1) + w_1(v_1) + w_{12}(v_1-v_2) \tag{9.8}$$

$$-\frac{dE_h}{dt} = v_2 f_2(v_2) + w_2(v_2) + w_{21}(v_2-v_1), \tag{9.9}$$

assuming that electric field $\mathbf{E} = (E,0,0)$ is along x-direction. In these equations, E_e and E_h are the internal energies (per layer) for the relative-electron and hole systems, and v_1 and v_2 are, respectively, the average velocities of the centers of mass, i.e. the average drift velocities of the electron and hole systems. The total current is due to both types of carriers and can be expressed as

$$J = -N_e ev_1 + N_h ev_2. \tag{9.10}$$

In the above equations $f_j(v_j)$ are the forces (per layer) exerted on the electron c.m. ($j=1$) and on the hole c.m. ($j=2$), due to scatterings by impurities and phonons, and $w_j(v_j)$ are the (per layer) energy transfer rates for the electron ($j=1$) and for hole ($j=2$) systems due to carrier-phonon interactions. Their expressions are exactly the same as in type-I superattices (Section 5). Finally, $f_{ij}(v_i - v_j)$ and $w_{ij}(v_i - v_j)$ $(j = 1, 2)$ are forces and energy transfer rates due to electron-hole interactions. We note that

$$f_{21}(v_2 - v_1) = -f_{12}(v_1 - v_2) \tag{9.11}$$

$$w_{21}(v_2 - v_1) = (v_1 - v_2)f_{12}(v_1 - v_2) - w_{12}(v_1 - v_2) \tag{9.12}$$

and the expressions for $f_{12}(v_1 - v_2)$ and $w_{12}(v_1 - v_2)$ are as follows:

$$\begin{aligned} f_{12}(v_1 - v_2) = &- \left(\frac{e^2}{2\epsilon_0\kappa}\right)^2 \frac{d}{\pi^2} \int_0^{\pi/d} dq_z \sum_{\mathbf{q}_\parallel} \frac{q_x}{q_\parallel^2} \int_{-\infty}^{\infty} d\omega \left[n\left(\frac{\omega}{T_e}\right) - n\left(\frac{\omega - \omega_{12}}{T_h}\right)\right] \\ &\times I_1(q_\parallel)^2 I_2(q_\parallel)^2 S^{12}(\mathbf{q}_\parallel, q_z) \Pi_2^{(1)}(q_z, \mathbf{q}_\parallel, \omega) \Pi_2^{(2)}(q_z, -\mathbf{q}_\parallel, \omega_{12} - \omega) \end{aligned} \tag{9.13}$$

and

$$\begin{aligned} w_{12}(v_1 - v_2) = &- \left(\frac{e^2}{2\epsilon_0\kappa}\right)^2 \frac{d}{\pi^2} \int_0^{\pi/d} dq_z \sum_{\mathbf{q}_\parallel} \frac{1}{q_\parallel^2} \int_{-\infty}^{\infty} d\omega\, \omega \left[n\left(\frac{\omega}{T_e}\right) - n\left(\frac{\omega - \omega_{12}}{T_h}\right)\right] \\ &\times I_1(q_\parallel)^2 I_2(q_\parallel)^2 S^{12}(\mathbf{q}_\parallel, q_z) \Pi_2^{(1)}(q_z, \mathbf{q}_\parallel, \omega) \Pi_2^{(2)}(q_z, -\mathbf{q}_\parallel, \omega_{12} - \omega), \end{aligned} \tag{9.14}$$

in which the form factors $I_j(q_\parallel)$ $(j = 1, 2)$ are given by Eq.(5.41); the density-density correlation functions $\Pi_2^{(j)}(q_z, \mathbf{q}_\parallel, \omega)$ are given by (5.54) for the electron $(j = 1)$ and hole $(j = 2)$ subsuperlattices, respectively, and finally

$$\begin{aligned} S^{12}(\mathbf{q}_\parallel, q_z) = &\frac{\cosh(q_\parallel d)\cos(q_z d) - 1}{\left[1 - \cosh(q_\parallel d)\cos(q_z d)\right]^2 + \left[\sinh(q_\parallel d)\sin(q_z d)\right]^2} \\ &+ \frac{\cosh\left[q_\parallel(d - 2b + a_1 - a_2)\right]}{\cosh(q_\parallel d) - \cos(q_z d)} \end{aligned} \tag{9.15}$$

and

$$\omega_{12} \equiv q_x(v_1 - v_2). \tag{9.16}$$

9.1.2 Steady-state nonlinear dc transport

The nonlinear resistivity function (per layer) can be defined as $R(v_1, v_2) = E/J$, which, according to Eqs.(9.6) and (9.10), has the following expression for steady state (Lei, Cui and Horing, 1988)

$$R(v_1, v_2) = \frac{f_1(v_1) + f_{12}(v_1 - v_2)}{N_e^2 e^2 (N_h v_2/N_e - v_1)}. \tag{9.17}$$

In the weak field limit

$$\begin{aligned} R_0 &\equiv R(v_1 \to 0, v_2 \to 0) \\ &= -e^{-2} \frac{f_1'(0) f_2'(0) + f_{12}'(0)\left[f_1'(0) + f_2'(0)\right]}{N_h^2 f_1'(0) + N_e^2 f_2'(0) + (N_e - N_h)^2 f_{12}'(0)} \end{aligned} \tag{9.18}$$

and for an intrinsic system with $N_e = N_h = N_s$, this reduces to

$$R_0 = R_{e-h} + R_e R_h/(R_e + R_h). \tag{9.19}$$

Here $R_e = -f_1'(0)/N_s^2e^2$ and $R_h = -f_2'(0)/N_s^2e^2$ are individual resistivities associated with electrons and with holes, and $R_{e-h} = -f_{12}'(0)/N_s^2e^2$ arises from the electron-hole interaction. Eq.(9.19) illustrates that the Ohmic resistivity of a type-II superlattice with equal concentrations of electrons and holes can be thought of in terms of two parallel resistance systems (specifically, electron and hole type-I superlattices), taken jointly with a series-connected resistance due to coupling between electrons and holes.

Numerical solutions for nonlinear steady state dc transport were carried out from balance equations (9.6)-(9.9) for InAs-GaSb type-II superlattices (Lei, Cui and Horing, 1988). Similar calculations were separately performed by Zheng, Cai and Lax (1988) for an electron-hole plasma in a semiconductor quantum well, a special case of the type-II superlattice with $b = 0$. We refer the reader to their original paper for details. In an electric field holes generally drift more slowly and tend to stay cooler than electrons since $m_h > m_e$, and hole-phonon coupling is more effective than electron-phonon coupling. Usually, the electron-hole interaction makes a very small contribution to the total dc resistivity. The reason for this is the spatial separation of the electrons and holes in the case of a type-II superlattice. On the other hand, when electrons and holes are confined in the same layer, the attraction between them can be sufficiently strong to cause the minority carrier to drift in the wrong direction relative to the applied electric field (Höpfel et al., 1986; Cui, Lei and Horing, 1988; Cai et al., 1988). This will be discussed in more detail in the following subsection.

9.1.3 Negative minority-electron mobility in an electron-hole plasma

Höpfel et al. (1986) reported their interesting experimental observation of negative minority-carrier mobility in GaAs/AlGaAs quantum well structures and attributed it physically to electron-hole attraction. To analyze this phenomenon, Cui, Lei and Horing (1988b) employed the balance equation formulation, performing numerical calculation of carrier transport for a two-component plasma consisting of minority electrons and majority-holes ($N_e \ll N_h$) in a GaAs/AlGaAs quantum well. The coupled momentum and energy-balance equations (9.6)-(9.9) were solved assuming polar-optical phonon couplings and acoustic deformation potential couplings with electrons, and optic and acoustic deformation potential coupling with holes. This analysis took full account of temperature-dependent RPA dynamic screening, to determine the carrier drift velocities v_1 and v_2 and the carrier temperature T_1 and T_2 in terms of the applied electric field. The associated electron and hole mobilities are $\mu_e = -v_1/E$ and $\mu_h = v_2/E$. The possibility of negative carrier mobility can be readily recognized from the low velocity limit of the balance equations

$$-N_e E + v_1 f_1'(0) + (v_1 - v_2) f_{12}'(0) = 0 \quad (9.20)$$

$$N_h e E + v_2 f_2'(0) - (v_1 - v_2 f_{12}'(0) = 0 \quad (9.21)$$

which yield

$$\mu_e = -\frac{v_1}{E} = \frac{e\,[f_{12}'(0)(N_h - N_e) - f_2'(0)N_e]}{f_1'(0)f_2'(0) + f_2'(0)^2 + f_1'(0)f_{12}'(0)}. \quad (9.22)$$

In the case when $N_h \gg N_e$, the term $f_{12}'(0)N_h$ can be of the same order of magnitude as $f_2'(0)N_e$. Since $f_2'(0)$, which is due mainly to the phonon scattering, decreases more rapidly than $f_{12}'(0)$, μ_e can change sign and become negative at low temperature when $f_{12}'(0)N_h$ dominates. Numerical results for Ohmic mobilities as functions of lattice temperature, and for nonlinear mobilities as functions of applied electric field strength are shown in Figs.9.1 and 9.2 (Cui, Lei, and Horing, 1988b). They are in good agreement with the experimental data of Höpfel et al. (1986). This problem has also been investigated by Cai, Zheng and Lax (1988) with an essentially similar balance equation method.

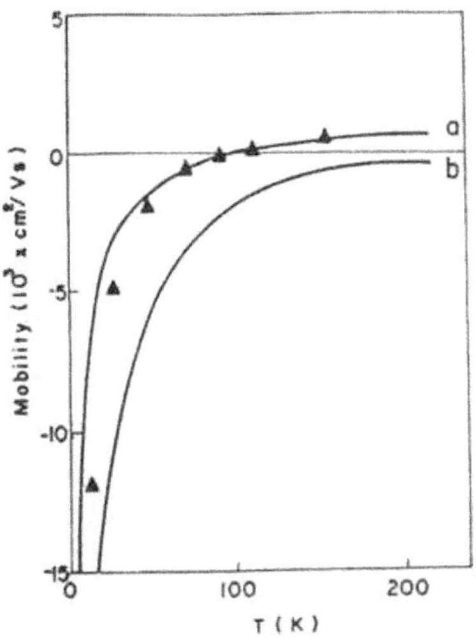

Fig.9.1 Carrier mobilities as functions of lattice temperature: curve a, minority electron mobility μ_e; b, majority hole mobility multiplied by −1. Solid triangles are experimental values of μ_e taken from Höpfel et al. (1986). (From Cui, Lei and Horing, 1988b)

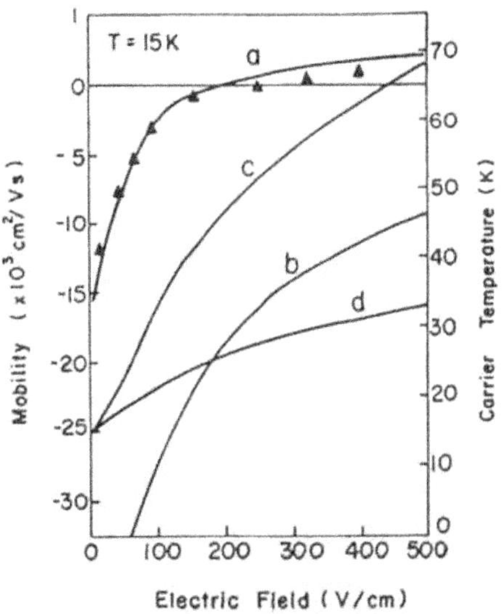

Fig.9.2 Carrier mobilities and temperatures as functions of the applied electric field at lattice temperature 15 K: a, μ_e ; b, μ_h multiplied by −1; c, electron temperature; d, hole temperature. Solid triangles are experimental values of μ_e taken from Höpfel et al. (1986). (From Cui, Lei and Horing, 1988b)

9.1.4 Linear high-frequency transport

Linear high-frequency transport in a type-II superlattice has also been studied using balance equations (Lei, 1987a; Lei, Cui and Horing, 1988). Applying a small ac field of frequency ω

$$E(t) = E_m e^{-i\omega t} + E_m e^{i\omega t} \tag{9.23}$$

along the x-direction, after the transient process, the centers of mass of the electrons and holes will execute simple harmonic oscillations in the field direction at the frequency of the applied ac field:

$$v_j(t) = v_{jm} e^{-i\omega t} + v_{jm}^* e^{i\omega t} \tag{9.24}$$

(j=1,2). The equations of motion for the centers of mass of the electrons and holes yield

$$-\omega v_{1m} = -\frac{e}{m_e} E_m + v_{1m} M^{(1)}(\omega) + (v_{1m} - v_{2m}) \frac{F_{12}(\omega)}{N_e m_e \omega}, \tag{9.25}$$

$$-\omega v_{2m} = \frac{e}{m_h} E_m + v_{2m} M^{(2)}(\omega) + (v_{1m} - v_{2m}) \frac{F_{12}(\omega)}{N_h m_h \omega}. \tag{9.26}$$

The memory functions $M^{(j)}(\omega)$ for electron ($j = 1$) and for hole ($j = 2$) systems are given by (7.18) and (7.19) due to impurity and phonon scatterings, and the function $F_{12}(\omega)$ due to electron-hole interaction is given by

$$F_{12}(\omega) = -\left(\frac{e^2}{2\epsilon_0 \kappa}\right)^2 \frac{d}{\pi} \int_0^{\pi/d} dq_z \sum_{\mathbf{q}_\parallel} \frac{q_x^2}{q_\parallel^2} I_1(q)^2 I_2(q)^2 S^{12}(\mathbf{q}_\parallel, q_z) \left[\chi(q_z, \mathbf{q}_\parallel, 0) - \chi(q_z, \mathbf{q}_\parallel, \omega)\right] \tag{9.27}$$

with

$$\chi(q_z, \mathbf{q}_\parallel, \omega) = \frac{1}{\pi} \int_{-\infty}^{\infty} d\omega' n(\frac{\omega'}{T}) \; [\Pi_2^{(1)}(q_z, \mathbf{q}_\parallel, \omega) \Pi^{(2)}(-q_z, -\mathbf{q}_\parallel, \omega - \omega') - \Pi^{(1)}(q_z, \mathbf{q}_\parallel, \omega + \omega') \Pi_2^{(2)}(-q_z, -\mathbf{q}_\parallel, -\omega')]. \tag{9.28}$$

Eqs.(9.25) and (9.26) determine the two complex drift velocities v_{1m} and v_{2m}, and thus the complex current density per sheet

$$J_m = ie(N_e v_{1m} - N_h v_{2m}). \tag{9.29}$$

As v_{1m} and v_{2m} are linear in E_m, we may determine the complex resistivity as $R(\omega) = E_m / J_m$. For an ideal intrinsic system, the hole density is equal to the electron density $N_e = N_h = N_s$, and there are no impurities to contribute to the resistivity. When the temperature is low enough and frequency is not too high, the dominant contribution to the resistivity comes from electron-hole interaction. In this case we have

$$R(\omega) = -i \frac{\mu}{e^2 N_s} [\omega + M(\omega)] \tag{9.30}$$

with the memory function

$$M(\omega) = \frac{1}{\mu N_s \omega} F_{12}(\omega) \tag{9.31}$$

and

$$\mu = m_e m_h/(m_e + m_h). \tag{9.32}$$

Numerical calculations of linear high frequency transport were carried out for InAs-GaSb type-II superlattices (Lei, Cui and Horing, 1988), focusing on the electron-hole scattering mechanism. Plasmon contributions were taken into careful consideration in the calculation. The high-frequency resistivity depends strongly not only on the separation between the electrons and holes, but also on the hole to electron mass ratio m_h/m_e. At high values of m_h/m_e, electron-hole scattering resembles electron-ionized impurity scattering, which is more effective in limiting electron mobility than easily mobile holes.

9.2 Multivalley Semiconductors

Conventional semiconductors such as Ge, Si, GaAs, are multivalley systems. A simple example is the Γ–L band model for carrier transport along the crystal axes in n-type GaAs systems. In this, the central valley (Γ) has the lowest energy and is nearly spherically symmetric with effective mass $m_1 \simeq 0.067 m_e$ (m_e is the free electron mass). The next lowest ones are four equivalent valleys at L-points. They have energies at $\varepsilon_g = 0.27eV$ above the Γ-valley, and are treated as spherically symmetric about $\mathbf{k}_L$, with effective mass $m_2 = 0.23 m_e$. This system has two nonequivalent sets of isotropic valleys with different effective masses, i.e. it is a system composed of two species of carriers. (Lei et al., 1987; Xing et al., 1988)

9.2.1 Systems of two carrier species with particle interchange

The main difference of a multivalley semiconductor from a type-II superlattice is that the carriers can be scattered from one valley to the other. Therefore we assume that the numbers of different carrier species, N_1 and N_2, are variable, but the total number of carriers $N = N_1 + N_2$ is fixed. We introduce the coordinate $\mathbf{R}$ for the center of mass of the whole system, and $\mathbf{R}_1$ and $\mathbf{R}_2$ for subsystems 1 and 2 (Lei et al., 1987):

$$\mathbf{R} = \frac{1}{N}\left[\sum_i^1 \mathbf{r}_{1i} + \sum_i^2 \mathbf{r}_{2i}\right] = (N_1/N)\mathbf{R}_1 + (N_2/N)\mathbf{R}_2, \delta \tag{9.33}$$

and momenta $\mathbf{P}_1$ and $\mathbf{P}_2$ for subsystems 1 and 2 :

$$\mathbf{P}_1 = \sum_i^1 \mathbf{p}_{1i}, \quad \mathbf{P}_2 = \sum_i^2 \mathbf{p}_{2i}. \tag{9.34}$$

The relative coordinates and momenta for the carrier subsystems 1 and 2 are defined as

$$\mathbf{r'}_{1i} = \mathbf{r}_{1i} - \mathbf{R}_1, \quad \mathbf{r'}_{2i} = \mathbf{r}_{2i} - \mathbf{R}_2, \tag{9.35}$$

$$\mathbf{p'}_{1i} = \mathbf{p}_{1i} - m_1 \mathbf{V}_1, \quad \mathbf{p'}_{2i} = \mathbf{p}_{2i} - m_2 \mathbf{V}_2. \tag{9.36}$$

According to these definitions, $[N_1 R_{1\alpha}, P_{1\beta}] = iN_1\delta_{\alpha\beta}$. Therefore it is consistent to consider $\mathbf{R}_1$ and $\mathbf{P}_1$ as canonical c.m. variables satisfying

$$[R_{1\alpha}, P_{1\beta}] = i\delta_{\alpha\beta}, \tag{9.37}$$

and N_1, the particle number of subsystem 1, as a variable of the relative electron subsystem 1, which commutes with c.m. variables. It is readily verified that to order $1/N_1$, the relative electron variables $\mathbf{r}'_{1i}$ and $\mathbf{p}'_{1j}$ obey the commutation relation:

$$\left[r'^{\alpha}_{1i}, p'^{\beta}_{1j}\right] = i\delta_{\alpha\beta}\left[\delta_{ij} + O(\frac{1}{N_1})\right]. \tag{9.38}$$

In terms of these new variables the overall Hamiltonian of the system in the presence of a uniform electric field $\mathbf{E}$ can be written as follows:

$$H = H_E + H_T + H_{1e} + H_{2e} + H_{ph} + H_I, \tag{9.39}$$

$$H_I = H_{12} + H_{1ei} + H_{2ei} + H_{1ep} + H_{2ep} + H^{12}_{ep} + H^{21}_{ep}. \tag{9.40}$$

Here

$$H_E = -Ne\mathbf{E}\cdot\mathbf{R}, \tag{9.41}$$

and

$$H_T = \frac{1}{2}m_1 N_1 V_1^2 + \frac{1}{2}m_2 N_2 V_2^2 \tag{9.42}$$

are the center-of-mass parts of the Hamiltonian. Particle numbers N_1 and N_2 can be expressed in second quantized representation for the relative carrier subsystems 1 and 2 as

$$N_1 = \sum_{\mathbf{k},\sigma}^{1} c^{\dagger}_{1\mathbf{k}\sigma}c_{1\mathbf{k}\sigma}, \quad N_2 = \sum_{\mathbf{k},\sigma}^{2} c^{\dagger}_{2\mathbf{k}\sigma}c_{2\mathbf{k}\sigma}, \tag{9.43}$$

where $c^{\dagger}_{\alpha\mathbf{k}\sigma}(c_{\alpha\mathbf{k}\sigma})$ are creation (annihilation) operators of the αth ($\alpha = 1, 2$) species of carrier relative to their respective center of mass. The summation over $\mathbf{k}$ is carried out in the vicinity of individual valley centers. $H_{\alpha e}(\alpha = 1, 2)$ are the relative parts of the Hamiltonian of the αth carrier subsystem, including intravalley Coulomb interaction:

$$H_{\alpha e} = \sum_{\mathbf{k},\sigma} \varepsilon_{\alpha\mathbf{k}} c^{\dagger}_{\alpha\mathbf{k}\sigma}c_{\alpha\mathbf{k}\sigma} + \frac{1}{2}\sum_{\mathbf{k},\mathbf{k}',\mathbf{q},\sigma,\sigma'} \nu_c(q) c^{\dagger}_{\alpha\mathbf{k}+\mathbf{q}\sigma} c^{\dagger}_{\alpha\mathbf{k}'-\mathbf{q}\sigma'} c_{\alpha\mathbf{k}'\sigma'} c_{\alpha\mathbf{k}\sigma}, \tag{9.44}$$

where $\nu_c(q) = e^2/\epsilon_0\kappa q^2$ is the Coulomb potential. For convenience we choose the band bottom of the Γ valley as the energy zero reference, so that $\varepsilon_{1\mathbf{k}} = k^2/2m_1$, and $\varepsilon_{2\mathbf{k}} = (\mathbf{k}-\mathbf{k}_L)^2/2m_2 + \varepsilon_g$. In Eq.(9.39), H_{12} stands for the Coulomb interaction between different species of carriers (intervalley Coulomb interaction):

$$H_{12} = \sum_{\mathbf{q}} \nu_c(q) e^{i\mathbf{q}\cdot(\mathbf{R}_1-\mathbf{R}_2)} \rho_{1,\mathbf{q}}\rho_{2,-\mathbf{q}}. \tag{9.45}$$

$H_{\alpha ei}$ and $H_{\alpha ep}$ are electron-impurity and electron-phonon couplings within the αth ($\alpha = 1, 2$) carrier subsystem:

$$H_{\alpha ei} = \sum_{\mathbf{q},a} u_\alpha(\mathbf{q}) e^{i\mathbf{q}\cdot(\mathbf{R}_\alpha - \mathbf{R}_a)} \rho_{\alpha\mathbf{q}}, \tag{9.46}$$

$$H_{\alpha ep} = \sum_{\mathbf{q},\lambda} M_\alpha(\mathbf{q},\lambda) \phi_{\mathbf{q}\lambda} e^{i\mathbf{q}\cdot\mathbf{R}_\alpha} \rho_{\alpha\mathbf{q}}, \tag{9.47}$$

where

$$\rho_{\alpha\mathbf{q}} = \sum_{\mathbf{k}\sigma} c^\dagger_{\alpha\mathbf{k}+\mathbf{q}\sigma} c_{\alpha\mathbf{k}\sigma} \tag{9.48}$$

is the density operator of the αth species of carrier, $u_\alpha(\mathbf{q})$ and $M_\alpha(\mathbf{q},\lambda)$ are intravalley electron-impurity potentials and electron-phonon matrix elements, and $\phi_{\mathbf{q}\lambda} \equiv b_{\mathbf{q}\lambda} + b^\dagger_{-\mathbf{q}\lambda}$. The last two terms of Eq.(9.40) are intervalley electron-phonon interactions: ($\alpha \neq \beta$; $\alpha, \beta = 1, 2$)

$$H^{\alpha\beta}_{ep} = \sum_{\mathbf{q},\lambda} M_{\alpha\beta}(\mathbf{q},\lambda) \phi_{\mathbf{q}\lambda} e^{i\mathbf{q}\cdot\mathbf{R}_\alpha} \sum_{\mathbf{k},\sigma} e^{i\mathbf{k}\cdot(\mathbf{R}_\alpha - \mathbf{R}_\beta)} c^\dagger_{\alpha\mathbf{k}+\mathbf{q}\sigma} c_{\beta\mathbf{k}\sigma}, \tag{9.49}$$

with $M_{12}(\mathbf{q},\lambda) = M^*_{21}(-\mathbf{q},\lambda)$ being the matrix elements for intervalley electron-phonon scattering. Intervalley electron impurity scatterings are neglected. Note that H^{12}_{ep} and H^{21}_{ep} do not commute with the subsystem particle number operators N_1, N_2.

Based on Hamiltonian (9.41), we can determine the rates of change of the center-of-mass momenta $\mathbf{P}_1$ and $\mathbf{P}_2$, the relative electron energies H_{1e} and H_{2e}, the phonon energy H_{ph}, and the subsystem particle number N_1, from the general relation $\dot{O} = -i\,[O, H]$ for an arbitrary operator O. The results are given in Eqs.(A1)-(A4) in the Appendix A of Lei et al.(1987).

9.2.2 Balance equations and Langevin equations

The derivation of balance equations and Langevin equations proceeds in a manner similar to that of Sections 2, 3 and 8: i.e., treating the center-of-mass variables classically and treating H_I perturbatively to lowest order. The coordinates and velocities of the centers of mass enter the density matrix only through their time-dependent statistical averages. It should be noted, however, that there is an important difference in the present case of two carrier species from that discussed previously. The separation of centers of mass from the relative degrees of freedom is incomplete in that H_T includes the relative electron variable N_1 which does not commute with the interaction associated with particle exchange between these two subsystems. Therefore the Liouville equation for the statistical density matrix $\hat{\rho}$ of the relative electron and phonon system takes the form (Lei et al., 1987)

$$i\frac{\partial \hat{\rho}}{\partial t} = [H_T + H_{1e} + H_{2e} + H_{ph} + H_I, \hat{\rho}]. \tag{9.50}$$

From this it is clear that interaction picture should be defined such that for any operator O,

$$O(t) = \exp(iH_0t)O\exp(-iH_0t), \tag{9.51}$$

with

$$H_0 = H_T + H_{1e} + H_{2e} + H_{ph}. \tag{9.52}$$

To determine an appropriate initial condition, we imagine turning off all the electron-impurity, electron-phonon, and intersubsystem electron-electron Coulomb interactions, as well as the electric field, at time t. Then the two carrier systems, decoupled from each other, will approach their respective thermal equilibrium states with temperatures T_1 and T_2 and chemical potentials μ_1 and μ_2, which of course depend on time t. The initial condition for the density matrix ρ is thus chosen to be

$$\hat{\rho}_0 = \frac{1}{Z}\exp(-H_{ph}/T)\exp[-(H_{1e} - \mu_1 N_1)/T_1]\exp[-(H_{2e} - \mu_2 N_2)/T_2]. \tag{9.53}$$

One can proceed in the same way as in Section 8.1 for a single species of carrier. Treating the operator equations (A.1)-(A.4) in Appendix A of Lei et al. (1987) as Heisenberg equations of motion, and expanding the time evolution operator (8.3) to lowest order in H_I, such that for any Heisenberg operator we have an expression similar to Eq.(8.5), and then neglecting the fluctuating part in the second term of the right hand side of Eq(8.5), we have (Xing et al., 1988)

$$m_\alpha \frac{d}{dt}[N_\alpha(t)V_\alpha(t)] = N_\alpha eE + F_\alpha(t) + f_\alpha, \tag{9.54}$$

$$\frac{d}{dt}N_\alpha(t) = -\frac{d}{dt}N_\beta(t) = D_\alpha(t) + d_\alpha, \tag{9.55}$$

where $N_\alpha eE + F_\alpha(t)$ is the interaction picture representation of the right-hand side of Eq.(A1) of Lei et al. (1987), and $D_\alpha(t)$ is the interaction picture representation of the operator

$$D_\alpha = -i(H_{ep}^{\alpha\beta} - H_{ep}^{\beta\alpha}) \tag{9.56}$$

($\alpha \neq \beta,\ \alpha, \beta = 1, 2$). $F_\alpha(t)$ and $D_\alpha(t)$ are fluctuations of forces and rates of change of particle number due directly to couplings with impurities and phonons. f_α and d_α are obtained from the second term of the right hand side of Eq.(8.5) by taking the ensemble average over the relative electron-phonon space. For $\alpha = 1$

$$\begin{aligned} f_1 &= -in_i\sum_{\mathbf{k},\mathbf{q}}|u_1(\mathbf{q})|^2 q_x \int_{-\infty}^{\infty} dt' A_{11}(\mathbf{k},\mathbf{q};t,t')\Pi^{(1)}(\delta\mathbf{k},\mathbf{q},t-t') \\ &\quad -i\sum_{\mathbf{k},\mathbf{q},\lambda}|M_1(\mathbf{q},\lambda)|^2 q_x \int_{-\infty}^{\infty} dt' A_{11}(\mathbf{k},\mathbf{q};t,t')\Lambda^{(1)}(\mathbf{k},\mathbf{q},\lambda;t-t') \\ &\quad -i\left[\sum_{\mathbf{k},\mathbf{q},\lambda}|M_{12}(\mathbf{q},\lambda)|^2(k_x+q_x)\int_{-\infty}^{\infty} dt' A_{12}(\mathbf{k},\mathbf{q};t,t')\Lambda^{(12)}(\mathbf{k},\mathbf{q},\lambda;t-t') + (1\leftrightarrow 2)\right] \\ &\quad -i\sum_{\mathbf{kq}}\nu_c(\mathbf{q})q_x\int_{-\infty}^{\infty} dt' \exp\left[iq_x\int_{t'}^{t}(v_1(s)-v_2(s))ds\right] S^{(12)}(\mathbf{k},\mathbf{q};t-t'), \end{aligned} \tag{9.57}$$

and

$$d_1 = -i \sum_{\mathbf{k},\mathbf{q},\lambda} |M_{12}(\mathbf{q},\lambda)|^2 \int_{-\infty}^{\infty} dt' A_{12}(\mathbf{k},\mathbf{q};t,t')\Lambda^{(12)}(\mathbf{k},\mathbf{q};t-t'). \tag{9.58}$$

The expressions for f_2 and d_2 are obtained by exchanging indices $1 \leftrightarrow 2$ in the above equations. In these equations,

$$A_{\alpha\beta}(\mathbf{k},\mathbf{q};t-t') \equiv \exp\left[i(k_x+q_x)\int_{t'}^{t} v_\alpha(s)ds - ik_x\int_{t'}^{t} v_\beta(s)ds\right], \tag{9.59}$$

and the correlation functions are

$$\Gamma^{(1)}(\mathbf{k},\mathbf{q},t-t') = -i\theta(t-t')\sum_\sigma\langle\left[c^\dagger_{1\mathbf{k}+\mathbf{q}\sigma}(t)c_{1\mathbf{k}\sigma}(t), c^\dagger_{1\mathbf{k}\sigma}(t')c_{1\mathbf{k}+\mathbf{q}\sigma}(t')\right]\rangle_0 \tag{9.60}$$

$$\Lambda^{(1)}(\mathbf{k},\mathbf{q},\lambda;t-t') = -i\theta(t-t') \sum_\sigma\langle\left[\phi_{\mathbf{q}\lambda}(t)c^\dagger_{1\mathbf{k}+\mathbf{q}\sigma}(t)c_{1\mathbf{k}\sigma}(t), \phi_{-\mathbf{q}\lambda}(t')c^\dagger_{1\mathbf{k}\sigma}(t')c_{1\mathbf{k}+\mathbf{q}\sigma}(t')\right]\rangle_0 \tag{9.61}$$

$$\Lambda^{(12)}(\mathbf{k},\mathbf{q},\lambda;t-t') = -i\theta(t-t') \sum_\sigma\langle\left[\phi_{\mathbf{q}\lambda}(t)c^\dagger_{1\mathbf{k}+\mathbf{q}\sigma}(t)c_{2\mathbf{k}\sigma}(t), \phi_{-\mathbf{q}\lambda}(t')c^\dagger_{2\mathbf{k}\sigma}(t')c_{1\mathbf{k}+\mathbf{q}\sigma}(t)\right]\rangle_0 \tag{9.62}$$

$$S^{(12)}(\mathbf{k},\mathbf{q};t-t') = -i\theta(t-t') \sum_\sigma\langle\left[c^\dagger_{1\mathbf{k}+\mathbf{q}\sigma}(t)c_{1\mathbf{k}\sigma}(t)\rho_{2,-q}(t), c^\dagger_{1\mathbf{k}\sigma}(t')c_{1\mathbf{k}+\mathbf{q}\sigma}(t')\rho_{2,q}(t')\right]\rangle_0. \tag{9.63}$$

Both $V_\alpha(t)$ and $N_\alpha(t)$ are thought of as consisting of drift parts and small fluctuations

$$V_\alpha(t) = v_\alpha(t) + \delta V_\alpha(t) \tag{9.64}$$

$$N_\alpha(t) = n_\alpha(t) + \delta N_\alpha(t), \tag{9.65}$$

with $\langle\delta V_\alpha(t)\rangle = 0$ and $\langle\delta N_\alpha(t)\rangle = 0$. Substituting (9.64) and (9.65) into Eqs. (9.54) and (9.55) and retaining only the linear order of δV_α and δN_α, we can derive the coupled balance equations for the drift motion and the coupled Langevin-type equations for the fluctuations. Note that the electron temperatures T_α, as parameters defining the initial canonical ensembles of the relative electron systems, should be devoid of fluctuations. Thus, we have drift variables n_α, v_α, and T_α, which are to be determined by the balance equations

$$\frac{d}{dt}n_1 = d_1 \tag{9.66}$$

$$m_\alpha\frac{d}{dt}(n_\alpha v_\alpha) = n_\alpha eE + f_\alpha \tag{9.67}$$

$$\frac{d}{dt}E_\alpha = \langle\dot{H}_\alpha\rangle \tag{9.68}$$

$(\alpha = 1,2)$. Eqs.(9.66) and (9.67) are obtained by taking the ensemble average of Eqs. (9.54) and (9.55), and d_α and f_α are the same as given by (9.57) and (9.58) with V_α and N_α replaced by v_α and n_α. Eq.(9.68) is obtained by taking the ensemble average of

Eq.(A2) in Appendix A of Lei et al (1987). $\langle \dot{H}_\alpha \rangle$ is given by the expression of Eq.(9.57) but with the factor $k_x + q_x$ replaced by $\varepsilon_{1\mathbf{k}+\mathbf{q}}$ in the right hand side of the third term and with the factor q_x in other terms replaced by $\varepsilon_{1\mathbf{k}+\mathbf{q}} - \varepsilon_{1\mathbf{k}}$.

In addition, we have fluctuation variables δN_α and δV_α, which satisfy Langevin equations established from the drift parts of Eqs.(9.54) and (9.55). Neglecting memory effects, i.e. replacing $\int_{t'}^{t} v_\alpha(s)ds$ by $v(t)(t-t')$ in the factor $A_{\alpha\beta}(\mathbf{k},\mathbf{q},t-t')$ of Eq.(9.59), the balance equations (9.66)-(9.68) for the drift motion reduce to (Lei et al. 1987)

$$\frac{dn_1}{dt} = n(v_1, v_2) \tag{9.69}$$

$$\frac{d}{dt}(n_1 v_1) = n_1 eE + f_1(v_1) + f_p^{12}(v_1, v_2) + f_{12}(v_1 - v_2) \tag{9.70}$$

$$\frac{d}{dt}(n_2 v_2) = n_2 eE + f_2(v_2) + f_p^{21}(v_2, v_1) + f_{21}(v_2 - v_1) \tag{9.71}$$

$$-C_1 \frac{dT_{1e}}{dt} - \lambda_1 \frac{dn_1}{dt} = v_1 f_1(v_1) + w_1(v_1) + w_p^{12}(v_1, v_2) + w_{12}(v_1 - v_2) \tag{9.72}$$

$$-C_2 \frac{dT_{2e}}{dt} - \lambda_2 \frac{dn_2}{dt} = v_2 f_2(v_2) + w_2(v_2) + w_p^{21}(v_2, v_1) + w_{21}(v_2 - v_1), \tag{9.73}$$

in which $C_\alpha = (\partial E_\alpha / \partial T_\alpha)_{n_\alpha}$, $\lambda_\alpha = (\partial E_\alpha / \partial n_\alpha)_{T_\alpha}$, $f_{21}(v_2 - v_1) = -f_{12}(v_1 - v_2)$ and $w_{21}(v_2 - v_1) = (v_1 - v_2) f_{12}(v_1 - v_2) - w_{12}(v_1 - v_2)$. The expressions for $f_\alpha(v_\alpha)$, $w_\alpha(v_\alpha)$, $f_{12}(v_1 - v_2)$, $w_{12}(v_1 - v_2)$, $f_p^{\alpha\beta}(v_\alpha, v_\beta)$ and $w_p^{\alpha\beta}(v_\alpha, v_\beta)$ are given in Appendix B of Lei et al. (1987). Note that in addition to the explicit dependence on the drift velocities v_α, these expressions include the electron temperatures T_1 and T_2, as well as the chemical potentials μ_1, μ_2 as parameters. The latter are related to particle number densities n_α by

$$n_\alpha = \sum_{\mathbf{k},\sigma} f((\varepsilon_{\alpha\mathbf{k}} - \mu_\alpha)/T_\alpha). \tag{9.74}$$

Finally, E_α is the internal energy of the αth subsystem:

$$E_\alpha = \sum_{\mathbf{k},\sigma} \varepsilon_{\alpha\mathbf{k}} f((\varepsilon_{\alpha\mathbf{k}} - \mu_\alpha)/T_\alpha). \tag{9.75}$$

9.2.3 Results for dc steady state and transient transport

Using the two-valley (Γ and L) model, Lei et al. (1987) carried out numerical calculations for dc steady state transport in GaAs from the balance equations (9.69)-(9.73), taking all the time derivatives (the left-hand-sides of all the above equations) to vanish. Numerical analyses were also performed for Si (Liu et al., 1988) based on similar equations with the applied electric field parallel to $< 111 >$, $< 110 >$ and $< 100 >$ crystallographic directions. With the extension of Eqs. (9.69)-(9.73) to three-valley systems, the case of dc steady state (Xing, Liu and Ting, 1988) and of transient (Liu, Xing and Ting, 1989) transport in n-type GaAs were discussed further with $\Gamma - L - X$ band structure. Figs. 9.3-9.6 show the calculated results at $T = 300K$.

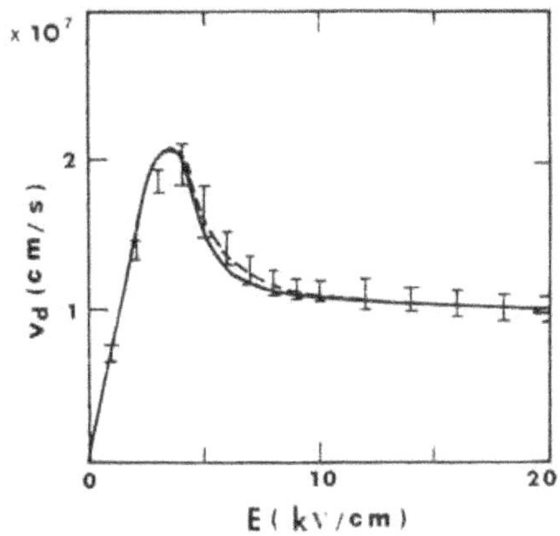

Fig.9.3 Drift velocity v_s vs electric field strength E for n-type GaAs using $\Gamma - L - X$ valley model at $T = 300$ K. The solid and dashed curves are derived from the balance-equation theory (solid curve for intervalley coupling constant $D_{\Gamma L} = 0.285 \times 10^9$ eV/cm, dashed curve for $D_{\Gamma L} = 0.36 \times 10^9$ eV/cm). The bars are compiled experimental data. (From Xing, Liu and Ting, 1988)

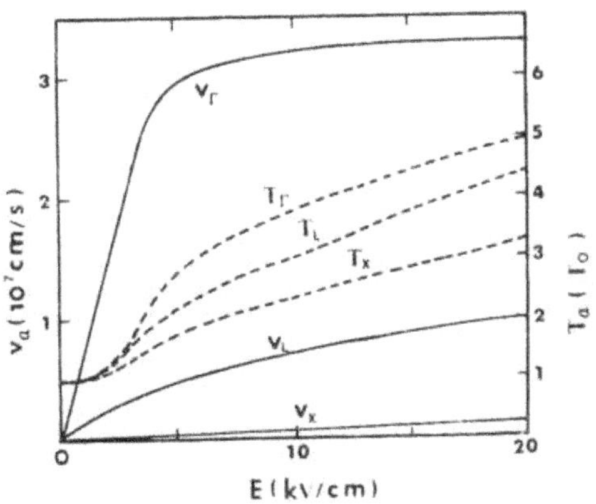

Fig.9.4 Mean velocities (solid lines) and temperatures (dashed lines) of electrons in Γ, L, and X valleys for GaAs vs electric field strength at $T = 300$ K. (From Xing, Liu and Ting, 1988)

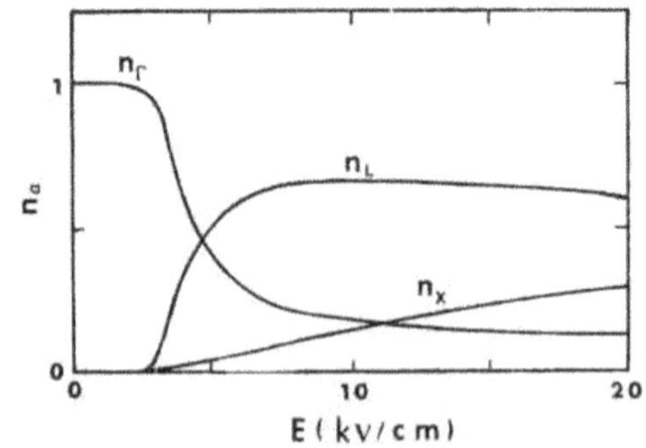

Fig.9.5 Fractions of electrons in Γ, L, and X valleys for GaAs vs electric field strength at T = 300 K. (From Xing, Liu and Ting, 1988)

Good agreement with experimental data and with the theoretical predictions of Monte Carlo simulation was obtained for both the cases of GaAs and Si, confirming the effectiveness of the balance equation approach in treating many valley semiconductors. More detail is available in the original papers.

9.2.4 Correlations of the velocity and particle-number fluctuations

The Langevin equations derived from Eqs. (9.54) and (9.55) are coupled equations for δV_α and δN_α. Xing, Liu and Ting (1988), and Xing, Liu, et al. (1988), using simplified forms which decouple the equations for δV_α from those for δN_α, carried out an analysis of the correlations of δV_α and δN_α: $\langle \delta V_\alpha(t)\delta N_\beta(0)\rangle$, $\langle \delta N_\alpha(t)\delta N_\alpha(0)\rangle$ and $\langle \delta V_\alpha(t)\delta N_\beta(0)\rangle$, and they calculated the diffusion coefficient in multivalley semiconductors. Fig.9.7 shows the longitudinal and transverse diffusion coefficients of GaAs as functions of the electric field, obtained from the $\Gamma - L - X$ three valley model (Xing, Liu and Ting, 1988), together with experimental data of Rush and Kino (1968) and Bareikis et al. (1980). Further detail may be found in the original papers.

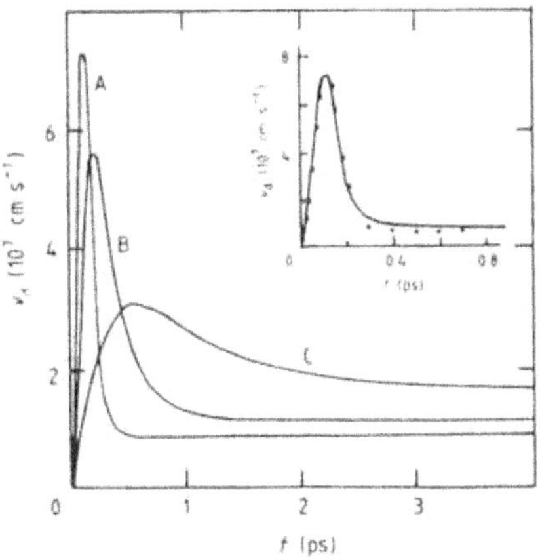

Fig.9.6 Transient drift velocity $v_s(t)$ against time when a time-step electric field is applied to a GaAs sample at $T = 300$ K: curve A, calculated results for $E = 40$ kV/cm; curve B, calculated results for $E = 15$ kV/cm; curve C, calculated results for 5 kV/cm. Comparison between curve A and the results obtained by Reggiani (1985) with a Monte Carlo procedure (black dots; Reggiani, 1985) is shown in the inset. (From Liu, Xing and Ting, 1989)

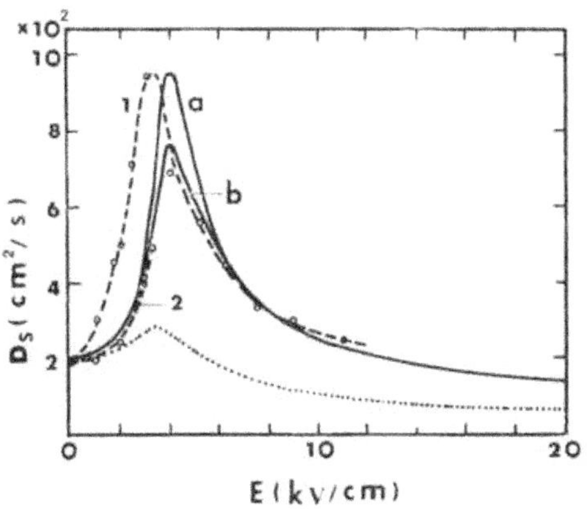

Fig.9.7 Longitudinal diffusion coefficient vs electric field strength. The solid curve is derived from the balance-equation theory (a, intervalley coupling constant $D_{\Gamma L} = 0.285 \times 10^9$ eV/cm; $b, D_{\Gamma L} = 0.36 \times 10^9$ eV/cm). The dashed curve 1, Rush and Kino (1968); curve 2, Bareikies et al. (1980)]. The dotted curve indicates the transverse diffusion coefficient obtained from the theory. (From Xing, Liu and Ting, 1988)

10. NONEQUILIBRIUM PHONON PHENOMENA

The observation of anomalously low carrier energy-loss rates by Shah et al. (1985) in multilayer GaAs-based quantum wells was attributed by them to the presence of a nonequilibrium population of optic phonons. This stimulated substantial study of the interplay between hot electrons and nonequilibrium phonons (Price, 1985; Kocevar, 1985; Cai, Marchetti and Lax, 1986; Lei and Horing, 1987a, 1988; Lugli and Goodnick, 1987; Marchetti, Cai and Lax, 1988; Ridley, 1989; Rieger et al.., 1989; Mickevicius and Reklaitis, 1989). Hot phonon dynamics have been extensively explored using the Boltzmann equation and the Monte-Carlo technique, as well as by means of the balance equation approach. It is particularly convenient to investigate the effects of a nonequilibrium phonon population on hot electron transport within the framework of balance equation theory. In the following we provide a brief description of this analysis based on Lei and Horing (1987a, 1988), and Lei, Cui and Horing (1987).

10.1 Nonequilibrium Phonons in 3D Bulk Systems

A system of electrons in transport may excite or absorb phonons, tending to drive the phonon system out of equilibrium. On the other hand, the phonon-phonon coupling H_{pp} tends to relax it toward equilibrium. Thus far, we have assumed that phonons relax much faster than the tendency to produce nonequilibrium phonons, such that the phonons are always essentially in a thermoequilibrium state with a heat bath at the "lattice temperature" T. Because of this the initial density matrix is chosen to be (7.30).

$$\hat{\rho}_0 = \frac{1}{Z} e^{-H_e/T_e - H_{ph}/T}. \tag{10.1}$$

While the electron temperature T_e is a parameter which may be time-dependent during slow transient processes, the lattice temperature is treated as a fixed external condition.

In considering that the phonon population $n_{\mathbf{q}\lambda}$ of mode $\mathbf{q}\lambda$ may deviate from the thermoequilibrium value at bath temperature T, $n(\Omega_{\mathbf{q}\lambda}/T)$, we assume instead that there is an equilibrium for each set of phonon modes such that we can describe the nonequilibrium phonon system by a mode-dependent phonon temperatures $T_{\mathbf{q}\lambda}$. The initial density matrix for the relative-electron-phonon system is chosen accordingly as

$$\hat{\rho}_0 = \frac{1}{Z_e} \exp(-H_e/T_e) \frac{1}{Z_p} \exp\left(-\sum_{\mathbf{q}\lambda} H_{\mathbf{q}\lambda}/T_{\mathbf{q}\lambda}\right), \tag{10.2}$$

where

$$H_{\mathbf{q}\lambda} \equiv \Omega_{\mathbf{q}\lambda} \hat{N}_{\mathbf{q}\lambda} \tag{10.3}$$

and

$$\hat{N}_{\mathbf{q}\lambda} \equiv b^{\dagger}_{\mathbf{q}\lambda} b_{\mathbf{q}\lambda} \tag{10.4}$$

are, respectively, the energy and number operators of phonons in mode $\mathbf{q}\lambda$. The rate of change of $\hat{N}_{\mathbf{q}\lambda}$ due to electron-phonon interaction is given by

$$\left(\frac{\partial \hat{N}_{\mathbf{q}\lambda}}{\partial t}\right)_{ep} = -i[\hat{N}_{\mathbf{q}\lambda}, H_{ep}]$$
$$= iM(\mathbf{q},\lambda)e^{i\mathbf{q}\cdot\mathbf{R}}b_{\mathbf{q}\lambda}\rho_{\mathbf{q}} - iM(-\mathbf{q},\lambda)e^{-i\mathbf{q}\cdot\mathbf{R}}b^{\dagger}_{\mathbf{q}\lambda}\rho_{-\mathbf{q}}. \tag{10.5}$$

The density matrix $\hat{\rho}$ for the relative electron and phonon system can be obtained from the Liouville equation and the initial condition (10.2) to first order in the interaction $H_I = H_{ep} + H_{ei} + H_{pp}$:

$$\hat{\rho}(t) = \hat{\rho}_0 + i\int_{-\infty}^{t} \exp[i\sum_{\mathbf{q}\lambda}(1-\alpha_{\mathbf{q}\lambda})H_{\mathbf{q}\lambda}(t'-t)]$$
$$[\hat{\rho}_0, H_I(t'-t)]\exp[-i\sum_{\mathbf{q},\lambda}(1-\alpha_{\mathbf{q}\lambda})H_{\mathbf{q}\lambda}(t'-t)]dt', \tag{10.6}$$

where

$$\alpha_{\mathbf{q}\lambda} \equiv T_e/T_{\mathbf{q}\lambda}, \tag{10.7}$$

$H(t) \equiv e^{iH_0t}He^{-iH_0t}$, and $H_0 \equiv H_e + H_{ph}$ is the Hamiltonian for relative electrons and noninteracting phonons.

The statistical average of the operator $(\partial\hat{N}_{\mathbf{q}\lambda}/\partial t)_{ep}$ can be evaluated in accordance with Eq.(2.26) to the lowest order in H_I, yielding

$$\left(\frac{\partial n_{\mathbf{q}\lambda}}{\partial t}\right)_{ep} \equiv \langle\left(\frac{\partial \hat{N}_{\mathbf{q}\lambda}}{\partial t}\right)_{ep}\rangle$$
$$= 2|M(\mathbf{q},\lambda)|^2\Pi_2(\mathbf{q},\Omega_{\mathbf{q}\lambda}-\mathbf{q}\cdot\mathbf{v})\left[n_{\mathbf{q}\lambda} - n\left(\frac{\Omega_{\mathbf{q}\lambda}-\mathbf{q}\cdot\mathbf{v}}{T_e}\right)\right] \tag{10.8}$$

Here, $n_{\mathbf{q}\lambda} = n(\Omega_{\mathbf{q}\lambda}/T_{\mathbf{q}\lambda})$ is the occupation number of phonons in mode $\mathbf{q}\lambda$ and $\mathbf{v}$ is the drift velocity of the electron system in transport under the influence of a uniform electric field. In writing this we have made use of the approximation (6.1) for $A(\mathbf{q},t,t')$, neglecting memory effects. The frictional force experienced by the center of mass of the electron system, $\mathbf{f} = \mathbf{f}_i + \mathbf{f}_p$ ($\mathbf{f}_i$ is due to impurities and $\mathbf{f}_p$ due to phonons), and the energy transfer rate from the electron system to phonon system, w, are readily determined in the same approximation. Since $\mathbf{f}_i$ is irrelevant to the phonon system, it has the same expression as Eq.(6.4) in Section 6, while $\mathbf{f}_p$ and w can be written in the physically transparent forms:

$$\mathbf{f}_p = -\sum_{\mathbf{q}\lambda}\mathbf{q}\left(\frac{\partial n_{\mathbf{q}\lambda}}{\partial t}\right)_{ep} \tag{10.9}$$

$$w = \sum_{\mathbf{q}\lambda}\Omega_{\mathbf{q}\lambda}\left(\frac{\partial n_{\mathbf{q}\lambda}}{\partial t}\right)_{ep} \tag{10.10}$$

involving quanta of phonon momentum and energy.

The determination of the rate of change of phonon occupation number due to the phonon-phonon interaction, $(\partial n_{\mathbf{q}\lambda}/\partial t)_{pp}$, can not be carried out without a detailed analysis of anharmonic couplings, as well as structure and boundary effects which are sample dependent. Therefore we represent them by an effective phonon relaxation time parameter τ_p:

$$\left(\frac{\partial n_{\mathbf{q}\lambda}}{\partial t}\right)_{pp} = -\frac{1}{\tau_p}\left[n_{\mathbf{q}\lambda} - n(\frac{\Omega_{\mathbf{q}\lambda}}{T})\right]. \tag{10.11}$$

The phonon relaxation time τ_p may be mode dependent, but in any case we treat it as phenomenological parameter(s).

The total rate of change of phonon occupation number can be written for each phonon mode as

$$\frac{d}{dt}n_{\mathbf{q}\lambda} = \left(\frac{\partial n_{\mathbf{q}\lambda}}{\partial t}\right)_{ep} - \frac{1}{\tau_p}\left[n_{\mathbf{q}\lambda} - n(\frac{\Omega_{\mathbf{q}\lambda}}{T})\right]. \tag{10.12}$$

These equations, taken jointly with the force- and energy-balance equations of the electron system,

$$nm\frac{d\mathbf{v}}{dt}n = e\mathbf{E}(t) + \mathbf{f}_i + \mathbf{f}_p \tag{10.13}$$

$$C_e\frac{dT_e}{dt} = -\mathbf{v}\cdot(\mathbf{f}_i + \mathbf{f}_p) - w, \tag{10.14}$$

determine all the parameters involved: $\mathbf{v}, T_e, n_{\mathbf{q}\lambda}$ (or $T_{\mathbf{q}\lambda}$), as functions of time t from a given initial state.

For a dc steady state, the occupation number for each phonon mode is constant, $dn_{\mathbf{q}\lambda}/dt = 0$, and Eqs.(10.12) and (10.8) lead to

$$n_{\mathbf{q}\lambda} = n\left(\frac{\Omega_{\mathbf{q}\lambda} - \mathbf{q}\cdot\mathbf{v}}{T_e}\right) + \left[1 + \frac{\tau_p}{\tau(\mathbf{q}\lambda, \Omega_{\mathbf{q}\lambda} - \mathbf{q}\cdot\mathbf{v})}\right]^{-1}\left[n\left(\frac{\Omega_{\mathbf{q}\lambda}}{T}\right) - n\left(\frac{\Omega_{\mathbf{q}\lambda} - \mathbf{q}\cdot\mathbf{v}}{T_e}\right)\right]. \tag{10.15}$$

Here,

$$\frac{1}{\tau(\mathbf{q}\lambda, \Omega_{\mathbf{q}\lambda} - \mathbf{q}\cdot\mathbf{v})} = -2|M(\mathbf{q},\lambda)|^2\Pi_2(\mathbf{q}, \Omega_{\mathbf{q}\lambda} - \mathbf{q}\cdot\mathbf{v}) \tag{10.16}$$

is a phonon-mode and electron-drift-velocity dependent inverse scattering time, reflecting the coupling strength between electrons and phonons of mode $(\mathbf{q}\lambda)$. If $\tau_p = 0$, Eq.(10.15) yields $n_{\mathbf{q}\lambda} = n(\Omega_{\mathbf{q}\lambda}/T)$ (i.e. equilibrium phonon occupation). For nonzero τ_p, the phonon occupation number deviates from the equilibrium value, such that the mode-dependent phonon temperature $T_{\mathbf{q}\lambda}$, which is determined by $n_{\mathbf{q}\lambda} \equiv n(\Omega_{\mathbf{q}\lambda}/T_{\mathbf{q}\lambda})$ and is strongly affected by the electron drift velocity, can be much higher than the lattice temperature T. This is so-called hot-phonon or "bottle-neck" phenomenon.

In a steady state, the phonon induced frictional force and the energy transfer rate may be written as

$$\mathbf{f}_p = \sum_{\mathbf{q},\lambda}\frac{\mathbf{q}}{\tau(\mathbf{q}\lambda, \Omega_{\mathbf{q}\lambda} + \mathbf{q}\cdot\mathbf{v}) + \tau_p}\left[n\left(\frac{\Omega_{\mathbf{q}\lambda} + \mathbf{q}\cdot\mathbf{v}}{T_e}\right) - n\left(\frac{\Omega_{\mathbf{q}\lambda}}{T}\right)\right], \tag{10.17}$$

$$w = \sum_{\mathbf{q}\lambda} \frac{\Omega_{\mathbf{q}\lambda}}{\tau(\mathbf{q}\lambda, \Omega_{\mathbf{q}\lambda} + \mathbf{q}\cdot\mathbf{v}) + \tau_p} \left[n\left(\frac{\Omega_{\mathbf{q}\lambda} + \mathbf{q}\cdot\mathbf{v}}{T_e}\right) - n\left(\frac{\Omega_{\mathbf{q}\lambda}}{T}\right)\right]. \tag{10.18}$$

Note that for this steady-state case, the nonequilibrium phonon occupation number $n_{\mathbf{q}\lambda}$ does not appear in the expressions (10.17) and (10.18) for $\mathbf{f}_p$ and w. Therefore the determination of steady state dc hot electron transport from the electron force- and energy-balance equations,

$$ne\mathbf{E} + \mathbf{f}_i + \mathbf{f}_p = 0 \tag{10.19}$$

$$\mathbf{v}\cdot(\mathbf{f}_i + \mathbf{f}_p) + w = 0, \tag{10.20}$$

involves only a knowledge of the phonon relaxation time τ_p. If the phonon relaxation is fast enough, $\tau_p \ll \tau$, Eqs. (10.17) and (10.18) become Eqs.(4.5) and (4.6). Otherwise the phonon-induced frictional force and the energy transfer rate for given values of T, T_e and v are evidently reduced.

10.2 Nonequilibrium Phonons in Quantum Wells and Superlattices

The bulk results obtained in the preceding subsection can easily be extended to superlattices if the phonons retain a 3-dimensional bulk character, i.e. extended plane waves are still the eigenstates of phonons. For simplicity, we consider the case where the electron sheet density is low and thermal excitation across subbands is negligible, such that only the lowest electron subband needs to be taken into account. In this case the superlattice counterparts (per layer) of the phonon-related quantities discussed in the preceding subsection may be obtained from the corresponding 3D formulae by replacing $\Pi(\mathbf{q},\omega)$ with $\Pi(q_z, \mathbf{q}_\parallel, \omega)$ and making the following substitutions:

$$\sum_{\mathbf{q}} \to \sum_{\mathbf{q}_\parallel} \frac{d}{2\pi}\int_{-\infty}^{\infty} dq_z \tag{10.21}$$

$$|M(\mathbf{q},\lambda)|^2 \to |M(\mathbf{q},\lambda)|^2 |I(iq_z)|^2 \tag{10.22}$$

in which the form factor $I(iq_z)$ and the correlation function $\Pi(q_z, \mathbf{q}_\parallel, \omega)$ are given by Eqs.(5.46) and (5.53), and d is the superlattice period. The result for a single quasi-2D sheet may be obtained in the limit $d \to \infty$.

The above results rely on the assumption that the phonon system quickly reaches a quasi-equilibrium for each plane wave mode $\mathbf{q}\lambda$, such that we have a temperature $T_{\mathbf{q}\lambda}$ for each $\mathbf{q}\lambda$, before it relaxes to the equilibrium state with unique lattice temperature in a finite time τ_p. It seems that this is not a good assumption for a quasi-2D electron system, in which the electron-phonon coupling is spatially inhomogeneous in nature due to the electron confinement in the z-direction. Emission and absorption of phonons due to electron-phonon interaction occur mainly in the small region where the electrons have substantial density. Such a spatial distribution is far from a single plane wave mode but a sum of many plane wave modes. To take account of this effect, an effective 2D description for polar optic phonon in terms of $A(\mathbf{q}_\parallel)$ and $A^\dagger(\mathbf{q}_\parallel)$ as creation and annihilation operators was introduced by Cai, Marchetti and Lax (1986) and reformulated by Lei and Horing (1987a). The effective electron-phonon interaction for a quasi-2D system can be written as

$$H_{ep} = \sum_{\mathbf{q}_\parallel} M(\mathbf{q}_\parallel) e^{i\mathbf{q}_\parallel\cdot\mathbf{R}}[A(\mathbf{q}_\parallel) + A^\dagger(-\mathbf{q}_\parallel)]\rho_{\mathbf{q}_\parallel}, \tag{10.23}$$

with

$$M(\mathbf{q}_\parallel)^2 = \frac{e^2\Omega}{2\epsilon_0}(\frac{1}{\kappa_\infty} - \frac{1}{\kappa})\frac{1}{q_\parallel}H(q_\parallel). \tag{10.24}$$

Here $\rho_{\mathbf{q}_\parallel}$ is the density operator for electrons in the lowest subband and $H(q_\parallel)$ is given by Eq.(5.42). The rate of change of the population number $n_{\mathbf{q}_\parallel} = \langle A^\dagger(\mathbf{q}_\parallel)A(\mathbf{q}_\parallel)\rangle$ of this effective phonon system can be written, in terms of a relaxation time τ_p, as

$$\frac{d}{dt}n_{\mathbf{q}_\parallel} = \left(\frac{\partial n_{\mathbf{q}_\parallel}}{\partial t}\right)_{ep} - \frac{1}{\tau_p}\left[n(\mathbf{q}_\parallel) - n(\frac{\Omega}{T})\right], \tag{10.25}$$

where $(\partial n_{\mathbf{q}_\parallel}/\partial t)_{ep}$ is the rate of charge of $n_{\mathbf{q}_\parallel}$ due to electron-phonon interaction H_{ep}:

$$\left(\frac{\partial n_{\mathbf{q}_\parallel}}{\partial t}\right)_{ep} = 2M(\mathbf{q}_\parallel)^2\Pi_2(\mathbf{q}_\parallel, \Omega - \mathbf{q}_\parallel\cdot\mathbf{v})\left[n_{\mathbf{q}_\parallel} - n(\frac{\Omega - \mathbf{q}_\parallel\cdot\mathbf{v}}{T_e})\right]. \tag{10.26}$$

The frictional force experienced by the electron center-of-mass due to electron-phonon interaction and the electron energy loss rate to phonons are given by

$$\mathbf{f}_p = -\sum_{\mathbf{q}_\parallel}\mathbf{q}_\parallel\left(\frac{\partial n_{\mathbf{q}_\parallel}}{\partial t}\right)_{ep}, \tag{10.27}$$

$$w = \sum_{\mathbf{q}_\parallel}\Omega\left(\frac{\partial n_{\mathbf{q}_\parallel}}{\partial t}\right)_{ep} \tag{10.28}$$

per layer. All the steady-state and time-dependent transport properties can be determined by the balance equations (10.25), (10.13) and (10.14).

10.3 Effects of Nonequilibrium Phonon Occupation on Hot-Electron Transport

On the basis of the formulae presented above, extensive calculations of hot carrier transport properties (dc and transient, mobility as well as electron and noise temperatures) were carried out in the presence of nonequilibrium phonon occupation for semiconductor heterostructures (Lei and Horing, 1987a,b, 1988; Lei, Cui, Horing, 1987). It was shown that the finite phonon relaxation time τ_p may have a significant impact on carrier transport in GaAs-based heterosystems and, in particular, it is important for the energy-transfer rate induced by the electron- polar-optic-phonon interaction. This effect can be understood from the fact that scatterings with small momentum transfer have a considerable weight in contributing to the electron energy loss, and in case of a Fröhlich coupling between electrons and polar optic phonons, the effective scattering matrix element $|M(\mathbf{q},\lambda)|^2$, and consequently $1/\tau(\mathbf{q}\lambda,\Omega+\mathbf{q}\cdot\mathbf{v}) \sim |M(\mathbf{q},\lambda)|^2\Pi_2(\mathbf{q},\Omega+\mathbf{q}\cdot\mathbf{v})$, are large at small q. The effects of such large values of $1/\tau$ are sharply reduced when taken in conjunction with the finite-phonon relaxation time. Of course, the phonon-induced frictional force is also modified by the finite τ_p. However, this effect is not as pronounced as that in the energy transfer rate because of the relatively small weight of small q scattering in the constitution of the force, in contrast to the energy loss.

The most striking effect of the finite-phonon relaxation time on hot carrier transport is that, for a given drift velocity, the electron temperature is much higher than in the case of $\tau_p = 0$, as can be seen from Fig.10.1, in which the nonlinear mobility μ (defined as the ratio of the drift velocity to electric field, $\mu = v/E$), normalized to Ohmic mobility μ_0, and the carrier temperature T_e, determined from steady-state balance equations (10.17)-(10.20) for an n-GaAs heterojunction at lattice temperature $T = 10K$, are shown as functions of drift velocity. Such an increase of electron temperature tends to increase the frictional force in a measure which outweighs the tendency toward diminution of it due to finite τ_p, and the normalized mobility μ/μ_0 is lower for finite τ_p than for $\tau_p = 0$ (Lei and Horing, 1987a).

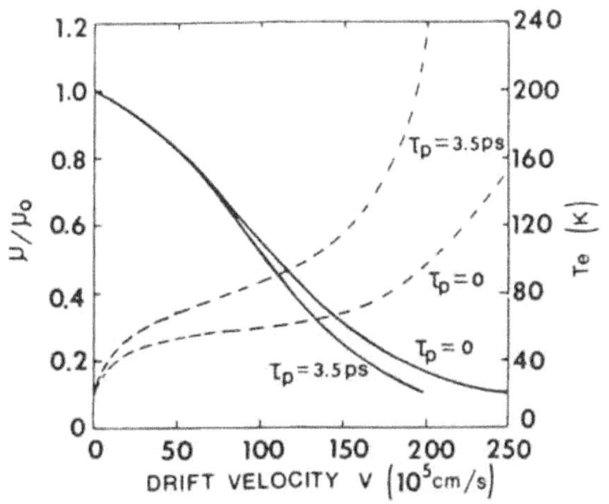

Fig.10.1 Nonlinear mobilities normalized to the Ohmic limit μ/μ_0 (solid curves) and electron temperatures T_e (dashed curves) are shown as functions of drift velocity v for phonon relaxation times $\tau_p = 0$ and $\tau_p = 3.5$ ps at lattice temperature $T = 10$ K for an n-GaAs heterojunction with carrier sheet density $N_s = 3.9 \times 10^{11}$ cm^{-2} and $\mu_0 = 2.2 \times 10^5$ cm^2/Vs. (From Lei and Horing, 1987a)

This effect is more pronounced at higher lattice temperature than at lower lattice temperature, as is seen in Fig.10.2, in which the normalized mobilities $\mu/\mu_0(T)$, calculated with $\tau_p = 0$ and $\tau_p = 7ps$, are shown as functions of electric field at lattice temperatures $T = 10$ K, 77 K and 100 K for a GaAs/AlGaAs heterojunction with carrier sheet density 4.0×10^{11} cm^{-2} and zero-temperature Ohmic mobility $\mu_0(0) = 2.2 \times 10^5$ cm$^2/Vs$. The experimental data from Drummond et al. (1981) are also shown in the figure for comparison. The inclusion of finite phonon relaxation time greatly improves the agreement between theory and experiments, especially at higher temperatures (Lei and Horing, 1988; Marchetti, Cai and Lax, 1988).

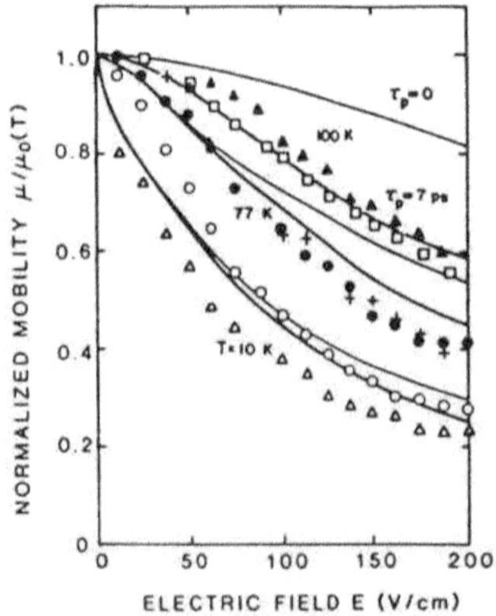

Fig.10.2 Normalized mobility μ/μ_0 vs electric field E. Solid curves are the theoretical results of the balance-equation approach for phonon relaxation times $\tau_p = 0$ and $\tau_p = 7$ ps at lattice temperatures $T = 10$, 77, and 100 K, respectively. The experimental data are taken from Drummond et al. (1981). (From Lei and Horing, 1988)

The effect of finite phonon relaxation time on transient response was shown to result in prolongation of the time required to approach the steady state (Lei, Cui and Horing, 1987), as is shown in Fig.6.3.

The role of nonequilibrium phonon occupation on the noise temperatures was discussed in Section 8.2 and shown in Figs.8.1 and 8.2.

10.4 Phonon Confinement in Semiconductor Quantum Wells

Raman scattering experiments have shown that optical phonons in GaAs-base quantum well superlattices are confined bulk modes and interface modes (Sood et al., 1985, Klein, 1986). Macroscopic theoretical models, based on dielectric continuum theory, including slab modes (Fuchs and Kliewer, 1965; Licari and Evard, 1977) and guided models (Sood et al., 1985; Babiker, 1986; Ridley, 1989b), and microscopic models (Huang and Zhu, 1988a,b) have been developed to describe these confined optic phonons. Interaction between electrons and such confined optic phonons is a problem of intense current interest. Unlike light scattering, transport properties depend on the totality of scatterings from all possible phonon modes, and may not be very sensitive to the detailed structure of the phonon modes. Nevertheless, the discrete nature of these confined phonon modes in quantum well superlattices may lead to substantial modifications of the electron power loss rate due to symmetry related selection rules at low temperature (Jain and Das Sarma, 1989).

It is straightforward to discuss the effect of these confined eigen-phonon modes on hot electron transport in a quasi-2D system, including nonequilibrium phonon occupations. Since these phonon eigenmodes do not spatially mismatch the electron density distribution, the assumption that there is a quasiequilibrium for each eigenmode before the whole phonon system relaxes (τ_p) to thermoequilibrium seems reasonable. With this, we can immediately write down the corresponding expressions from those in Subsection 10.1. For instance, the steady state force and energy transfer rate due to these LO phonon scatterings are given by

$$\mathbf{f}_p = \sum_{n',n,\mathbf{q}_\parallel,j} \frac{\mathbf{q}_\parallel}{\tau_{n'n}(\mathbf{q}_\parallel, j, \Omega_{\mathbf{q}_\parallel j} + \mathbf{q}_\parallel \cdot \mathbf{v}) + \tau_p} \left[n\left(\frac{\Omega_{\mathbf{q}_\parallel j} + \mathbf{q}_\parallel \cdot \mathbf{v}}{T_e}\right) - n\left(\frac{\Omega_{\mathbf{q}_\parallel j}}{T}\right)\right] \quad (10.29)$$

$$w = \sum_{n',n,\mathbf{q}_\parallel,j} \frac{\Omega_{\mathbf{q}_\parallel j}}{\tau_{n'n}(\mathbf{q}_\parallel, j, \Omega_{\mathbf{q}_\parallel j} + \mathbf{q}_\parallel \cdot \mathbf{v}) + \tau_p} \left[n\left(\frac{\Omega_{\mathbf{q}_\parallel j} + \mathbf{q}_\parallel \cdot \mathbf{v}}{T_e}\right) - n\left(\frac{\Omega_{\mathbf{q}_\parallel j}}{T}\right)\right], \quad (10.30)$$

with

$$\frac{1}{\tau_{n'n}(\mathbf{q}_\parallel, j, \omega)} = -2|M_{n'n}(\mathbf{q}, j)|^2 \Pi_2(n, n', \mathbf{q}_\parallel, \omega). \quad (10.31)$$

Here $\Omega_{\mathbf{q}_\parallel j}$ is the frequency of the eigen-phonon mode $(\mathbf{q}_\parallel, j)$, and $M_{n'n}(\mathbf{q}_\parallel, j)$ is the matrix element of electrons (in subbands n and n') and phonons (mode $\mathbf{q}_\parallel, j$).

Recently, Guillimot et al. (1990) carried out a high-field transport calculation for GaAs/AlGaAs heterostructures. They arrive at the same conclusion that finite phonon relaxation time greatly reduces the energy loss rate and increases the electron temperatures. As a result, the drift velocity is also reduced. By comparing the theoretical predictions with their experimental results they found that it would require $\tau_p = 5ps$ to give the correct energy loss rate if identifying T_e with photoluminescence-determined effective temperature, but this underestimates the high-field drift velocity. Careful calculation of the effective temperature from the distribution function in the balance equation theory can resolve the problem (Guillemot and Clérot, 1991, see Section 17.1).

11. BALANCE EQUATIONS FOR HOT-ELECTRON TRANSPORT IN THE PRESENCE OF A MAGNETIC FIELD

The balance equation theory discussed in the preceding sections for hot-carrier transport can be extended to the case in which both an electric field and a magnetic field are present (Lei, Cai and Ting, 1985; Cai, Lei and Ting, 1985; Cui, Lei and Horing, 1988a). We are going to give a brief introduction concerning the most fundamental problems in the crossed electric and magnetic fields. The balance-equation magnetotransport theory will be reviewed in detail in a separate paper in this volume.

11.1 Balance Equations in the Presence of Crossed Electric and Magnetic Fields

For an isotropic, parabolic band system described in Section 2, we consider a uniform magnetic field applied along the z-direction, $\mathbf{B} = B\hat{z}$, together with a uniform electric field $\mathbf{E}$ in x-y plane. The Hamiltonian for this system can be written in terms of center-of mass variables, $\mathbf{P}$, $\mathbf{R}$, and relative electron variables, $\mathbf{p}_i'$, $\mathbf{r}_i'$, as given by Eq.(2.9),

$$H = H_c + H_e + H_{ph} + H_{ei} + H_{ep}, \tag{11.1}$$

but with

$$H_c = \frac{1}{2Nm}(\mathbf{P} - Ne\mathbf{A}(\mathbf{R}))^2 - Ne\mathbf{E}\cdot\mathbf{R} \tag{11.2}$$

and

$$H_e = \sum_i \frac{1}{2m}(\mathbf{p}_i' - e\mathbf{A}(\mathbf{r}_i'))^2 + g\mu_B B\sum_i \sigma_i + \sum_{i<j}\frac{e^2}{4\pi\epsilon_0|\mathbf{r}_i' - \mathbf{r}_j'|}. \tag{11.3}$$

In this, $\mathbf{A}(\mathbf{r}) = (0, Bx, 0)$ is the vector potential function of the magnetic field, σ_i is the spin of the ith electron ($\sigma_i = \pm 1$), μ_B is the Bohr magneton and g is the effective g factor of electrons in the system. H_{ph}, H_{ei} and H_{ep} have exactly the same expressions as given in Eqs.(2.13)-(2.15), with $\rho_{\mathbf{q}}$ being the Fourier representation of the relative electron density operator given by Eq.(2.16). It can be seen from Eqs.(11.2) and (11.3) that while a uniform electric field acts only on the center of mass, and the relative electrons do not experience the electric field directly, the magnetic field appears both in H_c and in H_e.

In accordance with the Hamiltonian (11.1), the center-of-mass velocity and acceleration operators can be written as

$$\mathbf{V} = -i[\mathbf{R}, H] = \frac{1}{Nm}(\mathbf{P} - Ne\mathbf{A}(\mathbf{R})), \tag{11.4}$$

$$\dot{\mathbf{V}} = -i[\mathbf{V}, H] = \frac{e}{m}(\mathbf{V}\times\mathbf{B}) + \frac{e}{m}\mathbf{E} + \frac{1}{Nm}\mathbf{F}_i + \frac{1}{Nm}\mathbf{F}_p. \tag{11.5}$$

Here $\mathbf{F}_i$, $\mathbf{F}_p$, as well as the rate of change of the phonon energy, $W \equiv \dot{H}_{ph}$, have the same expressions as given in Eqs.(2.21), (2.22) and (2.23). The determination of the statistical averages of these quantities to lowest order in the scattering interaction, $H_I = H_{ei} + H_{ep}$, follows the same procedures as in Section 3 using the initial density matrix $\hat{\rho}_0$ given by (2.30) and identifying $\langle \mathbf{V} \rangle = \mathbf{v}$ as the drift velocity of the electron system. Balance equations for the drift or average quantities, and Langevin equations for the small fluctuating or random quantities, are obtained. The expressions for the frictional forces, energy loss rate, memory functions, fluctuating-force correlation functions, noise temperatures and diffusion coefficients are exactly the same as those without a magnetic field, when they are written in the plane-wave representation, except that there is also a Lorentz force term in the c.m. equation of motion. Thus, the steady-state force- and energy-balance equations are given by (Lei, Cai and Ting, 1985)

$$ne\mathbf{E} + ne\mathbf{v} \times \mathbf{B} + \mathbf{f}(\mathbf{v}) = 0, \tag{11.6}$$

$$\mathbf{v} \cdot \mathbf{f}(\mathbf{v}) + w(\mathbf{v}) = 0, \tag{11.7}$$

where the frictional force $\mathbf{f}(\mathbf{v})$ consists of contributions from impurity and phonon scatterings:

$$\begin{aligned}\mathbf{f}(\mathbf{v}) &= n_i \sum_{\mathbf{q}} |u(\mathbf{q})|^2 \mathbf{q} \Pi_2(\mathbf{q}, \omega_0) \\ &\quad + 2 \sum_{\mathbf{q}\lambda} |M(\mathbf{q}, \lambda)|^2 \mathbf{q} \Pi_2(\mathbf{q}, \Omega_{\mathbf{q}\lambda} + \omega_0) \left[n\left(\frac{\Omega_{\mathbf{q}\lambda}}{T}\right) - n\left(\frac{\Omega_{\mathbf{q}\lambda} + \omega_0}{T_e}\right) \right]\end{aligned} \tag{11.8}$$

with $\omega_0 \equiv \mathbf{q} \cdot \mathbf{v}$, and the energy transfer rate $w(\mathbf{v})$ is due to phonons:

$$w(\mathbf{v}) = 2 \sum_{\mathbf{q}\lambda} |M(\mathbf{q}, \lambda)|^2 \Omega_{\mathbf{q}\lambda} \Pi_2(\mathbf{q}, \Omega_{\mathbf{q}\lambda} + \omega_0) \left[n\left(\frac{\Omega_{\mathbf{q}\lambda}}{T}\right) - n\left(\frac{\Omega_{\mathbf{q}\lambda} + \omega_0}{T_e}\right) \right]. \tag{11.9}$$

Here $\Pi_2(\mathbf{q}, \omega)$ is the imaginary part of the relative electron density-density correlation function

$$\Pi(\mathbf{q}, t) = -i\theta(t) \langle [\rho_{\mathbf{q}}(t), \rho_{-\mathbf{q}}(0)] \rangle_0 \tag{11.10}$$

with the average $\langle \quad \rangle_0$ being in reference to $\hat{\rho}_0$ of Eq.(2.30), in which H_e represents the relative electrons, with Coulomb interaction among themselves, moving in the magnetic field (but with no electric field). As in the case without a magnetic field, the Coulomb interaction can be included in random phase approximation, resulting in

$$\Pi(\mathbf{q}, \omega) = \frac{\Pi_0(\mathbf{q}, \omega)}{1 - \nu_c(\mathbf{q}) \Pi_0(\mathbf{q}, \omega)} \tag{11.11}$$

where $\Pi_0(\mathbf{q}, \omega)$ is the density-density correlation function for noninteracting electrons in the presence of the magnetic field. It can be written in the Landau representation (Mermin and Canel, 1964; Horing 1965; Ting, Ying and Quinn 1977) as

$$\Pi_0(\mathbf{q}, \omega) = \frac{|e|B}{2\pi} \sum_{n_1, n_2} C_{n_1, n_2}(x) \, \Pi_0(n_1, n_2, q_z, \omega) \tag{11.12}$$

where $x \equiv q_\parallel^2/2|e|B$, $q_\parallel^2 \equiv q_x^2 + q_y^2$ and

$$C_{n,n+l}(x) = \frac{n!}{(n+l)!} x^l e^{-x} [L_n^l(x)]^2, \tag{11.13}$$

$L_n^l(x)$ being the associate Laguerre polynomials:

$$L_n^l(x) = \sum_m^n (-1)^m \frac{(n+l)! x^m}{(l+m)!(n-m)!m!}, \tag{11.14}$$

in which the summation over m begins from $m = -l$ if $l < 0$, otherwise from $m = 0$. The correlation function $\Pi_0(n_1, n_2, q_z, \omega)$ is given by

$$\Pi_0(n_1, n_2, q_z, \omega) = \sum_{k_z, \sigma} \frac{f(\varepsilon(n_2, k_z + q_z, \sigma), T_e) - f(\varepsilon(n_1, k_z, \sigma), T_e)}{\omega + \varepsilon(n_2, k_z + q_z, \sigma) - \varepsilon(n_1, k_z, \sigma) + i\delta}, \tag{11.15}$$

where $\varepsilon(n, k_z, \sigma) = (n + \frac{1}{2} + \frac{\beta}{2}\sigma)\omega_c + k_z^2/2m$ with $\omega_c = |e|B/m$ and $\beta = gm/m_e$, m_e being the free electron mass.

11.2 Hot Electron Magnetoresistivity

The balance equations (11.6) and (11.7), which accommodate electric field and magnetic field of arbitrary strength, facilitate the calculation of longitudinal and transverse nonlinear magnetoresistivities (Lei, Cai and Ting, 1985). For an isotropic system discussed above, the frictional force and energy transfer rate are functions of $v \equiv |\mathbf{v}|$, such that we can define a resistivity function

$$\rho(v) \equiv -\frac{f(v)}{n^2 e^2 v}, \tag{11.16}$$

and an equivalent energy dissipation resistivity function

$$\rho_E(v) = \frac{w(v)}{n^2 e^2 v^2}. \tag{11.17}$$

The energy balance equation (11.7) then becomes

$$\rho(v) - \rho_E(v) = 0 \tag{11.18}$$

which determines the electron temperature T_e as a function of v for given magnetic field B and lattice temperature T. The force-balance equation (11.6) can be written in x, y and z directions as

$$neE_x + neBv_y + f_x(v) = 0 \tag{11.19}$$

$$neE_y - neBv_x + f_y(v) = 0 \tag{11.20}$$

$$f_z(v) = 0. \tag{11.21}$$

Choosing the x-axis along the $\mathbf{v}$-direction we have

$$neE_x + f(v) = 0, \tag{11.22}$$

$$neE_y - neBv = 0, \tag{11.23}$$

and the current density is now only in x-direction

$$J_x = nev, \quad J_y = 0. \tag{11.24}$$

Therefore the nonlinear longitudinal and transverse resistivities are respectively

$$\rho_{xx} = E_x/J_x = \rho(v), \tag{11.25}$$

$$\rho_{yx} = E_y/J_x = B/ne. \tag{11.26}$$

On the other hand, if we choose the x-axis along the $\mathbf{E}$-direction and let φ be the angle between $\mathbf{v}$ and $\mathbf{E}$, Eqs.(11.19) and (11.20) become

$$neE + neBv\sin\varphi + f(v)\cos\varphi = 0, \tag{11.27}$$

$$-neBv\cos\varphi + f(v)\sin\varphi = 0. \tag{11.28}$$

This yields

$$E = [(nev\rho(v))^2 + B^2v^2]^{1/2} \tag{11.29}$$

and

$$\cot\varphi = -\frac{ne}{B}\rho(v). \tag{11.30}$$

The current density is

$$J_x = nev\cos\varphi, \quad J_y = nev\sin\varphi. \tag{11.31}$$

Then the nonlinear longitudinal and transverse conductivities are given by

$$\sigma_{xx} = \frac{J_x}{E} = \frac{\rho(v)}{\rho(v)^2 + (B/ne)^2}, \tag{11.32}$$

$$\sigma_{yx} = \frac{J_y}{E} = \frac{-B/ne}{\rho(v)^2 + (B/ne)^2}. \tag{11.33}$$

The rest resistivity and conductivity components can be determined by the relation between resistivity and conductivity tensors to be $\rho_{yy} = \rho_{xx}$, $\rho_{xy} = -\rho_{yx}$, $\sigma_{yy} = \sigma_{xx}$, and $\sigma_{xy} = \sigma_{yx}$. Note that the expression for the Hall resistivity, Eq.(11.26), or the Hall coefficient $R_H = 1/ne$, is exactly the same as that derived in linear theory for electrons with simple band structure. The present result shows that it is equally valid for the case of a strong electric field as long as the system has a simple band structure.

12. EFFECTS OF HIGHER ORDER SCATTERINGS

So far we have dealt with force- and energy-balance equations only to the lowest order in scattering interactions. Although Su, Chen and Birman (1987) gave a very general discussion of nonperturbative balance equations formulated in terms of path-integrals and nonequilibrium Green's functions, the only available practical calculations beyond the lowest order theory are based on Feynman diagrammatic techniques involving close-time-path-Green's functions (CTPG) (Lei and Cai, 1990a) or the Quantum Langevin equations of Hu and O'Connell (1987, 1988c). The former will be briefly discussed in this section.

12.1 CTPG Formulation of the Balance-Equation Approach

The force- and energy-balance equations for steady state,

$$ne\mathbf{E} + \mathbf{f} = 0, \tag{12.1}$$

$$\mathbf{v}_d \cdot \mathbf{f} + w = 0, \tag{12.2}$$

have extensive validity beyond lowest order in the scattering interaction, provideed that the frictional force $\mathbf{f}$ and the energy transfer rate w are written in the general forms:

$$\mathbf{f} = Tr\left[\hat{\rho}_0 \tilde{\mathbf{F}}(t)\right], \tag{12.3}$$

$$w = Tr\left[\hat{\rho}_0 \tilde{W}(t)\right]. \tag{12.4}$$

One can determine these statistical averages by the use of Feynman diagrammatic techniques by treating $H_I = H_{ei} + H_{ep}$ as a perturbation and writing, for instance, Eqs.(12.3) and (12.4) in terms of the Keldysh closed-time-path integration (Chou et al. 1985; Rammer and Smith, 1986),

$$\mathbf{f} = Tr\left\{T_p\left[\hat{\rho}_0 \mathbf{F}(t) S_p\right]\right\} \tag{12.5}$$

with

$$S_p = \exp\left[-i\int_p H_I(t_1)dt_1\right]. \tag{12.6}$$

Here $\mathbf{F}(t)$ and $H_I(t)$ are the interaction picture representations of $\mathbf{F}$ and H_I. The time path p is composed of a "+" branch (from $-\infty$ to $+\infty$) and a "$-$" branch (from $+\infty$ to $-\infty$), and T_p is the time-ordering operator on the path p. Note that the random impurity site average is implicitly included in the trace operations. The expression (12.3) can be recast into the form

$$\mathbf{f} = Tr\left\{T_p\left[\hat{\rho}_0 \mathbf{F} \exp\left(-i\int_p H_I(t_1 - t)dt_1\right)\right]\right\}. \tag{12.7}$$

which is manifestly time independent. By expanding the exponential part inside the trace of Eq.(12.7) we obtain a perturbative expansion. After averaging over the impurity sites and over the phonon variables, there remains only the configurations of momentum variables that are paired in each term of the perturbation. The trace operation with respect to the relative electron variables can be calculated by the use of the diagrammatic technique, based on Feynman-Keldysh rules as follows (Lei and Cai, 1990a). There is a Keldysh path branch index ($\alpha = +$ or $-$) with each vertex. The solid and wavy lines represent the components of the electron and phonon closed-time-path Green's functions, and dashed lines represent the impurity average pairing. All these lines are directional and carry wavevectors and frequencies. Furthermore, with each impurity (dashed) or phonon (wavy) line of wave-vector $\mathbf{q}$, there is an additional line of the same direction, which carries only frequency $\omega_{\mathbf{q}} = -\mathbf{q} \cdot \mathbf{v}_d$. Wavevectors and frequencies are conserved at each vertex, i.e. total wavevectors and total frequencies in are equal to those out. According to these rules we can easily construct the diagrams for the frictional force $\mathbf{f}$ and the energy loss rate w.

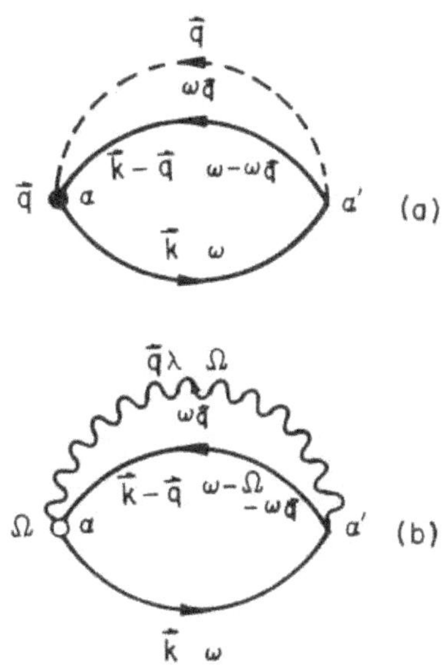

Fig.12.1 Feynman diagrams for (a) the frictional force due to impurities and (b) the electron energy-loss rate due to phonons.

In each frictional-force diagram we must add a factor of wavevector $\mathbf{q}$ at the primary vertex, which is indicated by a dot. The lowest order diagram for the impurity-induced frictional force is shown in Fig.12.1a. and it may be evaluated as

$$\mathbf{f}_i = n_i u^2 \sum_{\mathbf{q},\mathbf{k}} \int \frac{d\omega}{2\pi} \mathbf{q} \xi_\alpha \eta_{\alpha'} G_{\alpha\alpha'}(\mathbf{k}-\mathbf{q}, \omega + \mathbf{q}\cdot\mathbf{v}_d) G_{\alpha'\alpha}(\mathbf{k}, \omega). \tag{12.8}$$

Here two symbols ξ_α and η_β are defined by ξ_+ and $\xi_- = 1$, $\eta_+ = 1$ and $\eta_- = -1$, and the Einstein summation rule is implied for the repeated Greek-letter scripts. In writing Eq.(12.8) we have assumed that the impurity potential is short-ranged, such that

$|u(\mathbf{q})|^2 = u^2$. Employing the relation between the components of the closed-time-path Green's function $G_{\alpha\beta}$ and the retarded, advanced and correlation Green's functions G^r, G^a and G^c:

$$G_{\alpha\beta} = \frac{1}{2}(\xi_\alpha \eta_\beta G^r + \eta_\alpha \xi_\beta G^a + \xi_\alpha \xi_\beta G^c), \tag{12.9}$$

we can rewrite (12.8) as

$$\mathbf{f}_i = -n_i u^2 \int \frac{d\omega}{2\pi} \sum_{\mathbf{q},\mathbf{k}} \mathbf{q}$$
$$[G^r(\mathbf{k}-\mathbf{q}, \omega + \mathbf{q}\cdot\mathbf{v}_d) G^c(\mathbf{k},\omega) + G^c(\mathbf{k}-\mathbf{q}, \omega + \mathbf{q}\cdot\mathbf{v}_d) G^a(\mathbf{k},\omega)]. \tag{12.10}$$

These Green's functions are defined with respect to the initial density matrix $\hat{\rho}_0$, Eq.(2.30), i.e. a thermal equilibrium ensemble with Hamiltonian H_e at the electron temperature T_e. Therefore

$$G^c(\mathbf{k},\omega) = [1 - 2f(\omega)]\,[G^r(\mathbf{k},\omega) - G^a(\mathbf{k},\omega)], \tag{12.11}$$

where $f(\omega)$ is the Fermi function of the electron system at electron temperature T_e.

The effects of both the elastic (impurity) and inelastic (electron-electron and electron-phonon) scatterings on the self energy can be taken into account by introducing a life-time in the retarded and advanced electron Green's functions:

$$G^{r(a)}(\mathbf{k},\omega) = \frac{1}{\omega - \varepsilon_k \pm i/\tau_e} \tag{12.12}$$

Both elastic scattering and inelastic scattering contribute to $1/\tau_e$, such that

$$\frac{1}{\tau_e} = \frac{1}{\tau} + \frac{1}{\tau_{in}} \tag{12.13}$$

where τ and τ_{in} are, respectively, elastic and inelastic scattering times. Substituting expression (12.12) into Eq.(12.10) and noticing that terms with G^rG^r and G^aG^a vanish after ω-integration, we are left with

$$\mathbf{f}_i = -in_i u^2 \sum_{\mathbf{q}} \mathbf{q}\,\Pi(\mathbf{q}, \mathbf{q}\cdot\mathbf{v}_d) = n_i u^2 \sum_{\mathbf{q}} \mathbf{q}\,\Pi_2(\mathbf{q}, \mathbf{q}\cdot\mathbf{v}_d). \tag{12.14}$$

Here the correlation function

$$\Pi(\mathbf{q},\omega_1) = -i \int \frac{d\omega}{\pi} \sum_{\mathbf{k}} [f(\omega) - f(\omega+\omega_1)]\, G^r(\mathbf{k}-\mathbf{q}, \omega+\omega_1) G^a(\mathbf{k},\omega) \tag{12.15}$$

satisfies $\Pi(\mathbf{q},\omega) = \Pi(-\mathbf{q},-\omega)^*$, and in the case of small $1/\tau_e$ its imaginary part can be written as

$$\Pi_2(\mathbf{q},\omega_1) = 2\pi \sum_{\mathbf{k}} [f(\varepsilon_{\mathbf{k}}) - f(\varepsilon_{\mathbf{k}+\mathbf{q}})]\delta(\varepsilon_{\mathbf{k}+\mathbf{q}} - \varepsilon_{\mathbf{k}} + \omega_1). \tag{12.16}$$

Eq.(12.14) is the expression for impurity-induced frictional force given in Section 5.

The diagrams for the electron energy-loss rate can be constructed in the same way. The time derivative of the phonon field operator $\dot{\phi}_{\mathbf{q}\lambda}$ results in an additional frequency factor at the primary vertex in the ω-representation, which is indicated by a circle. The lowest order diagram for the electron energy-loss rate is shown in Fig.12.1b and it gives

$$w = i\int \frac{d\omega d\Omega}{(2\pi)^2} \sum_{\mathbf{k},\mathbf{q},\lambda} |M(\mathbf{q},\lambda)|^2 \Omega\, \xi_\alpha \eta_{\alpha'}$$
$$G_{\alpha\alpha'}(\mathbf{k}-\mathbf{q},\omega-\Omega+\mathbf{q}\cdot\mathbf{v}_d) G_{\alpha'\alpha}(\mathbf{k},\omega) D_{\alpha\alpha'}(\mathbf{q}\lambda,\Omega). \tag{12.17}$$

Here the components of the phonon closed-time-path Green's function $D_{\alpha\beta}$ are connected with the retarded, advanced and correlation functions as (12.9)

$$D_{\alpha\beta} = \frac{1}{2}\left(\xi_\alpha \eta_\beta D^r + \eta_\alpha \xi_\beta D^a + \xi_\alpha \xi_\beta D^c\right). \tag{12.18}$$

In the absence of a phonon life time, we have

$$\begin{aligned} D^r(\mathbf{q}\lambda,\Omega) &= (\Omega-\Omega_{\mathbf{q}\lambda}+i\delta)^{-1} - (\Omega+\Omega_{\mathbf{q}\lambda}+i\delta)^{-1} \\ D^a(\mathbf{q}\lambda,\Omega) &= -(\Omega-\Omega_{\mathbf{q}\lambda}-i\delta)^{-1} + (\Omega+\Omega_{\mathbf{q}\lambda}-i\delta)^{-1} \\ D^c(\mathbf{q}\lambda,\Omega) &= -i2\pi(1+2n(\Omega_{\mathbf{q}\lambda}/T))\left[\delta(\Omega-\Omega_{\mathbf{q}\lambda}) + \delta(\Omega+\Omega_{\mathbf{q}\lambda})\right], \end{aligned} \tag{12.19}$$

where $n(x) = 1/(e^x-1)$ is the Bose function and $\delta = 0^+$. We can calculate the expression (12.17) for w by reducing it to

$$\begin{aligned} w \;=\; & \frac{i}{2}\int \frac{d\omega d\Omega}{(2\pi)^2} \sum_{\mathbf{k},\mathbf{q},\lambda} |M(\mathbf{q},\lambda)|^2 \Omega \\ & \{[G^r(\mathbf{k}-\mathbf{q},\omega-\Omega+\mathbf{q}\cdot\mathbf{v}_d)G^c(\mathbf{k},\omega) + \\ & G^c(\mathbf{k}-\mathbf{q},\omega-\Omega+\mathbf{q}\cdot\mathbf{v}_d)G^a(\mathbf{k},\omega)]D^c(\mathbf{q}\lambda,\Omega) + \\ & [G^r(\mathbf{k}-\mathbf{q},\omega-\Omega+\mathbf{q}\cdot\mathbf{v}_d)G^a(\mathbf{k},\omega) + \\ & G^c(\mathbf{k}-\mathbf{q},\omega-\Omega+\mathbf{q}\cdot\mathbf{v}_d)G^c(\mathbf{k},\omega)]D^r(\mathbf{q}\lambda,\Omega)\}. \end{aligned} \tag{12.20}$$

The final result is

$$w = 2\sum_{\mathbf{q}\lambda} |M(\mathbf{q},\lambda)|^2 \Omega_{\mathbf{q}\lambda} \Pi_2(\mathbf{q},\Omega_{\mathbf{q}\lambda}+\mathbf{q}\cdot\mathbf{v}_d)\left[n(\frac{\Omega_{\mathbf{q}\lambda}}{T}) - n(\frac{\Omega_{\mathbf{q}\lambda}+\mathbf{q}\cdot\mathbf{v}_d}{T_e})\right], \tag{12.21}$$

as expected from Section 5.

Note that the phonon contribution to the frictional force $\mathbf{f}_p$ can also be represented by a diagram shown in Fig.12.1b if the circle vertex Ω is replaced by a dot vertex $\mathbf{q}$. The expression for $\mathbf{f}_p$ is obtained if the factor $\Omega_{\mathbf{q}\lambda}$ in Eq.(12.21) is replaced by $\mathbf{q}$.

12.2 Electric Field Effects on Weak Localization

We assume that impurity scattering dominates momentum relaxation such that the phonon contribution to the frictional force is negligible. Phonons, however, still play an important role in dissipating energy from the electron system. It is well known that an important quantum correction to conductivity arises from maximally crossed diagrams for elastic scattering. In the balance equation theory the contribution of the n-th order maximally crossed diagram to the frictional force is shown in Fig.12.2,

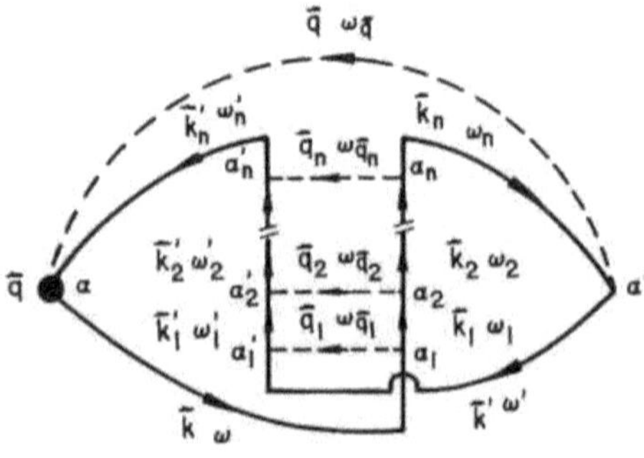

Fig.12.2 Contribution of the nth-order maximally crossed diagram to the frictional force.

and is given by

$$\delta \mathbf{f}_n = (n_i u^2)^{n+1} \sum_{\mathbf{q},\mathbf{k},\mathbf{q}_1\cdots\mathbf{q}_n} \mathbf{q} \int \frac{d\omega}{2\pi} \xi_\alpha \eta_{\alpha'} \eta_{\alpha_1} \eta_{\alpha_1'} \cdots \eta_{\alpha_n} \eta_{\alpha_n'}$$
$$G_{\alpha\alpha_n'}(\mathbf{k}_n', \omega_n') G_{\alpha_n'\alpha_{n-1}'}(\mathbf{k}_{n-1}', \omega_{n-1}') \cdots G_{\alpha_1'\alpha'}(\mathbf{k}', \omega') G_{\alpha'\alpha_n}(\mathbf{k}_n, \omega_n)$$
$$G_{\alpha_n\alpha_{n-1}}(\mathbf{k}_{n-1}, \omega_{n-1}) \cdots G_{\alpha_1\alpha}(\mathbf{k}, \omega), \tag{12.22}$$

where $\mathbf{k}_i = \mathbf{k} - \mathbf{q}_1 - \mathbf{q}_2 \cdots - \mathbf{q}_i$, $\omega_i = \omega + (\mathbf{q}_1 + \mathbf{q}_2 \cdots + \mathbf{q}_i) \cdot \mathbf{v}_d$, $\mathbf{k}'_i = \mathbf{k} - \mathbf{q}_{i+1} \cdots - \mathbf{q}_n - \mathbf{q}$, $\omega_i' = \omega + (\mathbf{q}_{i+1} \cdots + \mathbf{q}_n + \mathbf{q}) \cdot \mathbf{v}_d$ $(i = 1, \cdots n)$ and $\mathbf{k}_n = \mathbf{k}' + \mathbf{q}$, $\omega_n = \omega' - \mathbf{q} \cdot \mathbf{v}_d$, $\mathbf{k}'_n = \mathbf{k} - \mathbf{q}$, $\omega_n' = \omega + \mathbf{q} \cdot \mathbf{v}_d$. Using relations (12.9) and (12.11) we can express the above combinations of the components of the closed-time-path Green's function as the sum of terms composed of various numbers of retarded and advanced Green's function factors. The only terms that survive the summation over $\mathbf{q}_1, \mathbf{q}_2 \cdots \mathbf{q}_n$ are those with equal number of retarded and advanced Green's functions:

$$2\,[f(\omega_n) - f(\omega')]\, G^r(\mathbf{k}_n', \omega_n') G^r(\mathbf{k}_{n-1}', \omega_{n-1}') \cdots$$
$$G^r(\mathbf{k}', \omega') G^a(\mathbf{k}_n, \omega_n) G^a(\mathbf{k}_{n-1}, \omega_{n-1}) G^a(\mathbf{k}, \omega)$$

and the calculation is straightforward (Lei and Cai, 1990).

The sum of all orders of maximally crossed diagrams yields a total correction to the frictional force

$$\delta f = \sum_{n=1}^{\infty} \delta f_n$$

, and thus an addition to the steady state resistivity

$$\delta\rho = -\frac{\delta f}{n^2 e^2 v_d}$$

$$= -\frac{n_i u^2}{n^2 e^2 v_d^2} \sum_{\mathbf{q},\mathbf{k},\mathbf{k}'} \mathbf{q} \cdot \mathbf{v}_d \int \frac{d\omega}{\pi} [f(\omega' - \mathbf{q} \cdot \mathbf{v}_d) - f(\omega')]$$
$$G^r(\mathbf{k} - \mathbf{q}, \omega + \mathbf{q} \cdot \mathbf{v}_d) G^r(\mathbf{k}', \omega') G^a(\mathbf{k} + \mathbf{q}, \omega' - \mathbf{q} \cdot \mathbf{v}_d) G^a(\mathbf{k}, \omega) C(K_v) \quad (12.23)$$

where

$$C(K_v) = \frac{n_i u^2 \tau_e^{-1}}{1/\tau_{in} + (\tau_e/\tau)^2 D K_v^2}, \quad (12.24)$$

with $K_v = |\mathbf{k} + \mathbf{k}' + 2m\mathbf{v}_d|$, and $D = v_F^2 \tau / d$ is the diffusion coefficient due to impurity scattering in d dimensions. If $\tau_{in} \gg \tau$, or $\tau_e \simeq \tau$, which is the case for weak localization, we can write $C(K_v)$ approximately as

$$C(K_v) = \frac{1}{2\pi N(0)\tau^2} \frac{1}{D K_v^2 + 1/\tau_{in}}. \quad (12.25)$$

This is the well-known expression for the particle-particle diffusion propagator (Lee and Ramakrishnan, 1985). The existence of the inverse inelastic scattering time $1/\tau_{in}$ in Eq.(12.25) implies a cutoff of the quantum interference effect beyond the Thouless length $L_{Th} = (D\tau_{in})^{1/2}$ (Thouless, 1977). Such a cutoff was phenomenologically introduced previously (Anderson et al, 1979). The present derivation suggests a simple way to achieve this inelastic-scattering cutoff of the localization effect. Furthermore, this expression for $C(K_v)$ in the presence of a finite drift velocity ($v_d \neq 0$) indicates that a dc electric field produces no direct cutoff related to it. This is in agreement with Altshuler et al. (1981, 1982), who pointed out that a dc electric field does not break the time-reversal invariance and should not affect quantum interference. However, the drift velocity v_d, hence the dc electric field, does enter in the other factors of Eq.(12.23) and affects the quantum correction to the resistivity contributed by maximally crossed diagrams.

To calculate $\delta\rho$ from Eq.(12.23) we proceed by changing the summation variables $\mathbf{k}$ and $\mathbf{k}'$ to $\mathbf{k}_v = \mathbf{k}' - \mathbf{k}$ and $\mathbf{K}_v = \mathbf{k} + \mathbf{k}' + 2m\mathbf{v}_d$. The expression (12.25) of $C(K_v)$ indicates that the dominant contribution to the resistivity correction $\delta\rho$ comes from the small K_v region, such that in other factors of the integrand, we can neglect K_v in comparison with k_v, where the dominant contribution region is of order of k_F. With these considerations, the summation over k_v can be carried out, yielding at low temperature T_e

$$\delta\rho = \frac{2m^2}{\pi n^2 e^2 \tau^2} S_d \, g_d(2k_F v_d \tau), \quad (12.26)$$

where

$$S_d = \sum_{\mathbf{K}_v} \frac{1}{K_v^2 + L_{Th}^{-2}} \quad (12.27)$$

is the conventional dimension-dependent scale factor. Note that all the above formulations are written per unit volume and we assume the system size to be much larger than L_{Th}, (L_{Th} being the length cutoff for interference). The upper cutoff for K_v is generally taken to be ℓ^{-1}, $\ell = v_F\tau$ being the electron mean free path related to elastic scattering. We have,

$$S_3 = \frac{1}{2\pi^2}\left(\frac{1}{\ell} - \frac{\pi}{2L_{Th}}\right) \tag{12.28}$$

$$S_2 = \frac{1}{2\pi}\ln\left(\frac{L_{Th}}{\ell}\right) \tag{12.29}$$

$$S_1 = \frac{1}{2}L_{Th}. \tag{12.30}$$

These factors depend strongly on the electron temperature T_e because of the temperature dependence of the inelastic scattering time τ_{in}. The direct effect of a dc electric field is included in the function $g_d(2k_F v_d \tau)$:

$$g_3(x) = \frac{6}{x^2}\left[1 - \frac{1}{x}\arctan x + \frac{1}{x^2}\ln(1+x^2)\right] \simeq \begin{cases} 1 & x \ll 1 \\ 6/x^2 & x \gg 1 \end{cases} \tag{12.31}$$

$$g_2(x) = \frac{8}{\pi x^2}\int_0^1 \frac{d\xi}{\sqrt{1-\xi^2}}\left(1 - \frac{1}{\sqrt{1+x^2\xi^2}}\right) \simeq \begin{cases} 1 & x \ll 1 \\ 4/x^2 & x \gg 1 \end{cases} \tag{12.32}$$

$$g_1(x) = \frac{1}{1+x^2}. \tag{12.33}$$

The magnitude of $g_d(x)$ decreases with increasing x, indicating that localization effects are reduced when the electric field increases. (Lei and Cai, 1990a).

13. BALANCE EQUATIONS FOR WEAKLY NONUNIFORM SYSTEMS

Up to now we have dealt with spatially homogeneous case involving only a uniform applied electric field. Lei, Cai and Xie(1988) initiated the extension of the balance equation approach to a weakly nonuniform system.

13.1 Hamiltonian of the Small System Elements

Lei, Cai and Xie(1988) considered an interacting electron system under the influence of an inhomogeneous external electric field $\mathbf{E}_e(\mathbf{r}) = -\nabla\phi_e(\mathbf{r})$,

$$H = \sum_i \left[\frac{p_i^2}{2m} + \sum_{j\neq i}\frac{e^2}{8\pi\epsilon_0}\frac{1}{|\mathbf{r}_i - \mathbf{r}_j|} + e\phi_e(\mathbf{r}_i) + \Phi(\mathbf{r}_i)\right]. \tag{13.1}$$

$\mathbf{p}_i$ and $\mathbf{r}_i$ are the momentum and coordinate of ith electron, and

$$\Phi(\mathbf{r}_i) = \sum_a u(\mathbf{r}_i - \mathbf{R}_a) + \sum_l v_l(\mathbf{r}_i - \mathbf{R}_l) \tag{13.2}$$

is the scattering potential due to randomly distributed ($\mathbf{R}_a$) impurities and lattice vibrations ($\mathbf{R}_l$ stands for the lattice site). The whole system can be divided into many macroscopically small subsystems. A subsystem, which is termed a fluid element, or simply an element, is composed of electrons in a small volume $d\tau$ about a position $\mathbf{R}$. The Hamiltonian of the element $d\tau$ can be written as

$$\delta H = \sum_{i\in d\tau}\left[\frac{p_i^2}{2m} + \frac{e^2}{8\pi\epsilon_0}\sum_{j\neq i}\frac{1}{|\mathbf{r}_i - \mathbf{r}_j|} + e\phi_e(\mathbf{r}_i) + \Phi(\mathbf{r}_i)\right]. \tag{13.3}$$

Here, the second term in parentheses is the Coulomb interaction potential of the ith electron with all the other electrons. We divide it into two parts: contribution from electrons inside $d\tau$ and those from outside, and use a mean-field treatment for the latter. For small $d\tau$ it becomes the macroscopic induced potential $\phi_i(\mathbf{r}_i)$ caused by the electron charge distribution (plus positive charge background), such that

$$\delta H = \sum_{i\in d\tau}\left[\frac{p_i^2}{2m} + \sum_{\substack{j\neq i\\ j\in d\tau}}\frac{e^2}{8\pi\epsilon_0}\frac{1}{|\mathbf{r}_i - \mathbf{r}_j|} + e\phi(\mathbf{r}_i) + \Phi(\mathbf{r}_i)\right], \tag{13.4}$$

with $\phi(\mathbf{r}) = \phi_e(\mathbf{r}) + \phi_i(\mathbf{r})$ being the total potential. In the limit of small $d\tau$ the restricted sum over the particles inside $d\tau$ around $\mathbf{R}$ can be conveniently represented by a δ-function.

$$\sum_{i\in d\tau} \sim d\tau\sum_i \delta(\mathbf{r}_i - \mathbf{R}), \tag{13.5}$$

such that we can write

$$\delta H = d\tau H(\mathbf{R}) \quad \text{and} \quad H = \int d\tau H(\mathbf{R}). \tag{13.6}$$

Here,

$$H(\mathbf{R}) = H_e(\mathbf{R}) + \sum_i [e\phi(\mathbf{r}_i) + \Phi(\mathbf{r}_i)]\delta(\mathbf{r}_i - \mathbf{R}), \tag{13.7}$$

and

$$H_e(\mathbf{R}) = \sum_i \left[\frac{p_i^2}{2m} + \frac{e^2}{8\pi\epsilon_0} \sum_{\substack{j\neq i \\ j\in d\tau}} \frac{1}{|\mathbf{r}_i - \mathbf{r}_j|} \right] \delta(\mathbf{r}_i - \mathbf{R}). \tag{13.8}$$

$H_e(\mathbf{R})d\tau$ represents the kinetic and Coulombic energies of electrons in the element $d\tau$. Note that, under the present approximation, different elements consist of different sets of electrons, and the Coulomb coupling between electrons in different elements has been included in the total potential $\phi(\mathbf{r})$, which is a parametrized variable to be determined self-consistently by the resulting equations. Therefore, different subparts of H are independent and commutative. The number of electrons in element $d\tau$ is $\delta N = d\tau N(\mathbf{R})$ and

$$N(\mathbf{R}) = \sum_i \delta(\mathbf{r}_i - \mathbf{R}) \tag{13.9}$$

is the number density of electrons in the element around $\mathbf{R}$. It is convenient to identity $\mathbf{R}$ as the coordinate of the center of mass of the electrons in $d\tau$:

$$\mathbf{R} = \sum_{i\in d\tau} \mathbf{r}_i/\delta N. \tag{13.10}$$

The total momentum of the electrons in element $d\tau$ is $\delta\mathbf{P} = \sum_{i\in d\tau} \mathbf{p}_i = d\tau\mathbf{P}(\mathbf{R})$,

$$\mathbf{P}(\mathbf{R}) = \sum_i \mathbf{p}_i\delta(\mathbf{r}_i - \mathbf{R}) \tag{13.11}$$

being the momentum density. Letting $\mathbf{v}(\mathbf{R})$ be the average velocity of the electrons in element $d\tau$, which is a parameter to be determined self-consistently by the resulting balance equations, we can write the statistical expectation of the momentum density as

$$\langle\mathbf{P}(\mathbf{R})\rangle = mn(\mathbf{R})\mathbf{v}(\mathbf{R}), \tag{13.12}$$

where $n(\mathbf{R}) = \langle N(\mathbf{R})\rangle$ is the statistical average of the electron number density. It is convenient to use the relative electron variables $\mathbf{p}_i'$ and $\mathbf{r}_i'$, defined by

$$\mathbf{p}_i' = \mathbf{p} - m\mathbf{v}(\mathbf{R}), \qquad \mathbf{r}_i' = \mathbf{r}_i - \mathbf{R}. \tag{13.13}$$

They are momentum and coordinate of the ith electron relative to the center of mass of element $d\tau$. With these relative electron variables, the statistical average of $H_e(\mathbf{R})$ can be written as

$$\langle H_e(\mathbf{R})\rangle = u(\mathbf{R}) + \frac{1}{2}mn(\mathbf{R})v^2(\mathbf{R}), \tag{13.14}$$

where

$$u(\mathbf{R}) = \left\langle \sum_i \frac{p_i'^2}{2m}\delta(\mathbf{r}_i - \mathbf{R}) \right\rangle, \tag{13.15}$$

$u(\mathbf{R})d\tau$ being the average kinetic energy of the relative electrons in $d\tau$. The interaction potential energy between electrons inside $d\tau$, which is a higher-order quantity in $d\tau$, has been neglected.

13.2 Rates of Change of Particle Number, Momentum and Energy

We now consider the rates of charge of particle number, momentum and energy. First, the rate of change of particle number density $\dot{N}(\mathbf{R}) = -i[N(\mathbf{R}), H]$ can be easily calculated and this results (after taking statistical average) in the continuity equation:

$$\frac{\partial n}{\partial t} + \nabla \cdot (\mathbf{v}n) = 0. \tag{13.16}$$

Secondly, from Eq.(13.12), we have

$$\frac{\partial}{\partial t}\langle \mathbf{P}(\mathbf{R})\rangle = m\frac{\partial n}{\partial t}\mathbf{v} + mn\frac{\partial \mathbf{v}}{\partial t}. \tag{13.17}$$

On the other hand, we can calculate $\dot{\mathbf{P}}(\mathbf{R})$ according to $\dot{\mathbf{P}}(\mathbf{R}) = -i[\mathbf{P}(\mathbf{R}), H]$ to give

$$\dot{\mathbf{P}}(\mathbf{R}) = -\nabla_{\mathbf{R}} \cdot \left[\sum_i \frac{\mathbf{p}_i\mathbf{p}_i}{m}\delta(\mathbf{r}_i - \mathbf{R})\right] - \sum_i [\nabla\Phi(\mathbf{r}_i) + e\nabla\phi(\mathbf{r}_i)]\delta(\mathbf{r}_i - \mathbf{R}). \tag{13.18}$$

Changing to relative electron variables and taking the statistical average, we have

$$\langle\dot{\mathbf{P}}(\mathbf{R})\rangle = -\nabla[\frac{2}{3}u(\mathbf{R})] - \nabla \cdot [mn(\mathbf{R})\mathbf{v}(\mathbf{R})\mathbf{v}(\mathbf{R})] + en(\mathbf{R})\mathbf{E}(\mathbf{R}) + \mathbf{f}(\mathbf{R}), \tag{13.19}$$

where

$$\mathbf{f}(\mathbf{R}) = -\left\langle \sum_i \nabla\Phi(\mathbf{r}'_i + \mathbf{R})\delta(\mathbf{r}'_i)\right\rangle, \tag{13.20}$$

$\mathbf{f}(\mathbf{R})d\tau$ being the resistive force experienced by the fluid element due to impurity and phonon scatterings. Equating Eq.(13.19) to Eq.(13.17) and making use of the continuity equation (13.16), we obtain an Euler-type momentum balance equation:

$$\frac{\partial \mathbf{v}}{\partial t} + \mathbf{v}(\nabla \cdot \mathbf{v}) = -\frac{2}{3}\frac{\nabla u}{mn} + \frac{e}{m}\mathbf{E} + \frac{1}{mn}\mathbf{f}. \tag{13.21}$$

Thirdly, the rate of change of $\langle H_e(\mathbf{R})\rangle$ is calculated from Eq.(13.14) to be

$$\frac{\partial}{\partial t}\langle H_e(\mathbf{R})\rangle = \frac{du}{dt} + \frac{m}{2}\frac{dn}{dt}v^2 + mn\mathbf{v} \cdot \frac{d\mathbf{v}}{dt}, \tag{13.22}$$

which should be equated to that obtained from averaging the expression $\dot{H}_e(\mathbf{R}) = -i[H_e(\mathbf{R}), H]$:

$$\langle \dot{H}_e(\mathbf{R})\rangle = -\nabla \cdot \mathbf{v}(\mathbf{R})\frac{5}{3}u(\mathbf{R}) + \frac{m}{2}n(\mathbf{R})v^2(\mathbf{R}) + en(\mathbf{R})\mathbf{v}(\mathbf{R}) \cdot \mathbf{E}(\mathbf{R}) - w(\mathbf{R}), \tag{13.23}$$

where

$$w(\mathbf{R}) = \left\langle \sum_i \frac{\mathbf{p}_i'}{m} \cdot \nabla\Phi(\mathbf{r}_i' + \mathbf{R})\delta(\mathbf{r}_i') \right\rangle - \mathbf{v}(\mathbf{R}) \cdot \mathbf{f}(\mathbf{R}), \tag{13.24}$$

$w(\mathbf{R})d\tau$ being the energy-loss rate of electrons in element $d\tau$. Combining Eqs.(13.22) and (13.23) and considering Eqs. (13.16) and (13.21) we obtain the energy balance equation:

$$\frac{\partial u}{\partial t} + \mathbf{v} \cdot \nabla u = -\frac{5}{3}u(\nabla \cdot \mathbf{v}) - w - \mathbf{v} \cdot \mathbf{f}. \tag{13.25}$$

13.3 Hydrodynamic Balance Equations

To calculate the resistive force $\mathbf{f}$, the energy-loss rate w and other average quantities, we need the density matrix $\hat{\rho}$, which can be obtained by solving the Liouville equation with the appropriate initial condition. In the present model the interactions between different fluid elements are included approximately in the total potential with a mean-field treatment. Therefore different fluid elements, though geometrically correlative, are dynamically independent, and thus evolve separately from their own initial states.

In accordance with the precepts of the balance-equation approach, the unperturbed state is chosen to be described by a density matrix $\hat{\rho}_0$ of the local equilibrium form,

$$\hat{\rho}_0 = \frac{1}{Z} \exp[-\int [\delta H - \mu(\mathbf{R})\delta N]/T_e(\mathbf{R})] \exp(-H_{ph}/T), \tag{13.26}$$

where electron-temperature field $T_e(\mathbf{R})$ and chemical potential field $\mu(\mathbf{R})$ are parameters to be determined selfconsistently from the resulting balance equations, and H_{ph} is the phonon Hamiltonian.

The processes involved in deriving the perturbed density matrix $\hat{\rho} = \hat{\rho}_0 + \hat{\rho}_1$ to linear order in δH_I and in calculating the statistical expectations follow those of Section 3. To lowest nonzero order, the average local kinetic energy density of the relative electrons is determined by $\hat{\rho}_0$ only:

$$u = 2\sum_{\mathbf{k}} \varepsilon_{\mathbf{k}} f[(\varepsilon_{\mathbf{k}} - \mu)/T_e], \tag{13.27}$$

and the local chemical potential $\mu(\mathbf{R})$ is related to the local density $n(\mathbf{R})$ of electrons via the relation

$$n = 2\sum_{\mathbf{k}} f[(\varepsilon_{\mathbf{k}} - \mu)/T_e]. \tag{13.28}$$

Here $\varepsilon_{\mathbf{k}} = k^2/2m$ and $f(x) \equiv (e^x + 1)^{-1}$. The resistive force density $\mathbf{f}(\mathbf{R})$ and the energy loss rate $w(\mathbf{R})$ are found to be

$$\mathbf{f} = \mathbf{f}(n(\mathbf{R}), T_e(\mathbf{R}), \mathbf{v}(\mathbf{R})) = n_i \int \frac{d^3q}{(2\pi)^3} \mathbf{q} |u(\mathbf{q})|^2 \Pi_2(\mathbf{q}, \omega_0)$$
$$+2 \int \frac{d^3q}{(2\pi)^3} \mathbf{q} \sum_\lambda |M(\mathbf{q}, \lambda)|^2 \Pi_2(\mathbf{q}, \omega_0 + \Omega_{\mathbf{q}\lambda}) \left[n\left(\frac{\Omega_{\mathbf{q}\lambda}}{T}\right) - n\left(\frac{\Omega_{\mathbf{q}\lambda} + \omega_0}{T_e}\right) \right], \quad (13.29)$$
$$w = w(n(\mathbf{R}), T_e(\mathbf{R}), \mathbf{v}(\mathbf{R}))$$
$$= 2 \int \frac{d^3q}{(2\pi)^3} \sum_\lambda |M(\mathbf{q}, \lambda)|^2 \Omega_{\mathbf{q}\lambda} \Pi_2(\mathbf{q}, \omega_0 + \Omega_{\mathbf{q}\lambda}) \left[n\left(\frac{\Omega_{\mathbf{q}\lambda}}{T}\right) - n\left(\frac{\Omega_{\mathbf{q}\lambda} + \omega_0}{T_e}\right) \right]. \quad (13.30)$$

Here $\omega_0 \equiv \mathbf{q} \cdot \mathbf{v}(\mathbf{R})$, $n(x) \equiv (e^x - 1)^{-1}$, n_i is the impurity density around $\mathbf{R}$, $\Omega_{\mathbf{q}\lambda}$ is the phonon frequency, and $\Pi_2(\mathbf{q}, \omega)$ is the density-density correlation function of relative electrons in a unit volume with density $n(\mathbf{R})$ at temperature $T_e(\mathbf{R})$, including the effect of Coulomb interaction between carriers, which can be treated in RPA or beyond. We see that although the intercarrier Coulomb coupling inside the element gives a vanishing contribution to the average internal energy of the electron system, it plays an important role in screening.

These are six parametric field variables: $\mathbf{v}(\mathbf{R})$, $T_e(\mathbf{R})$, $u(\mathbf{R})$, $\mathrm{n}(\mathbf{R}), \mu(\mathbf{R})$, as well as the total electric potential $\phi(\mathbf{R})$ or field $\mathbf{E}(\mathbf{R})$. These variables may be time dependent on a macroscopic time scale for time-dependent processes. We have three hydrodynamic-type equations: continuity equation (13.16), momentum-balance equation (13.21), and energy-balance equation (13.25), supplemented by two relations (13.27) and (13.28) plus the Poisson equation relating electric charge density and potential (n^+ denotes the density of positive charge background):

$$\nabla^2 \phi = -\frac{e}{\epsilon_0}[n(\mathbf{R}) - n^+]. \quad (13.31)$$

All the space-time field variables can be determined for the given initial and boundary conditions.

14. ALTERNATIVE FORMULATIONS AND INTERPRETATIONS OF THE BALANCE EQUATION APPROACH

14.1 Dielectric Interpretation of the Impurity-Induced Force

One can gain insight into the physical processes associated with the balance equation description of transport dynamics through a reinterpretation of the resistive force on the c.m. due to impurity scattering (Horing, Lei and Cui, 1986). The impurity scattering potential in Eq. (1.2) can be written in terms of c.m. and relative electron coordinates as

$$H_{ei} = \sum_{i,a} u(\mathbf{r}'_i + \mathbf{R} - \mathbf{R}_a). \tag{14.1}$$

Considering that H_{ei} is the only mechanism which couples the c.m. and relative electron variables, one may take the view that the moving c.m. projects into space the single-particle potential $u(2)$, $(2 \equiv \mathbf{r}'_2, t_2)$:

$$u(2) = \sum_{a} u(\mathbf{r}'_2 + \mathbf{R}(t_2) - \mathbf{R}_a), \tag{14.2}$$

which polarizes the relative electrons, resulting in an effective electric potential $\phi(1)$, $(1 \equiv \mathbf{r}'_1, t_1;\;\; d2 \equiv d\mathbf{r}_2 dt_2)$

$$\phi(1) = \int d2 K(1,2) u(2)/e, \tag{14.3}$$

where $K(1,2)$ is the inverse dielectric function of the relative electron gas, representing its dynamic, nonlocal (and inhomogeneous) screening action in response to $u(2)$ impressed by the c.m. In this, the c.m. motion is taken to be classical, and for the steady state, $\mathbf{R}(t_2) = \mathbf{v}_d t_2$.

The charge density perturbation of the relative electron gas may be identify as $\rho(1) = -\epsilon_0 \nabla_1^2(\phi(1) - u(1)/e)$ (where we have separated the unperturbed density), and the effective electric field is given by $-\nabla_1\phi(1)$. Thus the force per unit volume of the relative electron gas is $-\rho(1)\nabla_1\phi(1)$, and by the law of action and reaction, the resistive force on the c.m. (integrating over the relative electron gas) is given by

$$\mathbf{F} = \frac{\epsilon_0}{e^2} \int d\mathbf{r}'_1 \int d2 \int d3 [\nabla_1 K(1,2) u(2)] \nabla_1^2 [K(1,3) - \delta(1,3)] u(3). \tag{14.4}$$

The integrand term involving the product $\nabla_1^2 K(1,3)$ may be rewritten in terms of the divergence of the electric stress tensor for the field $\nabla_1\phi(1)$, and consequently it does not contribute to the integrated resistive forces, whence

$$\mathbf{F} = \frac{\epsilon_0}{e^2} \int d\mathbf{r}'_1 \int d2 [\nabla_1 K(1,2) u(2)] \nabla_1^2 u(1). \tag{14.5}$$

To leading order in the impurity scattering potential, $\mathbf{F}$ is explicitly of order of u^2, and $K(1,2)$ is evaluated in the absence of u, and it therefore is decoupled from the electric field as well as impurity scattering. Such an evaluation of the inverse dielectric function nevertheless involves electron-electron interactions which can be treated within the framework of the random-phase approximation. Moreover, $K(1,2)$ is evaluated using an equilibrium ensemble for the relative electrons with a constant temperature. Since we focus here only on the resistive force due to impurity scattering, we may assume that there is a bath in contact with the system which dissipates the energy quickly enough to keep the system temperature at the lattice temperature while the bath itself contributes no resistive force. With this, $K(1,2)$ is translationally invariant in time and space for a homogeneous bulk medium, and it is convenient to Fourier analyze: $K(1-2) \to K(p) \equiv K(\mathbf{p},\omega)$. In these terms, the resistive force per unit volume may be written as

$$\mathbf{f} = n_i \frac{\epsilon_0}{e^2} \int \frac{d\mathbf{p}}{(2\pi)^3} \mathbf{p}|\mathbf{p}|^2 K_2(\mathbf{p}, \mathbf{p}\cdot\mathbf{v}_d)|u(\mathbf{p})|^2 \tag{14.6}$$

Here $\mathbf{f}$ represents the average of $\mathbf{F}$ over randomized impurity sites with impurity density n_i, and $K_2(\mathbf{p},\omega)$ is the imaginary part of $K(\mathbf{p},\omega) = K_1(\mathbf{p},\omega) + iK_2(\mathbf{p},\omega)$, and $u(\mathbf{p})$ is the impurity scattering potential in Fourier representation. Pertinent properties of inverse dielectric function $K(\mathbf{p},\omega) = 1/\epsilon(\mathbf{p},\omega)$ include $K(\mathbf{p},\omega) = K(-\mathbf{p},\omega)$, $K_1(\mathbf{p},\omega) = K_1(\mathbf{p},-\omega)$ and $K_2(\mathbf{p},\omega) = -K_2(\mathbf{p},-\omega)$ and its relation to the density-density correlation function $\Pi(\mathbf{p},\omega)$ is given by

$$K_2(\mathbf{p},\omega) = \frac{e^2}{\epsilon_0|\mathbf{p}|^2}\Pi_2(\mathbf{p},\omega). \tag{14.7}$$

Thus, (14.6) is exactly the impurity-induced frictional force given first by Lei and Ting (1984), Eq.(4.4).

The energy loss rate of the c.m. to the relative electron plasma, $\mathbf{f}\cdot\mathbf{v}_d$, may be likened to the energy loss rate of a passing particle to an isothermal plasma (held at a fixed temperature by a bath) (Klimontovich and Silin, 1960). In the latter case, the Coulomb interaction potential of the passing particle $u(\mathbf{r}-\mathbf{v}_d t)$ replaces the role of the impurity scattering potential (14.1) (and $\mathbf{r}$ replaces $\mathbf{r}'$). This facilitates recognition of the detailed correspondence between balance-equation transport theory and the well-known, classic, quantum many-body theoretic determination of energy loss of a passing particle to a solid state plasma. Indeed, the two are seen to be isomorphic (Horing, Lei and Cui, 1986), including energy transfer to plasmon and electron-hole pair excitations.

14.2 Drifted Electron-Temperature Model

Marchetti and Cai (1987) exhibited a connection between the balance equation approach and a generalized drifted electron-temperature model. They showed that the force- and energy-balance equations (6.2) and (6.3) for hot-electron transport in random-phase approximation, can be obtained formally from the Boltzmann equation for the electron distribution function $f(\mathbf{k})$:

$$\frac{\partial f(\mathbf{k})}{\partial t} - e\mathbf{E}\cdot\frac{\partial f(\mathbf{k})}{\partial \mathbf{k}} = \left[\frac{\partial f(\mathbf{k})}{\partial t}\right]_{ep} + \left[\frac{\partial f(\mathbf{k})}{\partial t}\right]_{ei} + \left[\frac{\partial f(\mathbf{k})}{\partial t}\right]_{ee}. \tag{14.8}$$

In this, $\mathbf{E}$ is the applied electric field and the right-hand side comprises the electron-phonon(ep), electron-impurity(ei) and electron-electron(ee) collision terms. The electron-phonon collision term, for instance, is given by

$$\left[\frac{\partial f(\mathbf{k})}{\partial t}\right]_{ep} = -\sum_{\mathbf{k}_1}\{W_{ep}(\mathbf{k}_1,\mathbf{k})f(\mathbf{k})[1-f(\mathbf{k}_1)] - W_{ep}(\mathbf{k},\mathbf{k}_1)f(\mathbf{k}_1)[1-f(\mathbf{k})]\}. \tag{14.9}$$

The transition probability of an electron from state $\mathbf{k}$ to $\mathbf{k}_1$ due to electron-phonon interaction, $W_{ep}(\mathbf{k}_1,\mathbf{k})$, takes the form

$$W_{ep}(\mathbf{k}_1,\mathbf{k}) = 2\pi\sum_{\mathbf{q}\lambda}|M(\mathbf{q},\lambda)|^2\sum_{\sigma=\pm1}\delta_{\mathbf{k}_1,\mathbf{k}+\mathbf{q}}\delta(\varepsilon_{\mathbf{k}} - \varepsilon_{\mathbf{k}_1} - \sigma\Omega_{\mathbf{q}\lambda})\left[n\left(\frac{\Omega_{\mathbf{q}\lambda}}{T}\right) + \frac{1}{2} + \frac{\sigma}{2}\right]. \tag{14.10}$$

They assumed that the role of the electron-electron collision term $[\partial f(\mathbf{k})/\partial t)]_{ee}$ is twofold: (a) to make the distribution function $f(\mathbf{k})$ take the form of a displaced Fermi-Dirac function at electron temperature T_e, having the wavevector shift $\mathbf{k}_d = m\mathbf{v}$, $\mathbf{v}$ being the drift velocity of the system:

$$f(\mathbf{k}) \to f(\varepsilon_{\mathbf{k}-\mathbf{k}_d}, T_e) \equiv \frac{1}{\exp[(\varepsilon_{\mathbf{k}-\mathbf{k}_d} - \mu)/T_e] + 1}, \tag{14.11}$$

and (b) to provide dynamic nonlocal screening of the electron-phonon matrix element and the electron-impurity potential, such that $M(\mathbf{q},\lambda)$ in (14.10) should be replaced by

$$\bar{M}(\mathbf{q},\lambda) = M(\mathbf{q},\lambda)/\epsilon_{RPA}(\mathbf{q},\Omega_{\mathbf{q}\lambda}) \tag{14.12}$$

and $u(\mathbf{q})$ by

$$\bar{u}(\mathbf{q}) = u(\mathbf{q})/\epsilon_{RPA}(\mathbf{q},0). \tag{14.13}$$

Here, $\Omega_{\mathbf{q}\lambda}$ is the frequency of the mode $\mathbf{q}\lambda$ phonon, and $\epsilon_{RPA}(\mathbf{q},\omega)$ is the "RPA" dielectric function given by

$$\epsilon_{RPA}(\mathbf{q},\omega) = 1 - 2v_c(q)\sum_{\mathbf{k}}\frac{f(\varepsilon_{\mathbf{k}+\mathbf{q}-\mathbf{k}_d},T_e) - f(\varepsilon_{\mathbf{k}-\mathbf{k}_d},T_e)}{\varepsilon_{\mathbf{k}+\mathbf{q}} - \varepsilon_{\mathbf{k}} + \omega + i\delta} = \epsilon(\mathbf{q},\omega + \mathbf{q}\cdot\mathbf{v}), \tag{14.14}$$

$\epsilon(\mathbf{q},\omega)$ being the dielectric function defined in Eq.(4.18). Multiplying Eq.(14.8) (without $[\partial f(\mathbf{k}/\partial t]_{ee}$ term) by $\mathbf{k}$, and by $(\mathbf{k}-\mathbf{k}_d)^2/2m$ respectively, summing over $\mathbf{k}$ and identifying

$$\mathbf{v} = \frac{2}{n}\sum_{\mathbf{k}}\frac{\mathbf{k}}{m}f(\mathbf{k}) \tag{14.15}$$

$$E_e = 2\sum_{\mathbf{k}}\frac{(\mathbf{k}-\mathbf{k}_d)^2}{2m}f(\mathbf{k}) \tag{14.16}$$

as average drift velocity and electron internal energy, with the replacements (14.12)-14.14), they obtain two coupled nonlinear equations

$$\frac{\partial}{\partial t}mn\mathbf{v} + en\mathbf{E} = \mathbf{f}(\mathbf{v}, T_e) \tag{14.17}$$

$$\frac{\partial}{\partial t}E_e = -\mathbf{v}\cdot\mathbf{f}(\mathbf{v}, T_e) - w(\mathbf{v}, T_e), \tag{14.18}$$

with $\mathbf{f} = \mathbf{f}_i + \mathbf{f}_p$ and w having the same expressions as given by Eqs.(4.4)-(4.6).

Since a displaced Fermi-Dirac function can not provide the solution to the dissipative Boltzmann equation in the presence of impurity and (or) phonon scattering(s), the above observation by Marchetti and Cai (1987) does not indicate the equivalence between the balance equation approach and the Boltzmann equation method (Lei and Ting, 1987).

It is also worth noting that a displaced equilibrium distribution function is not the distribution function of the balance equation theory. In principle, a system in transport is in a nonequilibrium state and can not be described by an equilibrium distribution function of any kind. To leading order of scattering interaction, $H_I = H_{ei} + H_{ep}$, the balance equation approach is a kind of linear response theory. The linear response of a system in the scattering interaction is determined by the properties of the system in the absence of the scattering interaction. Consequently, there is no need for detailed information about the distribution function in obtaining the bulk transport properties to leading order of H_I. This is one of the advantages of the balance-equation approach over the Boltzmann approach. Nevertheless, in balance-equation theory the distribution function can be defined for relative electrons as (Lei and Ting, 1987)

$$f(\mathbf{k}) = Tr(\hat{\rho}c_k^\dagger c_k). \tag{14.19}$$

According to Eq.(3.6), we can write $\hat{\rho} = \hat{\rho}_0 + \hat{\rho}_1$, $\hat{\rho}_0$ being the initial density matrix given by Eq.(2.30), whence

$$f(\mathbf{k}) = f_0(\mathbf{k}) + f_1(\mathbf{k}). \tag{14.20}$$

Here, $f_0(\mathbf{k}) = Tr(\rho_0 c_k^\dagger c_k)$ is an equilibrium distribution function, which may be thought of (disregarding the effects of the electron-electron interactions) as the Fermi-Dirac function (or the drifted Fermi-Dirac function in the fixed frame) at T_e. Moreover, $f_1(\mathbf{k})$ is the deviation from the equilibrium distribution function $f_0(\mathbf{k})$:

$$f_1(\mathbf{k}) = Tr(\hat{\rho}_1 c_k^\dagger c_k), \tag{14.21}$$

which is obviously not zero unless both $|u(\mathbf{q})|^2$ and $|M(\mathbf{q},\lambda)|^2$ vanish. This nonequilibrium distribution function has recently be calculated by Guillemot et al.(1991). The calculated distribution function resembles the experimental curve, which is markedly different from an equilibrium distribution function in the low energy range.

14.3 Generalized Quantum Langevin Equation Approach

Starting from the separation of c.m. variables from relative electron variables and the Hamiltonian of Eqs. (2.9)-(2.15), Hu and O'Connell (1987,1988a, 1989) developed a generalized quantum Langevin equation (GLE) approach to transport with the operator rates of change, $\dot{\mathbf{P}}$, $\dot{H}_{ph}$ and $\dot{H}_e$, given by Eqs. (2.20)-(2.24). They wrote, for instance, Eq.(2.20) as

$$\dot{\mathbf{P}} = Ne\mathbf{E} - i\sum_{\mathbf{q}} \mathbf{q} U^i_{\mathbf{q}}(\mathbf{R})\rho_{\mathbf{q}} - i\sum_{\mathbf{q}\lambda} \mathbf{q} M_{\mathbf{q}\lambda}(\mathbf{R}) O_{\mathbf{q}\lambda} \tag{14.22}$$

with the notation:

$$U^i_{\mathbf{q}}(\mathbf{R}) \equiv \sum_a u(\mathbf{q}) e^{i\mathbf{q}\cdot(\mathbf{R}-\mathbf{R}_a)} \tag{14.23}$$

$$M_{\mathbf{q}\lambda}(\mathbf{R}) \equiv M(\mathbf{q},\lambda) e^{i\mathbf{q}\cdot\mathbf{R}} \tag{14.24}$$

$$\rho_{\mathbf{q}} \equiv \sum_{\mathbf{k}} \rho_{\mathbf{kq}}, \qquad \rho_{\mathbf{kq}} \equiv \sum_{\sigma} c^\dagger_{\mathbf{k}+\mathbf{q}\sigma} c_{\mathbf{k}\sigma} \tag{14.25}$$

$$O_{\mathbf{q}\lambda} \equiv (b_{\mathbf{q}\lambda} + b^\dagger_{-\mathbf{q}\lambda})\rho_{\mathbf{q}} \equiv O^A_{\mathbf{q}\lambda} + O^E_{\mathbf{q}\lambda} \tag{14.26}$$

$$O^A_{\mathbf{q}\lambda} \equiv \sum_{\mathbf{k}} O^A_{\mathbf{kq}\lambda}, \qquad O^A_{\mathbf{kq}\lambda} \equiv b_{\mathbf{q}\lambda}\rho_{\mathbf{kq}} \tag{14.27}$$

$$O^E_{\mathbf{q}\lambda} \equiv \sum_{\mathbf{k}} O^E_{\mathbf{kq}\lambda}, \qquad O^E_{\mathbf{kq}\lambda} \equiv b^\dagger_{-\mathbf{q}\lambda}\rho_{\mathbf{kq}}. \tag{14.28}$$

Treating all the operators, $\mathbf{P}$, $\mathbf{R}$, $\rho_{\mathbf{q}}$ and $O_{\mathbf{q}\lambda}$, as time-dependent Heisenberg operators (thus $\dot{\mathbf{P}} \equiv d\mathbf{P}/dt$, $\dot{\mathbf{R}} \equiv d\mathbf{R}/dt$), and Eq. (14.22) as the Heisenberg equation of motion, Hu and O'connell developed explicit expressions for these Heisenberg operators by formally solving the Heisenberg equations of motion for $\rho_{\mathbf{kq}}$, $O^A_{\mathbf{kq}\lambda}$ and $O^E_{\mathbf{kq}\lambda}$. For instance, the Heisenberg equations of motion $\dot{\rho}_{\mathbf{kq}} = -i[\rho_{\mathbf{kq}}, H_B + H_I]$ and $\ddot{\rho}_{\mathbf{kq}} = -i[\dot{\rho}_{\mathbf{kq}}, H_B + H_I]$ yield the following second order differential equation:

$$\ddot{\rho}_{\mathbf{kq}} + \omega^2_{\mathbf{kq}}\rho_{\mathbf{kq}} = -\left[[\rho_{\mathbf{kq}}, H_I], H_B + H_I\right], \tag{14.29}$$

where $\omega_{\mathbf{kq}} \equiv \varepsilon_{\mathbf{k}+\mathbf{q}} - \varepsilon_{\mathbf{k}} = \mathbf{k}\cdot\mathbf{q}/m$. Assuming that as $t \to -\infty$, $H_I = 0$ and the initial condition of the relative electron system is an isotropic free electron gas (which is in agreement with our initial density matrix in Section 2.3), they obtain the solution of the differential equation (14.29) as (retaining only lowest order terms in electron-impurity and electron-phonon interactions and neglecting electron-electron interactions)

$$\begin{aligned}\rho_{\mathbf{kq}}(t) = e^{-i\omega_{\mathbf{kq}}t}\rho^0_{\mathbf{kq}} - i\int_{-\infty}^{t} dt' e^{-i\omega_{\mathbf{kq}}(t-t')}(n_{\mathbf{k}+\mathbf{q}} - n_{\mathbf{k}}) \\ \times \left[U^i_q(\mathbf{R}) + \sum_{\lambda} M_{\mathbf{q}\lambda}(\mathbf{R})(b_{\mathbf{q}\lambda} + b^\dagger_{-\mathbf{q}\lambda})\right].\end{aligned} \tag{14.30}$$

Here $\rho_{\mathbf{kq}}$ denotes the relative electron density operator in the absence of H_I. Similar expressions are obtained for $O^A_{\mathbf{kq}\lambda}(t)$ and $O^E_{\mathbf{kq}\lambda}(t)$. With these results Hu and O'Connell write Eq. (14.22) in the form

$$M\ddot{\mathbf{R}} = Ne\mathbf{E}(t) + \mathbf{F}(t) - \int_{-\infty}^{t} \mu(\dot{\mathbf{R}}; t, t') M\dot{\mathbf{R}} \tag{14.31}$$

which serves as momentum-GLE, in which $M = Nm$ and

$$\mathbf{F}(t) = -i\sum_{\mathbf{k},\mathbf{q}} \mathbf{q}U^i_{\mathbf{q}}(\mathbf{R})e^{-i\omega_{\mathbf{kq}}t}\rho^0_{\mathbf{kq}} -i\sum_{\mathbf{k},\mathbf{q}} \mathbf{q}M_{\mathbf{q}\lambda}(\mathbf{R})\left[e^{-i(\omega_{\mathbf{kq}}+\Omega_{\mathbf{q}\lambda})t}O^{0A}_{\mathbf{kq}\lambda} + e^{-i(\omega_{\mathbf{kq}}-\Omega_{\mathbf{q}\lambda})t}O^{0E}_{\mathbf{kq}\lambda}\right] \quad (14.32)$$

is the random force. The last term of the right-hand-side of (14.31) is the frictional force, which contains a memory function (Hu and O'Connell, 1988a, 1989)

$$\mu(\dot{\mathbf{R}};t,t') \equiv \sum_{\mathbf{k},\mathbf{q},s} d^s_{\mathbf{kq}}e^{i\mathbf{q}\cdot[\mathbf{R}(t)-\mathbf{R}(t')]-i\omega^s_{\mathbf{kq}}(t-t')}. \quad (14.33)$$

The definitions of $d^s_{\mathbf{kq}}$ and $\omega^s_{\mathbf{kq}}$ are detailed in their paper. It is easily seen that the random force $\mathbf{F}(t)$, Eq.(14.32), is equivalent to our fluctuation force operator given in Eq.(8.6). Furthermore, in obtaining Eq.(14.33) they have used the approximation

$$e^{i\mathbf{q}\cdot\mathbf{R}(t)}e^{-i\mathbf{q}\cdot\mathbf{R}(t')} = e^{-i\mathbf{q}\cdot[\mathbf{R}(t)-\mathbf{R}(t')]}, \quad (14.34)$$

neglecting the noncommutivity of the Heisenberg operators $\mathbf{R}(t)$ at differing times. This is reasonable for large N, but is obviously equivalent to treating the center-of-mass as a classical particle (Section 2.2).

The energy GLE of Hu and O'Connell (1988a) may be interpreted as

$$\frac{d}{dt}\left[\frac{1}{2}Nm\dot{\mathbf{R}}(t)^2\right] + \dot{H}_e(t) + \dot{H}_{ph}(t) = Ne\mathbf{E}\cdot\dot{\mathbf{R}}(t), \quad (14.35)$$

which is written in a form similar to Eq.(14.32). This equation is equivalent to our operator equation (2.26).

It is worth noting that the GLE's of Hu and O'connell, which correspond to our operator equations in the Heisenberg picture, do not incorporate statistical information. To establish balance equations and Langevin equations for transport, one must average them over the (initial) statistical ensemble. Although Hu and O'Connell did not clearly state what ensemble to use for the average, they adopted the idea of an electron temperature T_e in relation to the average $\langle\dot{H}_e\rangle$, indicating that they employed the same initial density matrix $\hat{\rho}_0$ as (2.30). Furthermore, the establishment of an energy balance equation involves an identification of the kind

$$\frac{d}{dt}\langle H_e\rangle = \langle\dot{H}_e\rangle. \quad (14.36)$$

By this means the electron temperature enters the theory as a time-dependent parameter.

Hu and O'Connell (1988b-f, 1989b-c) have used these equations in many investigations, particularly in problems related to c.m. velocity fluctuations.

14.4 Nonequilibrium Statistical Operator Method

The nonequlibrium statistical operator method, developed by Zubarev (1961, 1974), McLennan (1961) and Peletminskii and Yatsenko (1967), is based on Bogolyubov's idea (see Bogolyubov, 1962) that a physical system has a hierarchy of distinct relaxation times. If one considers the evolution of the system on a time scale which is larger than the characteristic time, τ_0, of micro-relaxation in the system, it is possible to use a set of macroscopic parameters to describe the system such that the time-dependence of these variables determines the time evolution of the statistical ensemble.

We assume that the system Hamiltonian can be broken up into two parts $H = H_0 + H_I$, where H_0 is the main Hamiltonian and H_I is a relatively weak interaction. A short micro-relaxation time τ_0 inherent in H_0 makes it possible to describe the system in a time scale larger than τ_0 by a set of macroscopic parameters $\gamma = \{\gamma_m\}$, which is the average value of the set of dynamic operators $\hat{\gamma} = \{\hat{\gamma}_m\}$, satisfying the commutation relations: $[H_0, \hat{\gamma}_m] = \sum_n \alpha_{mn}\hat{\gamma}_n$, α_{mn} being c-numbers. The density matrix, or the nonequilibrium statistical operator of the system depends on t only via the time t-dependence of the parameters $\gamma_m(t)$: $\rho = \hat{\rho}(t) = \hat{\rho}(\gamma)$. To determine such a density matrix satisfying the Liouville equation

$$\frac{d\rho}{dt} = i[\hat{\rho}, H], \tag{14.37}$$

a set of time-dependent parameters $\{X_m\}$ (which can be understood as the thermodynamic variables conjugate to $\{\gamma_m\}$), and a quasi-equilibrium statistical operator $\hat{\rho}^0 = \hat{\rho}^0(t) = \hat{\rho}^0(\gamma)$,

$$\hat{\rho}^0 = \frac{1}{Z}\exp[-\sum_m X_m\hat{\gamma}_m], \tag{14.38}$$

are introduced in such a way that the equality

$$\gamma_m \equiv Tr\{\hat{\rho}\hat{\gamma}_m\} = Tr\{\rho^0\hat{\gamma}_m\} \tag{14.39}$$

is satisfied. Eq.(14.39) gives the conditions for the determination of the parameters X_m. $\hat{\rho}^0(t)$ has the following meaning: if one turns off the interaction H_I at time t, the system, starting from $\hat{\rho}(t)$ and following the Liouville equation of Hamiltonian H_0 for a time longer than τ_0, would approach $\hat{\rho}^0(t)$. With these conditions the general form of the nonequilibrium statistical operator $\rho(t)$ can be derived (Zubarev, 1974, Peletmiskii and Yatsenko, 1967). To linear order in H_I, it can be written as

$$\hat{\rho} = \hat{\rho}^0 + i\int_{-\infty}^{0} d\tau e^{\epsilon\tau}[\hat{\rho}^0, H_I(\tau)] + i\int_{-\infty}^{0} d\tau e^{\epsilon\tau}\sum_m \frac{\partial\hat{\rho}^0}{\partial\gamma_m} Tr\{\hat{\rho}^0[\hat{\gamma}_m, H_I(\tau)]\}, \tag{14.40}$$

where $\epsilon \to 0^+$ and $H_I(\tau) \equiv e^{iH_0\tau}H_Ie^{-iH_0\tau}$. If $[\hat{\gamma}_m, \hat{\gamma}_n] = 0$, the last term of the right-hand-side of the above equation vanishes, leading to

$$\hat{\rho} = \hat{\rho}^0 + i\int_{-\infty}^{0} d\tau e^{\epsilon\tau}[\hat{\rho}^0, H_I(\tau)]. \tag{14.41}$$

With the help of these explicit expressions for the nonequilibrium statistical operator, the expectation value of any physical quantity can be calculated on a time scale larger than τ_0 under the nonequilibrium conditions subject to the Hamiltonian $H_0 + H_I$.

The nonequilibrium statistical operator method was first applied by Kalashnikov (1970), and then used by other authors (Ferry, 1981; Niez and Ferry 1983) in the study of warm-electron transport. Xing, Hu and Ting (1987) employed this method to discuss steady-state hot-electron transport in a strong electric field. They chose the total electron momentum $\mathbf{p}_e = \sum_{\mathbf{k}\sigma} \mathbf{k} c^{\dagger}_{\mathbf{k}\sigma} c_{\mathbf{k}\sigma}$, energy $H_e = \sum_{\mathbf{k}\sigma} \varepsilon_{\mathbf{k}} c^{\dagger}_{\mathbf{k}\sigma} c_{\mathbf{k}\sigma}$, electron number $N = \sum_{\mathbf{k}\sigma} c^{\dagger}_{\mathbf{k}\sigma} c_{\mathbf{k}\sigma}$, and the phonon energy H_{ph}, to form the parameter set. Nonlinear force and energy balance equations were derived to lowest order in the electron-impurity and electron-phonon interactions. These equations are exactly the same as the balance equations of Lei and Ting (1985b). Liu et al. (1988) applied the nonequilibrium statistical operator method to steady state dc transport in many-valley semiconductors, obtaining balance equations identical to those discussed in Section 9.2 with an appropriate selection of the parameter set. The nonequilibrium statistical operator method has also been applied to nonequilibrium phonons (Cai, Marchetti and Lax, 1986). Further detail may be found in the original papers and a separate review article by Xing and Liu in this volume.

15. ISOTHERMAL TRANSPORT AND BALANCE-EQUATION APPROACH

In Section 2, we emphasized the significance of choosing the initial state as one in which the electrons exhibit a thermal equilibrium distribution with respect to their center of mass. This choice enables us to establish a high-field transport theory for systems with strong intercarrier couplings. We will explore this further jointly with other related issues.

15.1 Full Quantum-Mechanical Treatment

In the preceding sections, our discussion started with writing Hamiltonian (2.1) in terms of center-of-mass variables and relative electron variables defined by Eqs.(2.2) and (2.3). They satisfy the relations (2.4), (2.5) and (2.7) and the constraint (2.6). We have made two assumptions: (a) the center of mass can be treated as a classical particle; (b) the $1/N$ terms in the exact noncanonical commutation relation (2.7) can be neglected, such that relative electron momentum $\mathbf{p}'_j$ and coordinate $\mathbf{r}'_j$ are regarded as canonically conjugate variables. Although these assumptions are valid in obtaining correct results to lowest order in resistive scattering in the presence of strong electron-electron interactions, they should not be used indiscriminately. For instance, by summing Eq.(2.7) over j, both the left hand side (LHS) and the right hand side (RHS) vanish. This consistency, $LHS = RHS$, would be seriously violated if one were to neglect the $1/N$ terms in Eq.(2.7). In this context of summing over j, the neglect of the $1/N$ terms incurs an error of order $N^0 = 1$ rather than $1/N$. This fact shows that the relative electron momenta, which satisfy constraint (2.6), do not provide the requisite translation generating qualities and that using them as translation generators may lead to unphysical results (Argyres, 1989).

An alternative discussion has been given by Lei (1990a) using the complete Hamiltonian (2.1), with a full quantum mechanical treatment. We rewrite the Hamiltonian here for the reader's convenience:

$$H = H_{el} + H_E + H_I + H_{ph}, \tag{15.1}$$

$$H_{el} = \sum_i \frac{p_i^2}{2m} + H_{ee} = \frac{P^2}{2Nm} + H_e \tag{15.2}$$

$$H_e = \sum_i \frac{p_i'^2}{2m} + H_{ee} \tag{15.3}$$

$$H_{ee} = \frac{1}{8\pi\epsilon_0}\sum_{i\neq j}\frac{1}{|\mathbf{r}_i - \mathbf{r}_j|} = \frac{1}{8\pi\epsilon_0}\sum_{i\neq j}\frac{1}{|\mathbf{r}'_i - \mathbf{r}'_j|} \tag{15.4}$$

$$H_E = -e\mathbf{E}\cdot\mathbf{R} \tag{15.5}$$

$$H_I = H_{ei} + H_{ep} \tag{15.6}$$

$$H_{ei} = \sum_{\mathbf{q},a} u(\mathbf{q})e^{-i\mathbf{q}\cdot\mathbf{R}_a}\rho_{\mathbf{q}} = \sum_{\mathbf{q},a} u(\mathbf{q})e^{i\mathbf{q}\cdot(\mathbf{R}-\mathbf{R}_a)}\bar{\rho}_{\mathbf{q}} \tag{15.7}$$

$$H_{ep} = \sum_{\mathbf{q},\lambda} M(\mathbf{q},\lambda)\phi_{\mathbf{q}\lambda}\rho_{\mathbf{q}} = \sum_{\mathbf{q},\lambda} M(\mathbf{q},\lambda)e^{i\mathbf{q}\cdot\mathbf{R}}\phi_{\mathbf{q}\lambda}\bar{\rho}_{\mathbf{q}} \tag{15.8}$$

$$H_{ph} = \sum_{\mathbf{q},\lambda}\Omega_{\mathbf{q}\lambda}b^{\dagger}_{\mathbf{q}\lambda}b_{\mathbf{q}\lambda}. \tag{15.9}$$

In this, $\mathbf{p}_i$, $\mathbf{r}_i$, $\mathbf{p}'_i$, $\mathbf{r}'_i$, $\mathbf{P}$, $\mathbf{R}$ and the other notations have the same meaning as given in Section 2, except that

$$\rho_{\mathbf{q}} = \sum_j e^{i\mathbf{q}\cdot\mathbf{r}_j} \tag{15.10}$$

now denotes the electron density operator in the laboratory reference frame, and

$$\bar{\rho}_{\mathbf{q}} = \sum_j e^{i\mathbf{q}\cdot\mathbf{r}'_j} = e^{-i\mathbf{q}\cdot\mathbf{R}}\rho_{\mathbf{q}} \tag{15.11}$$

represents the density operator for relative electrons.

For a full quantum-mechanical treatment the density matrix of the system, $\hat{\rho}$, obeys the Liouville equation

$$i\frac{d\hat{\rho}}{dt} = [H, \hat{\rho}], \tag{15.12}$$

with the complete Hamiltonian H.

To proceed we assume that both H_I and H_E are turned on at time $t = 0$. Thus for $t < 0$, the Hamiltonian is H_{el} with e-e interactions, plus H_{ph} of the independent phonons. As pointed out in Section 2 (Lei, 1990a), the balance equation approach is characterized by an initial state selected to be one in which electrons exhibit a thermal equilibrium distribution at the electron temperature with respect to their center-of-mass, which is moving at a velocity equal to the average drift velocity in the presence of an electric field (i.e. drifted thermal equilibrium state), plus an thermal equilibrium phonon system. The density matrix of this initial state, given in Eq.(2.30), can be written as

$$\hat{\rho}_0 = \frac{1}{Z}e^{-H_e/T_e}e^{-H_{ph}/T}|\mathbf{P}_d\rangle\langle\mathbf{P}_d|, \tag{15.13}$$

explicitly showing that c.m., as a quantum particle, is in a momentum eigenstate: $\mathbf{P}|\mathbf{P}_d\rangle = Nm\mathbf{v}_d$. Here $\mathbf{v}_d$ is the drift velocity of the final state, which, like the electron temperature T_e, is also a parameter to be determined by the resulting balance equations. Since the final steady state bears the same overall velocity $\mathbf{v}_d$ as the initial state (15.13), the evolution from the initial state to the final state is solely a thermalization process without the need for momentum relaxation. The time required for this evolution, τ_{th}, is the thermalization time of the relative electron system, and is limited by the relaxation time τ_{ee} determined by the intercarrier interactions H_{ee}. The Liouville equation for the density matrix together with the initial condition $\hat{\rho}(t = 0) = \hat{\rho}_0$, can be written in a form of integral equation:

$$\hat{\rho}(t) = \hat{\rho}_0 - i\int_0^t dt_1[H_I(t_1) + H_E(t_1), \hat{\rho}(t_1)]. \tag{15.14}$$

The statistical expectation of any dynamical variable A should be calculated according to

$$\langle A\rangle = Tr[\hat{\rho}(t)A(t)]. \tag{15.15}$$

Here, the interaction picture is defined as

$$A(t) = e^{i(H_{el}+H_{ph})t} A e^{-i(H_{el}+H_{ph})t}. \tag{15.16}$$

Our purpose is to calculate statistical expectation values $\mathbf{f} = Tr[\hat{\rho}(t)\mathbf{F}(t)]$ and $w = Tr[\hat{\rho}(t)W(t)]$, of the force operator

$$\mathbf{F} = -i\sum_{\mathbf{q},a} \mathbf{q}u(\mathbf{q})e^{-i\mathbf{q}\cdot\mathbf{R}_a}\rho_{\mathbf{q}} - i\sum_{\mathbf{q},\lambda}\mathbf{q}M(\mathbf{q},\lambda)\phi_{\mathbf{q}\lambda}\rho_{\mathbf{q}}, \tag{15.17}$$

and the energy transfer rate operator

$$W = -\sum_{\mathbf{q}\lambda} M(\mathbf{q},\lambda)\dot{\phi}_{\mathbf{q}\lambda}\rho_{\mathbf{q}}, \tag{15.18}$$

to obtain the balance equations for steady state:

$$Ne\mathbf{E} + \mathbf{f} = 0, \tag{15.19}$$

$$\mathbf{v}_d\cdot\mathbf{f} + w = 0. \tag{15.20}$$

Of course this requires a knowledge of density matrix.

15.2 Perturbative Expansion and Steady-State Balance Equations

Unfortunately, we cannot solve Eq.(15.12) or Eq.(15.14) to obtain $\hat{\rho}(t)$ exactly, and thus take recourse to a perturbation expansion as usual. A formal iteration of Eq.(15.14) yields

$$\hat{\rho}(t) = \hat{\rho}_0 + \sum_{n=1}(-i)^n\int_0^t dt_1\int_0^{t_1}dt_2\cdots\int_0^{t_{n-1}}dt_n[H_{IE}(t_1),[H_{IE}(t_2),\cdots[H_{IE}(t_n),\hat{\rho}_0]\cdots]], \tag{15.21}$$

where $H_{IE}\equiv H_I+H_E$. With this $\hat{\rho}(t)$, we can write the statistical average of any operator as an expansion in powers of both the electric field H_E and the scattering potential H_I jointly. For instance, the frictional force is given by (Lei, 1990a)

$$\mathbf{f}(t) = \sum_{n=1}^{\infty} i^n\int_0^t dt_1\int_0^{t_1}dt_2\cdots\int_0^{t_{n-1}}dt_n Tr\{\hat{\rho}_0[H_{IE}(t_{n-1}),\cdots[H_{IE}(t_1),\mathbf{F}(t)]\cdots]]\}. \tag{15.22}$$

We can regroup the terms on the right-hand side of Eq.(15.22) according to the respective powers of H_E and H_I:

$$\mathbf{f}(t) = \sum_{n=0}^{\infty}\sum_{m=1}^{\infty}\mathbf{f}_n^{(2m)}(t), \tag{15.23}$$

Here, $\mathbf{f}_n^{(2m)}(t)$ stands for all the terms with n factors of H_E and $(2m-1)$ factors of H_I in the perturbative expansion of the force.

A serious problem associated with this double expansion is that except for a few lowest order terms, the higher order terms are divergent when $t\to\infty$ for nonzero H_I and H_E, no matter how small they are, (Lei and Horing, 1989; Lei, 1990a). Thus the expansion series is divergent at large t. This divergence, however, does not invalidate the balance equation approach. Since the system, starting from the initial state $\hat{\rho}_0$, certainly reaches a unique steady state after the thermalization time τ_{th}, the divergence of the perturbation series is only an indication of the uselessness of the perturbation series itself for describing the evolution of the system at large t, and is irrelevant to the behavior of transport. Such perturbative series may still be useful within the finite time-ranges in which they converge.

We can easily determine the expansion parameters for these perturbation series. In the frictional force series Eq.(15.23), the expansion parameter with respect to H_I is seen to be t/τ_m, and that with respect to H_E is t/t_E. Here τ_m is estimated to be of order of the average momentum relaxation time due to impurity and phonon scatterings, and t_E is a characteristic time associated with the electric field: $t_E \sim (m/e\bar{q}E)^{1/2}$, with $\bar{q}$ being an average wavevector contributing in the momentum relaxation integration. Thus, up to time $t < t_E$, the perturbation expansion for the frictional force $\mathbf{f(t)}$, Eq.(15.23), is convergent and thus useful in describing the system transport. Since at time $t \sim \tau_{th}$ the system has essentially reached the steady state, the perturbative series Eq.(15.23) for $\mathbf{f}(t)$ can be used for the determination of steady state properties of transport, if the conditions

$$\tau_{th} \ll \tau_m \tag{15.24}$$

and

$$\tau_{th} \ll t_E \tag{15.25}$$

are satisfied. For $t < \tau_{th}$, the leading term in the force series is

$$\mathbf{f}_0^{(2)}(t) = -n_i \sum_{\mathbf{q}} |u(\mathbf{q})|^2 \mathbf{q} \int_0^t dt_1 Tr\{\hat{\rho}_0[\rho_{-\mathbf{q}}(t_1), \rho_{\mathbf{q}}(t)]\}$$
$$- \sum_{\mathbf{q},\lambda} |M(\mathbf{q},\lambda)|^2 \mathbf{q} \int_0^t dt_1 Tr\{\hat{\rho}_0[\phi_{-\mathbf{q}}(t_1)\rho_{-\mathbf{q}}(t_1), \phi_{\mathbf{q}}(t)\rho_{\mathbf{q}}(t)]\}. \tag{15.26}$$

Direct evaluation from the definition of $\rho_{\mathbf{q}}$ shows that

$$\rho_{-\mathbf{q}}(t_1)\rho_{\mathbf{q}}(t) = \exp\left[i\mathbf{q}\cdot\mathbf{P}(t-t_1)/Nm\right]\bar{\rho}_{-\mathbf{q}}(t_1)\bar{\rho}_{\mathbf{q}}(t), \tag{15.27}$$

$$[\rho_{-\mathbf{q}}(t_1), \rho_{\mathbf{q}}(t)] = \exp\left[i\mathbf{q}\cdot\mathbf{P}(t-t_1)/Nm\right][\bar{\rho}_{-\mathbf{q}}(t_1), \bar{\rho}_{\mathbf{q}}(t)]. \tag{15.28}$$

Here, the interaction picture quantity $\bar{\rho}_{\mathbf{q}}(t) \equiv \exp(iH_e t)\bar{\rho}_{\mathbf{q}}\exp(-iH_e t)$ and the commutator $[\bar{\rho}_{-\mathbf{q}}(t_1), \bar{\rho}_{\mathbf{q}}(t)]$ should be evaluated as if $\mathbf{r}'_i$, $\mathbf{p}'_j$ satisfy an exactly canonical commutation relation $[r'_{i\alpha}, p'_{j\beta}] = i\delta_{ij}\delta_{\alpha\beta}$. Therefore we can write $\mathbf{f}_0^{(2)}(t)$ as

$$\mathbf{f}_0^{(2)}(t) = n_i \sum_{\mathbf{q}} |u(\mathbf{q})|^2 \mathbf{q} \int_0^t ds \exp(i\omega_0 s)\langle[\bar{\rho}_{-\mathbf{q}}, \bar{\rho}_{\mathbf{q}}(s)]\rangle$$
$$+ \sum_{\mathbf{q},\lambda} |M(\mathbf{q},\lambda)|^2 \mathbf{q} \int_0^t ds \exp(i\omega_0 s)\cos(\Omega_{\mathbf{q}\lambda}s)\langle[\bar{\rho}_{-\mathbf{q}}, \bar{\rho}_{\mathbf{q}}(s)]\rangle 2n\left(\frac{\Omega_{\mathbf{q}\lambda}}{T}\right)$$
$$+ \sum_{\mathbf{q},\lambda} |M(\mathbf{q},\lambda)|^2 \mathbf{q} \int_{-t}^t ds \exp[i(\omega_0 - \Omega_{\mathbf{q}\lambda})s]\langle\bar{\rho}_{-\mathbf{q}}\bar{\rho}_{\mathbf{q}}(s)\rangle, \tag{15.29}$$

where $\omega_0 \equiv \mathbf{q} \cdot \mathbf{v}_d$, $n(x) \equiv (e^x - 1)^{-1}$, and the correlation functions of the density operators are calculated with respect to a canonical ensemble of Hamiltonian H_e at the electron temperature T_e. The density-density correlation function of the system $\langle \rho_{\mathbf{q}}(t)\rho_{-\mathbf{q}} \rangle$ generally differs appreciably from zero only in a limited time range $-\tau_{cr} < t < \tau_{cr}$, where τ_{cr} is a correlation time. When time t is large than the correlation time τ_{cr}, $\mathbf{f}_0^{(2)}$ approaches a finite limit value

$$\mathbf{f}_0^{(2)} = n_i \sum_{\mathbf{q}} |u(\mathbf{q})|^2 \mathbf{q} \Pi_2(\mathbf{q}, \omega_0) + \sum_{\mathbf{q},\lambda} |M(\mathbf{q},\lambda)|^2 \mathbf{q} \Pi_2(\mathbf{q}, \Omega_{\mathbf{q}\lambda} + \omega_0) \left[n\left(\frac{\Omega_{\mathbf{q}\lambda}}{T}\right) - n\left(\frac{\Omega_{\mathbf{q}\lambda} + \omega_0}{T_e}\right) \right]. \tag{15.30}$$

Therefore, the force balance equation, valid at time t of the order of, or larger than, τ_{th} and τ_{cr}, takes the form

$$Ne\mathbf{E} + \mathbf{f}_0^{(2)} = 0. \tag{15.31}$$

This is exactly the force balance equation obtained in Section 4 for the steady state.

The energy transfer rate $w(t)$ can also be expanded in powers of H_I and H_E:

$$w(t) = \sum_{n=0}^{\infty} \sum_{m=0}^{\infty} w_n^{(2m)}(t). \tag{15.32}$$

The lowest-order nonzero term $w_0^{(2)}(t)$ can be calculated in a similar fashion with the steady-state value

$$w_0^{(2)} = 2 \sum_{\mathbf{q},\lambda} |M(\mathbf{q},\lambda)|^2 \Omega_{\mathbf{q}\lambda} \Pi_2(\mathbf{q}, \Omega_{\mathbf{q}\lambda} + \omega_0) \left[n\left(\frac{\Omega_{\mathbf{q}\lambda}}{T}\right) - n\left(\frac{\Omega_{\mathbf{q}\lambda} + \omega_0}{T_e}\right) \right]. \tag{15.33}$$

The energy balance equation, which may be valid at time of the order of, or larger than, τ_{th} and τ_{cr}, is

$$\mathbf{v}_d \cdot \mathbf{f}_0^{(2)} + w_0^{(2)} = 0. \tag{15.34}$$

The correlation time τ_{cr} can be estimated roughly from the frequency range $\Delta\omega$ in which $\Pi_2(\mathbf{q}, \omega)$ differs appreciably from zero:

$$\tau_{cr} \approx 1/\Delta\omega. \tag{15.35}$$

As a conservative estimate we use the zero-temperature expression Eq.(4.11) for the correlation function in the absence of electron-electron interactions, yielding $1/\tau_{cr} \sim [q(2k_F + q)/m]$, where k_F is the Fermi wavevector. In the degenerate case the dominant contribution comes from wavevector transfer q of order of k_F, such that $\tau_{cr} \sim (4\varepsilon_F)^{-1}$. For a typical degenerate semiconductor of $\varepsilon_F \sim 1000K$, $\tau_{cr} \sim 10^{-14}$sec. In the presence of electron-electron interactions or at finite temperature, we anticipate an even larger $\Delta\omega$ and thus a shorter τ_{cr}.

The condition (15.25) imposes a restriction on the strength of the applied electric field $\mathbf{E}$. Fortunately, this restriction can be released on the proviso that the electric field is not treated as a perturbation, but is included in the unperturbed Hamiltonian (Lei and Horing, 1989). Then, effects due to acceleration by the electric field during the correlation time τ_{cr} (intracollisional field effect) will show up. We will discuss this matter in the next Section.

The thermalization among relative electrons due to e-e couplings can be represented mathematically by a damping factor in the single relative-electron retarded Green's function

$$G_r(\mathbf{k},t) \sim e^{-i\epsilon_{\mathbf{k}}t-|t|/\tau_e}, \tag{15.36}$$

where the damping time τ_e is related to the inelastic or electron-electron scatterings. With this, and the application of Feynman diagrammatic technique jointly with closed-time-path Green's functions, Chen, Ting and Horing (1989, 1990) showed that the corresponding perturbation series is devoid of long-time divergences and that the higher-order terms are negligible in the case $\tau_e \ll \tau_m$, thus establishing the validity of the balance equation approach.

The thermalization time τ_{th} is a collective feature of all kinds of carrier-carrier interactions: Coulombic, phonon-mediated, etc. It depends on the carrier density and temperature, and varies over a wide range for various systems. Wingreen, Stanton and Wilkins (1986) phenomenologically introduced in the Boltzmann equation a relaxation time τ_{ee} to represent electron-electron collisions, effect of which is to drive the distribution toward a displaced Maxwellian. This τ_{ee} resembles our τ_{th}. For an n-GaAs system with the electron density of the order of 10^{17} cm^{-3}, they estimated an effective contribution of about 4×10^{-13} sec from Coulomb scattering to τ_{ee} at electron temperature $T_e = 100$ K. Recently thermalization of nonequilibrium electrons by electron-electron interactions in n-type GaAs quantum wells has been observed to occur in less than $10 fs$ (Knox et al., 1988). Therefore, in real systems, thermalization time may be significantly shorter than previous theoretical and experimental estimates. The important role of strong electron-electron interactions in balance equation transport theory has also been recognized by Sernelius and Söderström (1991a).

15.3 Isothermal and Adiabatic Resistivities

In principle, the force-balance theory, which expresses the electric field as a function of current, is equivalent to the dynamic response theory, which calculates the current as the nonlinear response to an applied electric field (Su and Sakita, 1986). The realization in practical calculations, however, usually relies on perturbation methods, such that the different choices of the initial state, which determine the time required to reach the steady state, make these two theories dramatically different (Lei, 1990a). As discussed above, the balance equation theory, starting from the initial state given by the density matrix (15.13) and featuring force- and energy-balance equations (15.31) and (15.34) taken to leading order in the scattering interaction H_I, describes isothermal transport, in that it requires a rapid thermalization of the relative electron system due to strong electron-electron interactions. In balance equation theory the impurity-induced Ohmic resistivity, given in Eq.(4.28), can be written in the form

$$\rho = \frac{m}{ne^2}\langle\frac{1}{\tau}\rangle, \tag{15.37}$$

where

$$\frac{1}{\tau} = \frac{n_i m}{4\pi k^3}\int_0^{2k} |\bar{u}(q)|^2 q^3 dq \tag{15.38}$$

is a wavevector-dependent inverse relaxation time for impurity scattering, and the bracket $\langle\ \rangle$ represents an average defined by

$$\langle A(\mathbf{k})\rangle = (4/3n)\sum_{\mathbf{k}}[-\partial f(\varepsilon_{\mathbf{k}},T)/\partial\varepsilon_{\mathbf{k}}]\,\varepsilon_{\mathbf{k}}A(\mathbf{k}). \tag{15.39}$$

This result, known as isothermal resistivity, is expected valid for systems with strong e-e interactions.

It has been pointed out (Lei, 1990a) that the situation is quite different in the case of Kubo linear-response theory, where the total momentum of the initial state is zero. The initial state in Kubo theory is a thermal equilibrium state in the fixed reference frame:

$$\hat{\rho}_0 = \frac{1}{Z}e^{-H_{el}/T}e^{-H_{ph}/T}. \tag{15.40}$$

In order for the system to evolve from this initial state to the final transport state, a momentum relaxation mechanism is necessary. The electron-electron interaction, which induces no momentum relaxation, has no primary role in this process. The time required for the system to go from this initial state to the final state is mainly determined by τ_m, the momentum relaxation time, and is essentially not affected by the intercarrier interaction (save for shielding). When the strength of the impurity (and phonon) scattering gets weaker, this time becomes longer. This is the physical reason why one has to sum an infinite series of divergent terms in order to obtain the resistivity even in the weak scattering limit in Kubo linear response theory. The resistivity, thus derived in the absence of e-e interaction, has the form

$$\rho = \frac{m}{ne^2}\frac{1}{\langle\tau\rangle}. \tag{15.41}$$

This formula, known as adiabatic resistivity, can also be obtained from Boltzmann equation without e-e scattering.

The conflict of the resistivity formulas (15.37) and (15.41), i.e. so-called $\langle 1/\tau\rangle$ *vs* $1/\langle\tau\rangle$ dispute, has been in the literature for many years (see Huberman and Chester, 1985, and the references therein). Similar conflict also exists for phonon-induced linear resistivities, e.g. so-called $\frac{3}{2}(kT/\hbar\omega_{LO})$ problem in polaron mobility theory (see Peeters and Devreese, 1984, and the references therein). It is now sensed that these two types of formulas represent two extreme cases of the electron-electron interaction strength in the system (Sernelius and Söderström, 1991a).

The derivation of isothermal resistivity formula (15.37) from Kubo linear response theory by considering strong e-e interactions is an open problem. Inclusion of e-e interaction jointly with impurity scattering in the Boltzmann theory has recently been attempted by Sernelius and Söderström (1991a), treating e-e interaction in the relaxation time approximation. They showed that inclusion of electron-electron scattering in the solution to the Boltzmann equation brings the result closer to that from the force-balance type theories: very strong electron-electron scattering results in perfect agreement. Numerical calculation for doped GaAs suggested that for a real system the correct result probably lies somewhere in between these two extreme cases (Sernelius and Söderström, 1991b).

There has recently been much discussion about the possibilities of deriving adiabatic resistivity from balance-equation theory (Fishman and Mahan, 1989; Fishman, 1989; Chen, Ting and Horing, 1989, 1990a,b; Lei, 1990a). That is to explore what happens in the balance-equation theory if the e-e interaction is weak such that $\tau_{th} \sim \tau_m$. In this case we have to deal with times greater than the momentum relaxation time, $t > \tau_m$, and the perturbation series is divergent. Thus far the problem has been approached by summing the divergent series. Unfortunately, the sum of a divergent series depends on how one truncates it and groups its terms. If one retains in the frictional force expansion (15.23) only terms of zeroth and linear orders in E and formally performs the divergent partial sums $\mathbf{f}_0 = \sum_n \mathbf{f}_0^{(2n)}$ and $\mathbf{f}_1 = \sum_n \mathbf{f}_1^{(2n)}$, the resulting force balance equation for $t \to \infty$ yields a resistivity of the form Eq.(15.41) (Lei, 1990a). However, terms of higher order in E are more strongly divergent in the limit of $t \to \infty$. Therefore one may question the validity of using the balance equation without $f_n (n \geq 2)$ and taking $t \to \infty$ limit to obtain the resistivity (15.41). On the other hand, if one first performs the partial summations over all the orders of E but with fixed power of H_I: $\mathbf{f}^{2m} = \sum_n \mathbf{f}_n^{(2m)}$, we find that all $\mathbf{f}^{2m}$ approach zero when $t \to \infty$, such that one might expect a zero resistivity.

The situation becomes even more complicated if one performs the calculation in terms of c.m. and relative electron variables. The noncommutability of $\mathbf{P}, \mathbf{R}$ as quantum mechanical variables and the noncanonical $1/N$ terms in the commutation relation (2.7), produce large numbers of new terms, which are also divergent when $t \to \infty$. Although each of these terms may be of order of $1/N$, the number of these terms is of order of N, such that we have to do a careful accounting of them to avoid missing important contributions. In the literature, Fishman and Mahan (1989) and Fishman (1989) obtained Eq.(15.41) by treating relative electron variables as exactly canonical variables, but taking only partial account of the effects of the quantum mechanical aspect of the c.m. Chen, Ting and Horing (1990) pointed out that if the full quantum mechanical noncommutability of $\mathbf{P}$ and $\mathbf{R}$ is taken into account, contributions from the additional infinite set of divergent terms fully cancel those picked up by Fishman and Mahan (1989).

Terms arising solely from the noncanonical relative-electron commutation relation were also discussed by them and shown to produce resistivity of the form (15.41). However, these complicated consideration may be spurious. Lei (1990c) pointed out that the contribution from the noncanonical $1/N$ terms in the commutation relation (2.17) of the relative electron variables is exactly cancelled by that arising from the noncommutability of the center-of-mass coordinates $\mathbf{R}(t)$ at different times. Physically this should be the case. The separation of the center of mass from the relative electrons is only a mathematical convenience, introducing no new physics. Returning to the original electron variables is of course equivalent to taking both effects (noncommutability of $\mathbf{P}$ and $\mathbf{R}$ and noncanonical $1/N$ terms in (2.7)) into full consideration, just as we did in this section: no trace of the divergent terms of both effects emerges, but the long-time divergence of the perturbation series (15.24) remains.

Summing the divergent series to obtain the adiabatic resistivity (in the absence of e-e interaction) remains an open problem in the balance equation theory, as well as in the Kubo linear response theory (Lei and Cai, 1991).

15.4 Balance-Equation Theories for a Single Electron System

The use of quantum balance equations in transport was pioneered by Thornber and Feynman (1970), who analyzed the steady-state drift velocity of a single electron interacting with longitudinal optic phonons in polar crystal under the influence of an applied electric field, by calculating the average rate of change of the electron momentum and assuming that at the initial time the electron is injected into the crystal with a velocity equal to the steady state constant velocity. For the case of a single electron they were able to carry out the calculation unperturbatively with the help of path integral technique, and the theory is expected valid for arbitrary strength of electron-phonon couplings. Unfortunately the result is inconsistent with the energy balance. The distribution function corresponding to the Thornber-Feynman theory was shown to be a drifted Maxwellian with an effective electron temperature which is independent of electric field strength and the electron average velocity (Peeters and Devreese, 1981), a result which is physically questionable.

Motivated by Thornber-Feynman theory, Su, Chen and Ting (1988) recently advanced a consistent treatment for a single electron in a thermal crystal with an applied electric field, assuring that both the momentum balance and the energy balance are satisfied. The result, expected to be valid for noninteracting electrons, is of course not equivalent to that of the Lei-Ting balance equation approach, which is anticipated to describe many electron systems with strong intercarrier couplings. The theory of Su, Chen and Ting (1988) is more fully expounded in a separate review article in this volume.

16. BALANCE EQUATIONS FOR GENERAL SYSTEMS WITH INTRACOLLISIONAL FIELD EFFECT

The role of the electric field in electron acceleration during the collision process, known as "intracollisional field effect" or the "interference effect", was first proposed by Argyres (1960) and later explored by Levinson (1969) and by Barker (1973). Much attention has been given to it in both linear and nonlinear transport theories (Argyres, 1960,1963; Horing and Argyres, 1962; Argyres and Kelley, 1964; Levinson, 1969; Barker, 1973, 1978; Thornber, 1978; Barker and Ferry, 1979; Calecki and Pottier, 1981; Arora, 1983), and is still an active subject in the literature (Sarker, 1986; Khan et al., 1987; Krieger and Iafrate, 1987; Abdolsalami and Khan, 1990; Lipavsky, 1991; Rossi and Jacoboni, 1991). Nevertheless, relatively few explicit calculations of the intracollisional field effect have been carried through to a full determination of its influence on transport.

The appearance of the intracollisional field effect in balance equation transport theory was pointed out by Xing and Ting (1987). In fact, carrier acceleration by the electric field during a collision is naturally incorporated when one treats the electric field in the unperturbed Hamiltonian (Lei and Horing, 1989; Lei, 1990c). In this section we provide a brief discussion of this matter. Moreover, the force and energy balance equations will be expressed in more general forms, equally valid for crystalline, amorphous and liquid systems (Lei, 1990c).

16.1 Force and Energy-Loss-Rate with Intracollisional Field Effect

To write the balance equations in a more general form we consider n interacting electrons moving in a medium composed of atoms (ions) under the influence of a uniform electric field $\mathbf{E}$. The Hamiltonian of this electron-atom system consists of an atom part H_A, an electron part H_{el}, and electron-atom interaction $H_{e\text{-}A}$. The electron Hamiltonian H_{el} can be separated into $H_{el} = H_c + H_e$, as given in Section 2.1. The atom Hamiltonian depends on the momenta and positions of all atoms but the effects of the electric field on the atom system are neglected. The electron-atom interaction is assumed to be

$$H_{e\text{-}A} = \sum_{i,l} v_l(\mathbf{r}_i - \mathbf{R}_l), \tag{16.1}$$

with $v_l(\mathbf{r}_i - \mathbf{R}_l)$ denoting the potential between i-th electron and l-th atom whose position is $\mathbf{R}_l$, and can be written in the form

$$H_{e\text{-}A} = \sum_{\mathbf{q}} \rho_{\mathbf{q}} \gamma_{-\mathbf{q}} e^{i\mathbf{q}\cdot\mathbf{R}}, \tag{16.2}$$

where

$$\rho_{\mathbf{q}} = \sum_j e^{i\mathbf{q}\cdot\mathbf{r}'_j} \tag{16.3}$$

is the electron density operator in the relative frame, and

$$\gamma_{\mathbf{q}} = \sum_l v_l(\mathbf{q}) e^{i\mathbf{q}\cdot\mathbf{R}_l} \tag{16.4}$$

is the atom scattering operator with

$$v_l(\mathbf{q}) = \int e^{i\mathbf{q}\cdot\mathbf{r}} v_l(\mathbf{r})\, d^3r \tag{16.5}$$

being the Fourier transform of the electron-atom potential. In accordance with this Hamiltonian, the force operator $\mathbf{F}$ due to electron-atom interaction takes the form

$$\mathbf{F} = -i \sum_{\mathbf{q}} \mathbf{q} e^{i\mathbf{q}\cdot\mathbf{R}} \rho_{\mathbf{q}} \gamma_{-\mathbf{q}}, \tag{16.6}$$

and the energy loss rate of the electron system W is equal to the energy increase rate of the atom system due to electron-atom interaction:

$$W = -i\,[H_A, H] = -\sum_{\mathbf{q}} e^{i\mathbf{q}\cdot\mathbf{R}} \rho_{\mathbf{q}} \dot{\gamma}_{-\mathbf{q}}, \tag{16.7}$$

where we have defined

$$\dot{\gamma}_{\mathbf{q}} = -i[\gamma_{\mathbf{q}},\ H_A]. \tag{16.8}$$

It has been pointed out in Section 15 that the perturbation expansion with respect to the electric field is convergent and useful for a steady state calculation if the strength of the electric field is limited by the condition $\tau_{th} \ll t_E$. Lei and Horing (1989) have shown that this condition can be relaxed if the electric field is not treated as a perturbation but is included in the unperturbed Hamiltonian. In other words, we assume that electron-impurity and electron-phonon interactions are absent before time $t = 0$, wherein the relative variables of the electron are decoupled from the center of mass. On the other hand, the electric field is considered to already be present, so that the system is subject to the Hamiltonian $H_c + H_e + H_A$ at $t \leq 0$. Therefore the initial density matrix is chosen as

$$\hat{\rho}_0 = \frac{1}{Z} e^{-H_e/T_e} e^{-H_A/T} \left|\Psi\right\rangle \left\langle\Psi\right| \tag{16.9}$$

where $|\Psi\rangle$ is the traveling eigenstate of the center-of-mass Hamiltonian H_c with average momentum of $\langle\Psi|\mathbf{P}|\Psi\rangle = nm\mathbf{v}$, $\mathbf{v}$ being the drift velocity, and T is the lattice atom temperature, which is kept unchanged by assuming the existence of an energy dissipation mechanism which quickly drains energy out of the atom system. The Liouville equation for the evolution of the density matrix may be written in a form of an integral equation:

$$\hat{\rho}(t) = \hat{\rho}_0 - \int_0^t dt_1 [H_{e\text{-}A}(t),\ \hat{\rho}(t_1)], \tag{16.10}$$

in which an interaction representation has been defined for any operator O as

$$O(t) = e^{i(H_{el}+H_A)t} O e^{-i(H_{el}+H_A)t}. \tag{16.11}$$

The steady state calculation of the statistical average of the frictional force $\mathbf{f} = Tr\{\hat{\rho}(t)\mathbf{F}(t)\}$ and the electron energy loss rate $w = Tr\{\hat{\rho}(t)W(t)\}$ becomes easy for an system with a strong electron-electron interactions. In such a system the time to approach the steady state from the initial state (16.9), τ_{th}, is determined mainly by the intercarrier coupling and is short in comparison with the relaxation time due to electron-atom scattering. The perturbative expansions of the frictional force $\mathbf{f}$ and the energy loss rate w in powers of the electron-atom scattering potential, obtained by using the formal iterative series of Eq.(16.10), converge up to time $t \sim \tau_{th}$. In the case of a weak electron-atom potential it is a good approximations to keep only leading terms, yielding (Lei, 1990c)

$$\mathbf{f} = -\sum_{\mathbf{q}} \int_0^t ds \langle [\rho_{\mathbf{q}}(s)\gamma_{-\mathbf{q}}(s),\ \rho_{-\mathbf{q}}\gamma_{\mathbf{q}}] \rangle \exp[i\mathbf{q}\cdot\mathbf{v}s + \frac{ie}{2m}\mathbf{q}\cdot\mathbf{E}s^2] \tag{16.12}$$

$$w = i\sum_{\mathbf{q}} \int_0^t ds \langle [\rho_{\mathbf{q}}(s)\dot{\gamma}_{-\mathbf{q}}(s),\ \rho_{-\mathbf{q}}\gamma_{\mathbf{q}}] \rangle \exp[i\mathbf{q}\cdot\mathbf{v}s + \frac{ie}{2m}\mathbf{q}\cdot\mathbf{E}s^2]. \tag{16.13}$$

Here, the average should be taken with respect to the initial density matrix (16.9). The s^2 term in the argument of the exponential function, which is due to the acceleration of the center-of-mass by the electric field:

$$\mathbf{R}(t) = \mathbf{R} + \frac{\mathbf{P}}{Nm}t + \frac{e\mathbf{E}}{2m}t^2, \tag{16.14}$$

gives rise to the intracollisional field effect. The dominant contributions to the integrations in Eqs. (16.12) and (16.13) come from the time range of s less than the characteristic time τ_{cr} of the correlation function. When time t is larger than the correlation times the integrations in Eqs. (16.12) and (16.13) will approach steady values and the upper limits of the time integrals can be extended to ∞.

The correlation time may be identified with the duration of a collision. The s^2 term may be expected relevant only when $|e|\mathbf{q}\cdot\mathbf{E}\tau_{cr}^2/2m \gtrsim 1$. If the electric field strength E is much smaller than the field

$$E_{cr} = \frac{2m}{|e|q\tau_{cr}^2}, \tag{16.15}$$

the intracollisional field term is obviously negligible. Employing the estimation $\tau_{cr} \sim (4\varepsilon_F)^{-1}$ we have from Eq. (16.15) $E_{cr} \sim 1 \times 10^6$V/cm for $m = 1.8 \times 10^{-31}$kg and $\varepsilon_F = 1000K$; and $E_{cr} \sim 1 \times 10^4$V/cm if $m = 0.6 \times 10^{-31}$kg and $\varepsilon_F = 70K$.

16.2 General System Formulations without Intracollisional Field Effect

Neglecting the intracollisional field terms, Eqs.(16.12) and (16.13) become ($\omega_0 \equiv \mathbf{q}\cdot\mathbf{v}$)

$$\mathbf{f} = -2\sum_{\mathbf{q}} \int \frac{d\omega}{2\pi} \mathbf{q}\Pi_2(\mathbf{q}, \omega + \omega_0)\Gamma(\mathbf{q}, \omega) n\left(\frac{\omega + \omega_0}{T_e}\right) \tag{16.16}$$

$$w = -2\sum_{\mathbf{q}} \int \frac{d\omega}{2\pi} \omega\Pi_2(\mathbf{q}, \omega + \omega_0)\Gamma(\mathbf{q}, \omega) n\left(\frac{\omega + \omega_0}{T_e}\right) \tag{16.17}$$

Here $n(x) \equiv 1/(e^x - 1)$ is the Bose function, and $\Gamma(\mathbf{q}, \omega)$ represents the dynamic scattering factor, the spectral function of the time correlation function $\Gamma(\mathbf{q}, t)$ of the atom system in interaction with electrons:

$$\Gamma(\mathbf{q}, \omega) = \int dt\, e^{i\omega t} \Gamma(\mathbf{q}, t), \tag{16.18}$$

$$\Gamma(\mathbf{q}, t) = \langle \gamma_{-\mathbf{q}}(t)\, \gamma_{\mathbf{q}} \rangle, \tag{16.19}$$

where the average is taken with respect to the initial state (16.9), which, as far as the atom system is concerned, is a thermoequilibrium state at temperature T. Because of this the spectral function has the general property:

$$\Gamma(-\mathbf{q}, -\omega) = e^{-\omega/T} \Gamma(\mathbf{q}, \omega). \tag{16.20}$$

Furthermore, in deriving Eq. (16.16) we have used the following relations:

$$\langle \dot{\gamma}_{-\mathbf{q}}(t)\, \gamma_{\mathbf{q}} \rangle = -\langle \gamma_{-\mathbf{q}}(t)\, \dot{\gamma}_{\mathbf{q}} \rangle = \frac{\partial}{\partial t} \Gamma(\mathbf{q}, t). \tag{16.21}$$

Note that in Eqs. (16.16) and (16.17) the Bose factor $n[(\omega + \omega_0)/T_e]$ can be replaced by

$$\left[\frac{e^{-\omega/T}}{1 - e^{(\omega+\omega_0)/T_e}} - 1 \right] \tag{16.22}$$

or by

$$\frac{1}{2} \left[\frac{1 - e^{-\omega/T}}{1 - e^{-(\omega+\omega_0)/T_e}} - 1 \right]. \tag{16.23}$$

The force and the energy balance equations can be written in terms of $\mathbf{f}$ and w, and the resistivity is defined from the frictional force as

$$\rho = -\frac{\mathbf{f} \cdot \mathbf{v}}{n^2 e^2 v^2}. \tag{16.24}$$

In weak current limit we expand Eq. (16.16) to linear order in $\mathbf{v}$ and make use of the approximate expression Eq.(4.34) for $\Pi_2(\mathbf{q}, \omega)$ in the degenerate case, to obtain the Ohmic resistivity from Eq.(16.24) as (Lei, 1990c):

$$\rho = \frac{12\pi}{e^2 v_F^2} \int_0^1 d\left(\frac{q}{2k_F}\right) \left(\frac{q}{2k_F}\right)^3 \int_{-\infty}^{\infty} \frac{d\omega}{2\pi} \frac{\omega/T}{e^{\omega/T} - 1} \Gamma(\mathbf{q}, \omega). \tag{16.25}$$

This is the generalized Ziman formula for linear resistivity of liquid and amorphous metals (Ziman 1961, 1967), as well as for crystalline systems if the elastic Bragg part due to the long-range ordered atoms of the lattice is substracted from the dynamic scattering factor $\Gamma(\mathbf{q}, \omega)$ (Lei, 1980). For solids it is convenient to write the atom position $\mathbf{R}_l$ as the sum of the static position and a small oscillatory deviation $\mathbf{u}_l$ from it. The dynamic scattering factor is then divided into an elastic part $2\pi\Gamma_0(\mathbf{q})\delta(\omega)$ and an inelastic part $\Gamma_{el}(\mathbf{q}, \omega)$:

$$\Gamma(\mathbf{q}, \omega) = 2\pi\Gamma_0(\mathbf{q})\delta(\omega) + \Gamma_{el}(\mathbf{q}, \omega) \tag{16.26}$$

The latter includes all multiphonon effects. Both the elastic and the inelastic parts contribute to the frictional force, but only the inelastic part contributes to the electron energy-loss rate. For crystalline solids only randomly distributed impurities contribute to the elastic part of the dynamic scattering factor, $\Gamma_0(\mathbf{q}) = n_i|u(\mathbf{q})|^2$, yielding the frictional force from Eq. (16.16)

$$\mathbf{f}_i = n_i \sum_{\mathbf{q}} |u(\mathbf{q})|^2 \mathbf{q} \Pi_2(\mathbf{q}, \omega_0) \tag{16.27}$$

due to elastic scattering. Here n_i is the impurity density and $u(\mathbf{q})$ is the difference of an impurity potential and a regular lattice atom potential $v(\mathbf{q})$. Retaining only the dynamic single-phonon process in the inelastic part of the scattering factor, we have the phonon-induced frictional force $\mathbf{f}_p$ and the energy loss rate w of the form

$$\mathbf{f}_p = 2\sum_{\mathbf{q},\lambda} |M(\mathbf{q},\lambda)|^2 \mathbf{q}\, \Pi_2(\mathbf{q}, \Omega_{\mathbf{q}\lambda} + \omega_0) \left[n\left(\frac{\Omega_{\mathbf{q}\lambda}}{T}\right) - n\left(\frac{\Omega_{\mathbf{q}\lambda} + \omega_0}{T_e}\right)\right] \tag{16.28}$$

$$w = 2\sum_{\mathbf{q},\lambda} |M(\mathbf{q},\lambda)|^2 \Omega_{\mathbf{q}\lambda} \Pi_2(\mathbf{q}, \Omega_{\mathbf{q}\lambda} + \omega_0) \left[n\left(\frac{\Omega_{\mathbf{q}\lambda}}{T}\right) - n\left(\frac{\Omega_{\mathbf{q}\lambda} + \omega_0}{T_e}\right)\right] \tag{16.29}$$

where

$$|M(\mathbf{q},\lambda)|^2 = \frac{N_L(\mathbf{q}\cdot \mathbf{e}_{\mathbf{q}\lambda})^2}{2M_L\Omega_{\mathbf{q}\lambda}} |v(q)|^2. \tag{16.30}$$

M_L and N_L being the mass and number density of the atom, $\Omega_{\mathbf{q}\lambda}$ and $\mathbf{e}_{\mathbf{q}\lambda}$ the phonon frequency and polarization vector.

16.3 Evaluation of the Intracollisional Field Effect

Although it has been shown that the intracollisional field terms in Eqs.(16.12) and (16.13) are negligible when the applied electric field $E \ll E_{cr}$, it is still not clear how important they are when $E \gtrsim E_{cr}$. To evaluate the effect of these intracollisional field terms on the high-field mobility for $E \gtrsim E_{cr}$, we rewrite Eqs. (16.12) and (16.13) as (Lei and Song, 1992)

$$\mathbf{f} = 2\sum_{\mathbf{q}} \mathbf{q} \iint_{-\infty}^{\infty} \frac{d\omega d\omega'}{(2\pi)^2} D(\omega_0 - \omega - \omega', \alpha) \Pi_2(\mathbf{q}, \omega) \Gamma(\mathbf{q}, \omega') \left(\frac{1 - e^{-\omega/T_e - \omega'/T}}{1 - e^{-\omega/T_e}} \right) \tag{16.31}$$

$$w = 2\sum_{\mathbf{q}} \iint_{-\infty}^{\infty} \frac{d\omega d\omega'}{(2\pi)^2} \omega' D(\omega_0 - \omega - \omega', \alpha) \Pi_2(\mathbf{q}, \omega) \Gamma(\mathbf{q}, \omega') \left(\frac{1 - e^{-\omega/T_e - \omega'/T}}{1 - e^{-\omega/T_e}} \right) \tag{16.32}$$

where

$$\alpha \equiv \frac{e}{2m} \mathbf{q}\cdot \mathbf{E} \tag{16.33}$$

and the function $D(\beta,\alpha)$ is defined by

$$D(\beta,\alpha) \equiv \int_0^\infty \cos(\beta t + \alpha t^2)dt. \tag{16.34}$$

Retaining contributions from the elastic part and the single phonon part of the dynamic scattering factor $\Gamma(\mathbf{q},\omega)$, we have the frictional forces $\mathbf{f}_i$ and $\mathbf{f}_p$, due to elastic and single phonon processes respectively, as

$$\mathbf{f}_i = \sum_{\mathbf{q}} \mathbf{q}\Gamma_0(\mathbf{q}) \int \frac{d\omega}{\pi} D(\omega_0 - \omega, \alpha)\Pi_2(\mathbf{q},\omega) \tag{16.35}$$

$$\mathbf{f}_p = \sum_{\mathbf{q},\lambda} \mathbf{q}|M(\mathbf{q},\lambda)|^2 \int \frac{d\omega}{\pi} \Pi_2(\mathbf{q},\omega) \left[n\left(\frac{\Omega_{\mathbf{q}\lambda}}{T}\right) - n\left(\frac{\omega}{T_e}\right)\right] D(\omega_0 - \omega + \Omega_{\mathbf{q}\lambda}, \alpha) \tag{16.36}$$

and the energy transfer rate w, due to the single phonon process, is given by

$$w = \sum_{\mathbf{q},\lambda} \Omega_{\mathbf{q}\lambda}|M(\mathbf{q},\lambda)|^2 \int \frac{d\omega}{\pi} \Pi_2(\mathbf{q},\omega) \left[n\left(\frac{\Omega_{\mathbf{q}\lambda}}{T}\right) - n\left(\frac{\omega}{T_e}\right)\right] D(\omega_0 - \omega + \Omega_{\mathbf{q}\lambda}, \alpha). \tag{16.37}$$

The function $D(\beta,\alpha)$ defined in Eq. (16.34), which was discussed by Thornber (1978) and by Barker (1979), can be written in terms of Fresnel integrals (Abramowitz and Stegun, 1965)

$$D(\beta,\alpha) = \left(\frac{\pi}{2|\alpha|}\right)^{1/2} \left\{\cos\left(\frac{\beta^2}{4|\alpha|}\right)\left[\frac{1}{2} - C\left(\frac{\alpha\beta}{2|\alpha|^{3/2}}\right)\right] + \sin\left(\frac{\beta^2}{4|\alpha|}\right)\left[\frac{1}{2} - S\left(\frac{\alpha\beta}{2|\alpha|^{3/2}}\right)\right]\right\} \tag{16.38}$$

with

$$C(x) = \left(\frac{2}{\pi}\right)^{1/2} \int_0^x \cos y^2 dy,$$

$$S(x) = \left(\frac{2}{\pi}\right)^{1/2} \int_0^x \sin y^2 dy.$$

It has delta-function behavior when $\alpha \to 0$:

$$D(\beta,\alpha) \to 2\pi\delta(\beta). \tag{16.39}$$

Thus, in the limit of $E \ll E_{cr}$, Eqs. (16.35)-(16.37) reduce to the corresponding expressions without intracollisional field terms in the preceding subsection. The effect of high electric field resides in the oscillatory function $D(\beta,\alpha)$. The complicated oscillatory behavior of the integrands in Eqs.(16.35)-(16.37) makes numerical calculation very difficult. In most previous investigations of the intracollisional field effect, the field-dependent oscillatory functions were replaced by nonoscillatory functions to facilitate calculation. Lei and Song (1992) performed numerical calculation directly from these equations by taking advantage of the Fresnel integral

$$C(x) = \frac{1}{2} + f(x)\sin(\frac{\pi}{2}x^2) - g(x)\cos(\frac{\pi}{2}x^2) \tag{16.40}$$

$$S(x) = \frac{1}{2} - f(x)\cos(\frac{\pi}{2}x^2) - g(x)\sin(\frac{\pi}{2}x^2) \tag{16.41}$$

expressed in terms of slowly varying auxiliary functions $f(x)$ and $g(x)$, which are tabulated (Abramowitz and Stegun, 1965). Using (16.40) and (16.41), one only needs to treat oscillatory integrations with the simple factor $\sin y^2$ A special interpolation was employed to carry out these oscillatory integrations semianalytically. In this way Lei and Song (1992) were able to solve the force- and energy-balance equations for Ge including the intracollisional field effect correction to first order, with the help of a microcomputer. They found that the field effect is unimportant even for very high field because it is always suppressed by an increasing of the carrier temperature which induces broadening in the density-density correlation function. This result is in conformity with most existing studies.

17. RECENT DEVELOPMENTS OF THE BALANCE-EQUATION THEORY

17.1 Distribution Function in the Balance-Equation Theory

It has been emphasized (Lei and Ting, 1987) that a displaced Fermi-Dirac (or Maxwell) function, as a Galilean-transformed equilibrium distribution function, can not be the distribution function of electrons in balance-equation transport theory. Although there is no need to explicitly determine the electron distribution function in the calculation of major bulk transport properties to leading order in the scattering interactions in the balance-equation theory, such information about the electron distribution is necessary in making connection with more detailed experimental measurements. In balance-equation theory, the electron temperature T_e, which is defined as the thermodynamic temperature of the relative electron system after turning-off the scattering interactions (Lei and Ting, 1985b) and serves as one of the major parameters, is not a directly experimentally measurable quantity (Lei and Horing, 1987b). One must exercise caution when attempting to identify T_e as an experimental quantity. For instance, photoluminescence experiments usually measure an effective temperature T_{eff} by fitting the luminescence intensity at high energy with the form $\sim \exp(-\omega/T_{eff})$. Identifying T_{eff} as the electron temperature T_e is inappropriate under conditions of high-field transport. A meaningful comparison requires the knowledge of the distribution function of electrons.

As discussed in Section 14.2, the distribution function of the relative electrons is defined in balance-equation transport as (Lei and Ting, 1987)

$$f(\mathbf{k}) = Tr(\hat{\rho} n_{\mathbf{k}}), \tag{17.1}$$

($n_{\mathbf{k}} \equiv \sum_\sigma c^\dagger_{\mathbf{k}\sigma} c_{\mathbf{k}\sigma}$), which can be written as the sum of an equilibrium distribution function at the electron temperature T_e,

$$f_0(\mathbf{k}) = Tr(\hat{\rho}_0 n_{\mathbf{k}}), \tag{17.2}$$

and a deviation from it:

$$f(\mathbf{k}) = f_0(\mathbf{k}) + f_1(\mathbf{k}). \tag{17.3}$$

$f_1(\mathbf{k})$ can be written in the interaction picture as

$$f_1(\mathbf{k}) = Tr[\hat{\rho}_1(t) n_{\mathbf{k}}(t)], \tag{17.4}$$

with $\hat{\rho}_1(t) \equiv \hat{\rho}(t) - \hat{\rho}_0$. A perturbative treatment for $f_1(\mathbf{k})$ is possible for a system with relatively strong carrier-carrier interactions. Calculation can proceed with the use of the iterative expansion of the density matrix $\rho_1(t)$; or by the use of Feynman diagrammatic techniques in the CTPG formulation (Sect. 12.1).

Recently Guillemot and Clérot (1991) did a calculation of the distribution function for an interacting relative electron gas to leading nonvanishing order in the scattering interactions. Their result is expressed as

$$f_0(\mathbf{k}) = \int d\omega A_{\mathbf{k}}(\omega) f_{\mathbf{k}}(\omega) \tag{17.5}$$

$$\begin{aligned} f_1(\mathbf{k}) = -\frac{2}{\pi} \int d\omega A_{\mathbf{k}}^2(\omega) \Bigg\{ & n_i \sum_{\mathbf{q}} |u(\mathbf{q}|^2 \Pi_2(\mathbf{k}, \mathbf{q}, \mathbf{q}\cdot\mathbf{v}_d) \\ & + \sum_{\mathbf{q}\lambda} |M(\mathbf{q},\lambda)|^2 \left[n\left(\frac{\Omega_{\mathbf{q}\lambda}}{T_{\mathbf{q}\lambda}}\right) - n\left(\frac{\Omega_{\mathbf{q}\lambda} - \mathbf{q}\cdot\mathbf{v}_d}{T_e}\right)\right] \\ & \times [\Pi_2(\mathbf{k}, \mathbf{q}, \mathbf{q}\cdot\mathbf{v}_d - \Omega_{\mathbf{q}\lambda} + \Pi_2(\mathbf{k}, -\mathbf{q}, -\mathbf{q}\cdot\mathbf{v}_d + \Omega_{\mathbf{q}\lambda})] \Bigg\}, \end{aligned} \tag{17.6}$$

where $A_{\mathbf{k}}(\omega)$ is the spectral density function of the interacting electron system, $\Pi_2(\mathbf{k}, \mathbf{q}, \omega)$ is the imaginary part of the Fourier spectrum of the three particle correlation function

$$\Pi(\mathbf{k}, \mathbf{q}, t_1 - t_2) = -i\theta(t_1 - t_2)\langle[[n_{\mathbf{k}}, \rho_{\mathbf{q}}(t_1)], \rho_{-\mathbf{q}}(t_2)]\rangle_0, \tag{17.7}$$

and $T_{\mathbf{q}\lambda}$ is the mode-dependent phonon temperature, taking account of nonequilibrium phonon occupation (Sect.10). Therefore, the first correction to the distribution function is given by the product of the finite lifetime of an electron in state $\mathbf{k}$ of the interacting electron gas in thermodynamic equilibrium at temperature T_e, $\int d\omega A_{\mathbf{k}}^2(\omega)$, with the drift-velocity modulated scattering rate due to impurities and phonons. This result is physically quite understandable: Coulomb interactions between electrons drive the relative electron gas toward an equilibrium distribution $f_0(\mathbf{k})$, while the electron-impurity and electron-phonon interactions tend to deviate the relative electron system from equilibrium. Guillemot and Clérot (1991) calculated $A_{\mathbf{k}}(\omega)$ and $\Pi_2(\mathbf{k}, \mathbf{q}, \omega)$ in the spirit of RPA. For a $161\mathring{A}$ wide GaAs/AlGaAs quantum well with carrier sheet density $N_s = 1.62 \times 10^{11} cm^{-2}$ subject to LO phonon scattering and an electric field of $300 V/cm$, they found that the distribution function is highly anisotropic, exhibiting dips at $\omega = \Omega_{LO} - \mathbf{q}\cdot\mathbf{v}_d$ in the direction of the drift velocity. Under the assumption of $\tau_p = 0$ (equilibrium phonon occupation), the drift velocity and electron temperature derived from the high-field balance equations for this system are $v_d = 1.5 \times 10^7 cm/s$ and $T_e = 83.5K$. However, by averaging the electron distribution function in all the directions they obtained an effective temperature $T_{eff} = 98K$ with $\tau_p = 0$, and a $T_{eff} = 115K$ assuming $\tau_p = 1ps$. The latter is not only in agreement with the experimental photoluminescence-determined effective temperature of 120K (Guillemot et al., 1990), thus satisfying energy-loss rate data, but it also gives the correct drift velocity. Previous comparison between theory and experiment by identifying T_e with the experimental of T_{eff}, required a phonon relaxation time of $\tau_p = 5ps$ to give correct energy loss rate, but at the same time it underestimated the high-field drift velocity (Guillemot et al., 1990).

17.2 Improved Balance-Equation Calculation for GaAs/AlGaAs Heterojunctions—A Step towards Device Simulation

Recently Magnus, Sala and De Meyer (1990) undertook the inclusion of some nonparabolic features of a real energy band in balance equation theory. They indicated that exactly the same set of balance equations given by Lei and Ting (1985b) can still be useful: to take account of nonparabolic effects, one need only evaluate the density-density correlation function using the nonparabolic dispersion of $\varepsilon_{\mathbf{k}}$. For a quasi-2D electron gas confined at a GaAs/AlGaAs heterojunction, to account for the nonparabolicity of band structure in GaAs they use a Kane-like expression modified by Tanimoto et al. (1988),

$$\varepsilon_{\alpha n}(\mathbf{k}_{\parallel}) = \varepsilon_{c_0} + \frac{\varepsilon_g}{2}\left\{-1 + \left[\left(1 + \frac{2W_{\alpha n}}{\varepsilon_g}\right)^2 + \frac{2k_{\parallel}^2}{m_{\alpha}^{*}\varepsilon_g}\right]^{1/2}\right\}, \tag{17.8}$$

as the energy dispersion with $W_{\alpha n}$ denoting the energy of the nth subband induced from the α-valley, ε_{c_0} the energy of the conduction band edge and ε_g the bandgap. For small $k_{\parallel}$ the dispersion relation is quadratic, while in the limit of large $k_{\parallel}$ it becomes linear. There is an upper bound v_m for the group velocity:

$$v_m = (\varepsilon_g/2m^{*})^{1/2}. \tag{17.9}$$

The imaginary parts of the intra- and inter-valley density-density correlation functions (5.27) of the 2D electron gas with dispersion given by (17.8) have been evaluated and incorporated in a computer program (Magnus, Sala and De Meyer, 1990). They considered remote impurity, intra- and inter-valley optic phonon scatterings and allowed for an arbitrary number of subbands to be included in the simulation by making use of an algorithm for the rapid evaluation of subband wavefunctions, energies and populations (Sala, Magnus and De Meyer, 1991). Although the results obtained have not yet provided an absolute quantitative comparison with experimental data because of the large uncertainty in the material parameters, they believe that the high accuracy/CPU cost ratio of this method will make it suitable for implementation in an all-purpose device simulator, allowing one to study transport in high electron mobility transistors in much greater detail than before.

17.3 Balance Equations for a Single Arbitrary Energy Band and Application to Superlattice Miniband Transport

The inclusion of nonparabolic features in the last subsection is mainly directed at wide energy bands, for which it is suffices to consider nonparabolic corrections to otherwise parabolic band systems. In the case of a narrow band, or an extremely nonparabolic band, such nonparabolicity corrections are no longer an appropriate vehicle for the determination of carrier transport.

Recently, Lei (1991) developed a balance-equation theory for hot electron transport in a single energy band of arbitrary structure, by distinguishing the center-of-mass momentum from its velocity. In this subsection we will give a brief discussion on this approach.

The Hamiltonian of the system has exactly the same form as given in Eqs.(15.1)-(15.10), but the electron Hamiltonian H_{el} consists of a single electron part and an electron-electron interaction,

$$H_{el} = \sum_j h_j + H_{ee}, \tag{17.10}$$

with h having eigenfunctions $\psi_{\mathbf{k}}(\mathbf{r})$ and eigenenergies $\varepsilon(\mathbf{k})$ for a single band: $h\psi_{\mathbf{k}}(\mathbf{r}) = \varepsilon(\mathbf{k})\psi_{\mathbf{k}}(\mathbf{r})$. It is now well-established that, limiting consideration to electrons moving within this single band, one can use the effective Hamiltonian

$$\varepsilon(-i\nabla_j) \equiv \varepsilon(\hat{\mathbf{p}}_j) \tag{17.11}$$

to replace h_j. Here $\hat{\mathbf{p}}_j \equiv -i\nabla_j$ is the momentum operator of the j-th electron. Employing a periodic zone scheme to describe the electron Bloch state, i.e. extending the wave vector $\mathbf{k}$ to the entire wavevector space in such a way that for any reciprocal vector $\mathbf{G}$, $\mathbf{k}$ and $\mathbf{k}+\mathbf{q}$ represent the same electron state, and writing the electron density operator $\rho_{\mathbf{q}} = \sum_j e^{i\mathbf{q}\cdot\mathbf{r}_j}$ in second quantization representation of the Bloch state, we have

$$\rho_{\mathbf{q}} = \sum_{\mathbf{k}} \rho_{\mathbf{kq}} \tag{17.12}$$

$$\rho_{\mathbf{kq}} = \sum_{\sigma} g(\mathbf{k},\mathbf{q}) c^{\dagger}_{\mathbf{k}+\mathbf{q}\sigma} c_{\mathbf{k}\sigma}, \tag{17.13}$$

$g(\mathbf{k},\mathbf{q})$ being a form factor.

With the center-of-mass position and momentum operators $\mathbf{R}$ and $\mathbf{P}$ defined by Eqs. (2.2)-(2.5), we can express the c.m. velocity operator as

$$\hat{\mathbf{v}} = -i[\mathbf{R}, H] = \frac{1}{N}\sum_j \mathbf{v}(\mathbf{P}_j) = \frac{1}{N}\sum_{\mathbf{k},\sigma} \mathbf{v}(\mathbf{k}) c^{\dagger}_{\mathbf{k}\sigma} c_{\mathbf{k}\sigma} \tag{17.14}$$

($\mathbf{v}(\mathbf{k}) \equiv \nabla\varepsilon(\mathbf{k})$), and obtain the acceleration in the form

$$\frac{d\hat{\mathbf{v}}}{dt} = -i[\hat{\mathbf{v}}, H] = e\mathbf{E}\cdot\hat{\mathcal{K}} + \hat{\mathbf{A}}_i + \hat{\mathbf{A}}_p, \tag{17.15}$$

and the rate of change of the electron energy as

$$\dot{H}_e = -i[H_e, H] = Ne\mathbf{E}\cdot\hat{\mathbf{v}} - \hat{W}. \tag{17.16}$$

Detailed expressions for the operators $\mathcal{K}$, $\hat{\mathbf{A}}_i$, $\hat{\mathbf{A}}_p$ and $\hat{W}$ are given by Lei (1991)

The essential idea of the balance equation theory is to choose a parameterized quasi-equilibrium initial state, from which the system evolves to the final state in a time of order of the system thermalization time τ_{th}. In the case of an arbitrary energy band the initial state is chosen to be

$$\hat{\rho}_0 = e^{-H_{er}/T_e}|\mathbf{P}_d\rangle\langle\mathbf{P}_d|. \tag{17.17}$$

Here $|\mathbf{P}_d\rangle$ is a single particle state bearing an average c.m. momentum $\mathbf{P}_d \equiv N\mathbf{p}_d$, in the presence of the electric field, and

$$H_{er} = \sum_{\mathbf{k},\sigma}\varepsilon(\mathbf{k}-\mathbf{p}_d)c^{\dagger}_{\mathbf{k}\sigma}c_{\mathbf{k}\sigma} + H_{ee} \equiv \sum_{\mathbf{k},\sigma}\bar{\varepsilon}(\mathbf{k})c^{\dagger}_{\mathbf{k}\sigma}c_{\mathbf{k}\sigma} + H_{ee}. \tag{17.18}$$

The implication of this density matrix is that the initial electron system obeys a canonical distribution in a reference frame in which the total momentum of the system vanishes.

The balance equations are obtained by taking the statistical average of the operator equations (17.15) and (17.16) and identifying $\langle d\hat{\mathbf{v}}/dt\rangle = d\mathbf{v}_d/dt$ and $\langle \dot{H}_e\rangle = d\langle H_e\rangle/dt$:

$$\frac{d\mathbf{v}_d}{dt} = e\mathbf{E}\cdot\mathcal{K} + \mathbf{A}_i + \mathbf{A}_p, \tag{17.19}$$

$$\frac{dh_e}{dt} = e\mathbf{E}\cdot\mathbf{v}_d - W. \tag{17.20}$$

Here $h_e \equiv \langle H_e\rangle/N$ is the system energy per carrier, $\mathbf{v}_d = \langle\hat{\mathbf{v}}\rangle$ is the average drift velocity and $\mathcal{K} = \langle\hat{\mathcal{K}}\rangle$ is an ensemble averaged inverse effective mass tensor. To the leading order in the scattering interactions, we have

$$h_e = \frac{2}{N}\sum_{\mathbf{k}}\varepsilon(\mathbf{k})f(\bar{\varepsilon}(\mathbf{k}),T_e) \tag{17.21}$$

$$\mathbf{v}_d = \frac{2}{N}\sum_{\mathbf{k}}\mathbf{v}(\mathbf{k})f(\bar{\varepsilon}(\mathbf{k}),T_e) \tag{17.22}$$

$$\mathcal{K} = \frac{2}{N}\sum_{\mathbf{k}}\nabla\nabla\varepsilon(\mathbf{k})f(\bar{\varepsilon}(\mathbf{k}),T_e), \tag{17.23}$$

where $f(\varepsilon,T_e) = 1/\exp[(\varepsilon-\mu)/T_e+1]$ is the Fermi function at T_e, and μ is the chemical potential, which is determined by the total number of electrons, N, according to

$$N = 2\sum_{\mathbf{k}}f(\bar{\varepsilon}(\mathbf{k}),T_e) = 2\sum_{\mathbf{k}}f(\varepsilon(\mathbf{k}),T_e). \tag{17.24}$$

In Eqs. (17.19) and (17.20), the frictional accelerations $\mathbf{A}_i$, $\mathbf{A}_p$ and the energy loss rate W are given by

$$\mathbf{A}_i = \frac{2\pi n_I}{N}\sum_{\mathbf{k},\mathbf{q}}|u(\mathbf{q})|^2|g(\mathbf{k},\mathbf{q})|^2\left[\mathbf{v}(\mathbf{k}+\mathbf{q})-\mathbf{v}(\mathbf{k})\right]\delta\left(\varepsilon(\mathbf{k}+\mathbf{q})-\varepsilon(\mathbf{k})\right) \times\frac{[f(\bar{\varepsilon}(\mathbf{k}),T_e)-f(\bar{\varepsilon}(\mathbf{k}+\mathbf{q}),T_e)]}{|\epsilon(\mathbf{q},\bar{\varepsilon}(\mathbf{k})-\bar{\varepsilon}(\mathbf{k}+\mathbf{q}))|^2}, \tag{17.25}$$

$$\mathbf{A}_p = \frac{4\pi}{N}\sum_{\mathbf{k},\mathbf{q},\lambda}|M(\mathbf{q},\lambda)|^2|g(\mathbf{k},\mathbf{q})|^2\left[\mathbf{v}(\mathbf{k}+\mathbf{q})-\mathbf{v}(\mathbf{k})\right]\delta\left(\varepsilon(\mathbf{k}+\mathbf{q})-\varepsilon(\mathbf{k})+\Omega_{\mathbf{q},\lambda}\right) \times\frac{[f(\bar{\varepsilon}(\mathbf{k}),T_e)-f(\bar{\varepsilon}(\mathbf{k}+\mathbf{q}),T_e)]}{|\epsilon(\mathbf{q},\bar{\varepsilon}(\mathbf{k})-\bar{\varepsilon}(\mathbf{k}+\mathbf{q}))|^2}\left[n\left(\frac{\Omega_{\mathbf{q},\lambda}}{T}\right)-n\left(\frac{\bar{\varepsilon}(\mathbf{k})-\bar{\varepsilon}(\mathbf{k}+\mathbf{q})}{T_e}\right)\right] \tag{17.26}$$

and

$$W = \frac{4\pi}{N} \sum_{\mathbf{k},\mathbf{q},\lambda} |M(\mathbf{q},\lambda)|^2 |g(\mathbf{k},\mathbf{q})|^2 \Omega_{\mathbf{q},\lambda} \delta\left(\varepsilon(\mathbf{k}+\mathbf{q}) - \varepsilon(\mathbf{k}) + \Omega_{\mathbf{q},\lambda}\right)$$
$$\times \frac{[f(\bar{\varepsilon}(\mathbf{k}),T_e) - f(\bar{\varepsilon}(\mathbf{k}+\mathbf{q}),T_e)]}{|\epsilon(\mathbf{q},\bar{\varepsilon}(\mathbf{k}) - \bar{\varepsilon}(\mathbf{k}+\mathbf{q}))|^2} \left[n\left(\frac{\Omega_{\mathbf{q},\lambda}}{T}\right) - n\left(\frac{\bar{\varepsilon}(\mathbf{k}) - \bar{\varepsilon}(\mathbf{k}+\mathbf{q})}{T_e}\right)\right]. \quad (17.27)$$

using the random-phase approximation. Here,

$$\epsilon(\mathbf{q},\omega) = 1 - \nu_c(q) \sum_{\mathbf{k}} \Pi^0(\mathbf{k},\mathbf{q},\omega) \quad (17.28)$$

and

$$\Pi^0(\mathbf{k},\mathbf{q},\omega) = 2|g(\mathbf{k},\mathbf{q})|^2 \frac{f(\bar{\varepsilon}(\mathbf{k}),T_e) - f(\bar{\varepsilon}(\mathbf{k}+\mathbf{q},T_e)}{\bar{\varepsilon}(\mathbf{k}+\mathbf{q}) - \bar{\varepsilon}(\mathbf{k}) + \omega - i\delta}. \quad (17.29)$$

These equations have recently been applied to the discussion of negative differential conductivity due to Bloch electron transport through a superlattice miniband by Lei, Horing and Cui (1991), using tight-binding sums for the envelope functions and

$$\varepsilon(\mathbf{k}) = \frac{k_{\parallel}^2}{2m} + \varepsilon_1(k_z), \quad (17.30)$$

$$\varepsilon_1(k_z) = \frac{\Delta}{2}(1 - \cos k_z d), \quad (17.31)$$

for the energy dispersion relation of the lowest miniband, Δ being the miniband width.

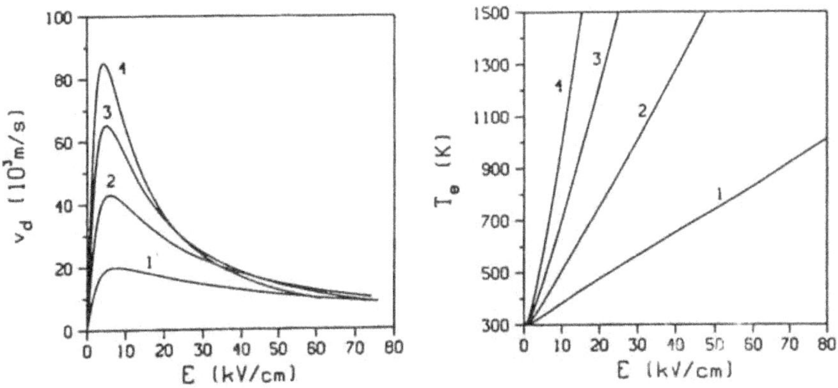

Fig.17.1 Calculated drift velocity and electron temperature as functions of electric field for a series of GaAs/AlAs superlattices with $N = 2 \times 10^{10} cm^{-2}$, $d = 57 Å$, having various miniband widths: curve 1, $\Delta = 300\,K$; curve 2, $\Delta = 500\,K$; curve 3, $\Delta = 700\,K$; curve 4, $\Delta = 900\,K$. The lattice temperature $T = 300\,K$.

The drift velocity v_d and the electron temperature T_e for steady state transport are obtained from the balance equations (17.19) and (17.20) as functions of the applied electric field under the conditions $dv_d/dt = 0$ and $dh_e/dt = 0$. The results are shown in Fig.17.1. In all the cases investigated the velocity-field curve exhibits negative differential mobility with a peak drift velocity v_p at the critical field E_c. The calculated results for v_p/d and E_cd as functions of miniband width Δ, shown in Fig.17.2, compare favorably with the experimental observation of Sibille et al. (1990).

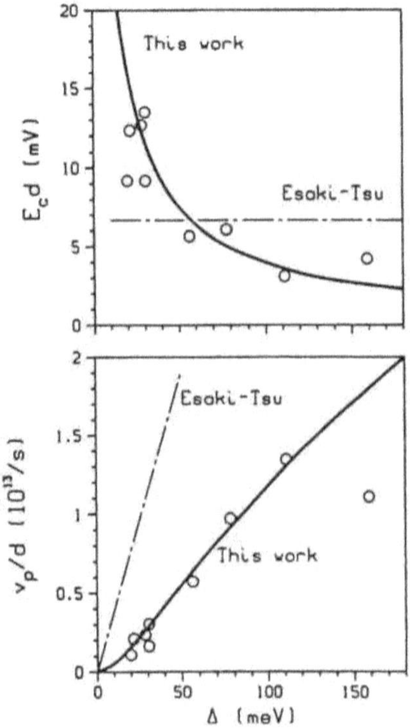

Fig.17.2 Dependencies of v_p/d and E_cd on miniband width Δ. Solid curves are calculated using balance equations at $T = 300\,K$ for GaAs/AlAs superlattices with electron sheet density $N_s = 2 \times 10^{10} cm^{-1}$ and a fixed impurity scattering rate. Experimental data (open circles) are taken from Sibille et al. (1990). Values of v_p/d and E_cd based on Esaki-Tsu theory (with $\tau = 0.1\,ps$) are also shown (as dot-dashed lines).

Miniband transport in the presence of a magnetic field and transient response to a step and/or impulsive electric field have also been carried out (Lei, Horing and Cui, 1991) for various superlattice systems, using the same set of balance equations (17.19) and (17.20) extended to incorporate magnetic fields.

ACKNOWLEDGMENTS

This work was supported by the National Natural Science Foundation of China. The authors wish to thank Mr. X.F. Wang and Mr. J. Cai for their help in preparing this manuscript. One of the authors (NJMH) wishes to thank P.N. Argyres for a proper introduction to the intricacies of quantum transport theory as a graduate student, and for many helpful discussions over the years following.

REFERENCES

Abdolsalami, F. and Khan, F.S., 1990, Phys. Rev. B**41**, 3494

Abramowitz, M. and Stegun, I.A., 1965, *Handbook of Mathematical Functions* (US GPU, Washington D.C.)

Altshuler, B.L., Aronov, A.G. and Khmelnitsky, D.E., 1981, Solid State Commun. **39**, 619

Altshuler, B.L., Aronov, A.G. and Khmelnitsky, D.E., 1982, J. Phys. C **15**, 7367

Anderson, P.W., Abrahams, E. and Ramakrishnan, T.V., 1979, Phys. Rev. Lett. **43**, 718

Ando, T., Fowler, A.B. and Stern, F., 1982, Rev. Mod. Phys. **54**, 437

Ando, T., 1976, Phys. Rev. Lett. **36**, 1383

Arai, M.R., 1983, Appl. Phys. Lett. **42**, 906

Argyres, P.N., 1960, Phys. Rev. **117**, 315

Argyres, P.N., 1963, Phys. Rev. **132**, 1527

Argyres, P.N. and Kelley, P.L., 1964, Phys. Rev. **134**, A98

Argyres, P.N., 1989, Phys. Rev. B**39**, 2982

Arora, V.K., 1983, J. Appl. Phys. **54**, 824

Bareikis, V., et al., 1980, Soviet. Phys.-Semicond. **14**, 847

Babiker, M, 1986, J. Phys. C **19**, 683

Barker, J.R., 1973, J. Phys. C **6**, 2663

Barker, J.R., 1978, Solid State Electron. **21**, 267

Barker, J.R., 1979, Solid State Commun. **32**, 1013

Bogolyubov, N.N., 1962, *Problems of Dynamic Theory in Statistical Physics* (North-Holland, Amsterdam)

Cai, J., 1989, Commun. Theor. Phys. **12**, 249

Cai, J. and Lei, X.L., 1990, Commun. Theor. Phys. **13**, 307

Cai, W. et al., 1989, Phys. Rev. B**40**, 7671

Cai, W., Lei, X.L. and Ting, C.S., 1985, Phys. Rev. B**31**, 4070

Cai, W., Marchetti, M.C. and Lax, M., 1986, Phys. Rev. B**34**, 8573

Cai, W., Zheng, T.F. and Lax, M., 1988, Phys. Rev. B**37**, 8205

Calecki, D. and Pottier, N., 1981, J. Phys. (Paris) Colloq. **42**, 67-271

Chen, L.Y., Ting, C.S. and Horing, N.J.M., 1990a, Phys. Rev. B**42**, 1129

Chen, L.Y., Ting, C.S. and Horing, N.J.M., 1990b, Solid State Commun. **73**, 437

Chou, K.C., Su, Z.B., Hao, B.L., 1985, Phys. Rep. **118**, 1

Cui, H.L., Lei, X.L. and Horing, N.J.M., 1988a, Phys. Stat. Sol. (b) **146**, 189

Cui, H.L., Lei, X.L. and Horing, N.J.M., 1988b, Phys. Rev. B**37**, 8223

Dreicer, H., 1959, Phys. Rev. **115**, 238

Drummond, T.J. et al., 1981, Electron Lett. **17**, 545

Fishman, R.S., 1989, Phys. Rev. B**39**, 2994

Fishman, R.S. and Mahan, G.D., 1989, Phys. Rev. B**39**, 2990

Fuchs, R. and Kliewer, K.L., 1965, Phys. Rev. **140**, A2076
Ganguly, A.K. and Ting, C.S., 1977, Phys. Rev. B**16**, 3541
Gerlach, E., 1986, J. Phys. C **19**, 4585
Götze, W. and Wölfle, P., 1972, Phys. Rev. B**6**, 122
Guillemot, C. et al., 1991, in *Proc. 7th Int. Conf. Hot Carriers in Semiconductors*, N
Japan, July 1-5, 1991
Guillemot, C. et al., 1990, Superlattices and Microstructures, **8**, 259
Guillemot, C. and Clérot, F., 1990, Superlattices and Microstructures, **8**, 268
Guillemot, C. and Clérot, F., 1991, preprint
Hirakawa, K. and Sakaki, H., 1988, J. Appl. Phys. **63**, 803
Höpfel, R.A. and Weimann, G., 1985, Appl. Phys. Lett. **46**, 291
Höpfel, R.A. et al., 1986, Phys. Rev. Lett. **25**, 2736
Horing, N.J.M., 1965, Ann. Phys. **31**, 1
Horing, N.J.M. and Argyres, P.N., 1962, in *Report of the Int. Conf. on the Physic Semicond.*, July 1962, Pub: Inst. of Phys. and Physical Society, London, (1962) p.58
Horing, N.J.M., Cui, H.L. and Lei, X.L., 1987, Phys. Rev. B**35**, 6438
Horing, N.J.M., Lei, X.L. and Cui, H.L., 1986, Phys. Rev. B**33**, 6929
Huang, K. and Zhu, B.F., 1988a, Phys. Rev. B**38**, 2183
Huang, K. and Zhu, B.F., 1988b, Phys. Rev. B**38**, 13377
Huberman, M. and Chester, G. M., 1975, Adv. Phys. **24**, 489
Hu, G.Y. and O'Connell, R.F., 1987, Phys. Rev. B**36**, 5798
Hu, G.Y. and O'Connell, R.F., 1988a, Physica A**149**, 1
Hu, G.Y. and O'Connell, R.F., 1988b, Physica A**151**, 33
Hu, G.Y. and O'Connell, R.F., 1988c, Physica A**153**, 114
Hu, G.Y. and O'Connell, R.F., 1988d, Solid State Commun. **65**, 33
Hu, G.Y. and O'Connell, R.F., 1988e, Phys. Rev. B**37**, 10391
Hu, G.Y. and O'Connell, R.F., 1988f, J. Phys. C **21**, 4325
Hu, G.Y. and O'Connell, R.F., 1989a, Phys. Rev. B**39**, 12717
Hu, G.Y. and O'Connell, R.F., 1989b, Phys. Rev. B**40**, 3600
Hu, G.Y. and O'Connell, R.F., 1989c, Superlattices and Microstructures, **5**, 515
Hu, P. and Ting, C.S., 1986, Phys. Rev. B**34**, 7003
Hu, P. and Ting, C.S., 1987a, Phys. Rev. B**35**, 4162
Hu, P. and Ting, C.S., 1987b, Phys. Rev. B**36**, 9671
Jain, J.K. and Das Sarma, S., 1989, Phys. Rev. Lett. **62**, 2305
Kalashnikov, V.P., 1970, Physica **48**, 93
Keldysh, L.V., Kirzhritz, D.A. and Maradudin, A.A., 1989, *The Dielectric Function Condensed Systems* (North-Holland, Amsterdam)
Khan, F.S., Davis, J.H. and Wilkins, J.W., 1987, Phys. Rev. B**36**, 2578
Klein, M.V., 1986, IEEE J. Quantum Electron. QE-**22**, 1760

Klimontovich, Yu.L., Silin, V.P., 1960, Usp. Fiz. Nauk, **70**, 247 (Sov. Phys.–Usp. **3**, 84)
Knox, W.H. et al., 1989, J. Phys. Cond. Matter, **1**, 9401
Kocevar, P., 1985, Physica **134B**, 155
Kohn, W. and Luttinger, J.M., 1957, Phys. Rev. **108**, 590
Krieger, J.B. and Iafrate, G.J., 1987, Phys. Rev. B**36**, 9644
Kubo, R., 1957, J.Phys. Soc. Jpn. **12**, 570
Lee, P.A. and Ramakrishnan, T.V., 1985, Rev. Mod. Phys. **57**, 287
Lei, X.L., 1980, Acta Phys. Sinica, **29**, 1395
Lei, X.L., 1985a, J. Phys. C **18**, L593
Lei, X.L., 1985b, J. Phys. C **18**, L731
Lei, X.L., 1987a, Commun. Theoret. Phys. **8**, 385
Lei, X.L., 1987b, Chinese Phys. Lett. **4**, 193
Lei, X.L., 1990a, Phys. Rev. B**41**, 8085
Lei, X.L., 1990b, Phys. Lett. A**148**, 384
Lei, X.L., 1990c, in *Lattice Dynamics and Semiconductor Physics*, edited by Xia, J.B. et al. (World Scientific, Singapore) p.507
Lei, X.L., 1991, Phys. Stat. Solidi (b) (to be published)
Lei, X.L., Birman, J.L. and Ting, C.S., 1985, J. Appl. Phys. **58**, 2270
Lei, X.L. and Cai, J., 1990a, Phys. Rev. B**42**, 1574
Lei, X.L. and Cai, J., 1990b, Phys. Lett. A**144**, 316
Lei, X.L. and Cai, J. and Xie, L.M., 1988, Phys. Rev. B**38**. 1529
Lei, X.L. and Cai, W. and Ting, C.S., 1985, J. Phys. C **18**, 4315
Lei, X.L., Cui, H.L. and Horing, N.J.M., 1987, J. Phys. C **20**, L287
Lei, X.L., Cui, H.L. and Horing, N.J.M., 1988, Phys. Rev. B**38**, 8230
Lei, X.L. et al., 1986, Phys. Rev. B**33**, 4382
Lei, X.L. et al., 1987, Phys. Rev. B**36**, 9134
Lei, X.L. and Horing, N.J.M., 1986, Solid State Commun. **58**, 235
Lei, X.L. and Horing, N.J.M., 1987a, Phys. Rev. B**35**, 6281
Lei, X.L. and Horing, N.J.M., 1987b, Phys. Rev. B**36**, 4238
Lei, X.L. and Horing, N.J.M., 1988, Solid State Electron. **31**, 531,
Lei, X.L. and Horing, N.J.M., 1989, Phys. Rev. B**40**, 3756
Lei, X.L., Horing, N.J.M. and Cui, H.L., 1991, Phys. Rev. Lett. **66**, 3277
Lei, X.L., Horing, N.J.M. and Cui, H.L., 1991, *Proc. 7th Int. Conf. Hot Carriers in Semiconductors*, Nara, Japan, July 1-5, 1991
Lei, X.L., Horing, N.J.M. and Zhang, J.Q., 1986a, Phys. Rev. B**33**, 2912
Lei, X.L., Horing, N.J.M. and Zhang, J.Q., 1986b, Phys. Rev. B**34**, 1139
Lei, X.L., Horing, N.J.M. and Zhang, J.Q., 1987, Phys. Rev. B**35**, 2834
Lei, X.L. and Song, W.D., 1992, Commun. Theoret. Phys. (to be published)
Lei, X.L. and Ting, C.S., 1984, Phys. Rev. B**30**, 4809

Lei, X.L. and Ting, C.S., 1985a, J. Phys. C **18**, 77
Lei, X.L. and Ting, C.S., 1985b, Phys. Rev. B**32**, 1112
Lei, X.L. and Ting, C.S., 1985c, Solid State Commun. **53**, 305
Lei, X.L. and Ting, C.S., 1985d, Acta Phys. Sinica, **8**, 983
Lei, X.L. and Ting, C.S., 1987, Phys. Rev. B**36**, 8162
Lei, X.L. and Zhang, J.Q., 1986, J. Phys. C **19**, L73
Levinson, I.B., 1969, Zh. Eksp. Teor. Fiz. **57**, 660 [Sov. Phys.-JETP, **30**, 362, (1970)]
Lifshitz, E.M. and Pitaevskii, L.P., 1981, *Physical Kinetics* (Pergamon, New York)
Lipavsky, P. et al., 1991, Phys. Rev. B**43**, 4885
Liu, M. et al. 1988, Phys. Rev. B**37**, 2997
Liu, M., Xing, D.Y. and Ting, C.S., 1989, J. Phys.: Condens. Matter **1**, 407
Lugli, P. and Goodnick, S., 1987, Phys. Rev. Lett. **59**, 716
Magnus, W., Sala, C. and De Meyer, K., 1990, Phys. Rev. B**41**, 5197
Mahan, G.D., 1981, *Many Particle Physics* (Plenum, New York), p.42
Mahan, G.D., 1972, in *Polarons in Ionic Crystals and Polar Semiconductors*, edited by J.T. Devreese (North-Holland, Amsterdam), P.533
Marchetti, M.C. and Cai, W., 1987, Phys. Rev. B**36**, 8159
Marchetti, M.C., Cai, W. and Lax, M., 1988, Solid State Electron. **31**, 677
McLennan jr., J.A., 1961, Phys. of Fluids **4**, 1319
Mermin, N.D. and Canal, E., 1964, Ann. Phys. **26**, 247
Nougier, J.P., 1980, in *Physics of Nonlinear Transport in Semiconductors*, edited by D.K. Ferry, J.R. Barker and C. Jacoboni (Plenum, New York) p.415
Nougier, J.P. and Rolland, M., 1973, Phys. Rev. B**8**, 5728
Peeters, F.M. and Devreese, J.T., 1981, Phys. Rev. B**23**, 1936
Peeters, F.M. and Devreese, J.T., 1984, in *Solid State Physics, v38*, edited by H. Ehrenreich and D. Turnbull (Academic Press, Orlando) p.82
Peletminskii, S. and Yatsenko, A., 1967, Zh. Eksp. Theor. Fiz. **53**, 1327 [Sov. Phys. -JETP **26**, 773 (1968)]
Price, P.J., 1985, Physica **134**B, 164
Rammer, J. and Smith, H., 1986, Rev. Mod. Phys., **58**, 323
Reggiani, L., 1985, in *Hot-Electron Transport in Semiconductors* (Springer-Verlag, Berlin)
Ridley, B.K., 1989b, Phys. Rev. B**39**, 5282
Ridley, B.K., 1989a, Semicond. Sci. Tech. **4**, 1142
Rieger, M. et al., 1989, Phys. Rev. B**39**, 7866
Rossi, F. and Jacoboni, C., 1991, in *Proc. 7th Int. Conf. Hot Carriers in Semiconductors*, Nara, Japan, July 1-5, 1991
Sala, C., Magnus, W. and De Meyer, K., 1991, J. Appl. Phys. **69**, 7689
Sarker, S.K., 1986, Phys. Rev. B**33**, 7263
Schmidt, G., 1979, *Physics of High Temperature Plasmas* (Academic Press, New York)

Seeger, K., 1982, *Semiconductor Physics* (Springer-Verlag, Berlin)
Sernelius, B.E. and Sőderstrőm, E., 1991a, J. Phys. Cond. Matter **3**, 1493
Sernelius, B.E. and Sőderstrőm, E., 1991b, J. Phys. Cond. Matter **3**, 8425
Shah, J. et al., 1984, Appl. Phys. Lett. **44**, 322
Shah, J. et al., 1985, Phys. Rev. Lett. **47**, 264
Sibille et al., 1990, Phys. Rev. Lett. **64**, 52
Sood, A.K. et al., 1985, Phys. Rev. Lett. **54**, 2111
Spitzer, L., 1956, *Physics of Fully Ionized Gases* (Interscience, New York)
Stern, F. and Howard, W.E., 1964, Phys. Rev. **163**, 816
Su, Z.B., Chen, L.Y. and Ting, C.S., 1988, Phys. Rev. Lett. **60**, 2323; Phys. Rev. B**38**, 3866
Su, Z.B., Chen, L.Y. and Birman, J.L., 1987, Phys. Rev. B**35**, 9744
Su, Z.B. and Sakita, B., 1989, Phys. Rev. B**40**, 9959
Tanimoto, H. et al., 1988, Jpn. J. appl. Phys. **27**, 563
Thornber, K.K. and Feynman, R.P., 1970, Phys. Rev. B**1**, 4099
Thornber, K.K., 1978, Solid State Electron. **21**, 259
Thornber, K.K., 1971, Phys. Rev. B**3**, 1929
Thouless, D.J., 1977, Phys. Rev. Lett. **39**, 1167
Ting, C.S. and Lei, X.L., 1984, Solid State Commun. **51**, 553
Ting, C.S. and Nee, T.W., 1986, Phys, Rev. B**33**, 7056
Ting, C.S., Ying, S.C. and Quinn, J.J., 1976a, Phys. Rev. B**14**, 4439
Ting, C.S., Ying, S.C. and Quinn, J.J., 1976b, Phys. Rev. Lett. **37**, 215
Ting, C.S., Ying, S.C. and Quinn, J.J., 1977, Phys. Rev. B**16**, 5394
Tzoar, N., Platzman, P.M. and Simons, A., 1976, Phys. Rev. Lett. **36**, 1200
Tzoar, N. and Zhang, C., 1985, Phys. Rev. B**32**, 1146
Vogl, P. 1981, in *Physics of Nonlinear Transport in Semiconductors*, edited by D.K. Ferry, J.R. Barker, and C. Jacoboni (Plenum, New York) p. 75
Wingreen, N.S., Stanton, C.J. and Wilkins, J.W., 1986, Phys. Rev. Lett. **57**, 1084
Xing, D.Y. Hu, P. and Ting, C.S., 1987, Phys. Rev. B**35**, 6379
Xing, D.Y. et al. 1988, J. Phys. C **21**, 2881
Xing, D.Y. Liu. M., Hu, P. and Ting, C.S., 1988, Phys. Rev. B**37**, 10283
Xing, D.Y. and Ting, C.S., 1987, Phys. Rev. B**35**, 3971
Yang, R.Q., Lei, X.L. and Tsai, C.H., 1990, Commun. Theor. Phys. **13**, 167
Yang, R.Q. et al., 1988, Surface Science **196**, 487
Zheng, T.F., Cai, W. and Lax, M., 1988, Phys. Rev. B**38**, 1406
Ziman, J.M., 1960, *Electrons and Phonons*, (Oxford University Press, Oxford)
Ziman, J.M., 1961, Phil. Mag. **6**, 1013
Ziman, J.M., 1967, Advan. Phys. **16**, 551

Ziman, J.M., 1972, *Prinicples of the Theory of Solids* (Cambridge University Press, Cambridge), Chapter 6

Zubarev, D.N., 1961, Dokl. Akad. Nauk SSSR, **144**, 92

Zubarev, D.N., 1974, *Nonequilibrium Statistical Theormodynamics* (Consultants Bureau, New York)

International Journal of Modern Physics B, Vol. 6, No. 7 (1992) 937–973

Recent Developments in Magnetotransport Theory

Norman J.M. Horing and H.L. Cui
Department of Physics and Engineering Physics
Stevens Institute of Technology, Hoboken, New Jersey 07030 USA
X.L. Lei
Department of Solid State Physics, Shanghai Institute of Metallurgy
Chinese Academy of Sciences, Shanghai 200050, China

A selection of topics recently examined in the literature of normal, dissipative magnetotransport is reviewed here, principally from the perspective of the Lei-Ting balance equation approach. We employ this isothermal formalism to revisit linear and nonlinear bulk 3D magnetoresistance, discussing longitudinal impurity limited conduction and magnetophonon resonance structure. For the transverse configuration, the characteristic linear impurity magnetoresistance log-divergence is seen to be removed by nonlinearity alone, with or without screening. At high magnetic fields, prominent hot-electron magnetophonon resonance structure is discussed for transverse transport as well as thermal noise. Another principal focus of concern here is the role of Landau quantization in linear and nonlinear transport in semiconductor microstructures, particularly quantum well superlattices (Type-I and Type-II) and heterostructures, including magnetoplasmon and cyclotron resonances as well as Shubnikov-de Haas oscillations. Electron-hole drag and negative absolute minority carrier mobility are discussed. Finally, we briefly discuss the density modulated 2DEG having magnetoresistance commensurability oscillations due to the interplay of the length scales of the cyclotron radius and the modulation period (superposed as an envelope upon Shubnikov-de Haas oscillations).

1 Introduction: Balance Equation Transport Theory in Magnetic Fields

The magnetic field has played a central role in the development of our understanding of the physics of all phases of matter. In particular, deep insight into the properties of solids has been generated by phenomena such as the Hall effect, de Haas-van Alphen-Shubnikov oscillations, cyclotron resonance, and numerous others. In the crucially important area of transport, key to a host of vital device applications, magnetic field phenomena abound and it is necessary to delimit the scope of any single review. Our focus will be upon normal, dissipative, magnetotransport in semiconductor microstructures, such as quantum well superlattices and heterostructures, with emphasis on the nonlinear regime, replete with Shubnikov-de Haas oscillations and magnetophonon effects, *etc.* In this choice, we do not address other very important areas of magnetotransport research which are well reviewed elsewhere, such as the quantum Hall effect and mesoscopics, as well as tunneling transport which has grown into a field unto itself. Other areas which have been abundantly reviewed and will be correspondingly de-emphasized here include linear bulk (3D) magneto-transport which has been reviewed in depth (Kubo, Hashitsume, and Miyake, 1965; Roth and Argyres, 1966) and linear two-dimensional (2D) magneto-transport, already extensively reviewed(Ando, Fowler, and Stern, 1982). Particular topics of special importance, such as cyclotron resonance, have also been thoroughly reviewed in the past.

Having clarified the general orientation of this review, it is appropriate to say that the boundaries between subjects treated and those omitted are rather porous, because of the perspective gained in our extensive experience with the Lei-Ting balance equation approach to transport theory (Lei and Ting, 1984; Lei and Horing, 1991). This method is uniquely significant for its inclusion of the role of relatively strong electron-electron interactions (either direct Coulomb or bath-phonon mediated electron-electron scattering) in promoting a strong tendency toward rapid thermalization about the drifted transport state. As well as producing interesting effects in transport in semiconductor microstructures, this phenomenology yields results that differ from the "standard" Boltzmann equation results for transport in bulk in the absence of electron-electron interactions, except at zero temperature, even in the linear limit (both with and without an ambient magnetic field). It is therefore appropriate to undertake a limited review of linear bulk magnetotransport from this point of view, within the more general Lei-Ting formulation of nonlinear balance equations for transport described in this volume (Lei and Horing, 1991) with two new features characterizing the ambient magnetic field $\mathbf{B}$: These features include a Lorentz force term ($\mathbf{v} \times \mathbf{B}$) in the center of mass (c.m.) force-balance equation, and installation of the proper magnetic field dependence into the structure of the density-density correlation function Π (or inverse dielectric function K)

for the system at hand. Thus the c.m. drift velocity $\mathbf{v}$ is governed by

$$nm\frac{d\mathbf{v}}{dt} = ne\mathbf{E} + ne(\mathbf{v}\times\mathbf{B}) + \mathbf{f}_i(\mathbf{v}) + \mathbf{f}_p(\mathbf{v}) \tag{1}$$

for time-dependent as well as dc transport in electric field $\mathbf{E}$, with the impurity and phonon induced frictional forces (per unit volume) given by

$$\mathbf{f}_i(\mathbf{v}) = n_i\sum_{\mathbf{q}}|u(\mathbf{q})|^2\mathbf{q}\,\Pi_2(\mathbf{q},\mathbf{q}\cdot\mathbf{v}), \tag{2}$$

$$\mathbf{f}_p(\mathbf{v}) = 2\sum_{\mathbf{q},\lambda}|M(\mathbf{q},\lambda)|^2\mathbf{q}\,\Pi_2(\mathbf{q},\Omega_{\mathbf{q}\lambda}+\mathbf{q}\cdot\mathbf{v})\left[n\left(\frac{\Omega_{\mathbf{q}\lambda}}{T}\right)-n\left(\frac{\Omega_{\mathbf{q}\lambda}+\mathbf{q}\cdot\mathbf{v}}{T_e}\right)\right]. \tag{3}$$

This describes resistive contributions due to impurity and phonon scatterings in terms of $\Pi_2(\mathbf{q},\omega)$ as before, except that $\Pi_2(\mathbf{q},\omega)$ incorporates both classical and quantum effects of the magnetic field in the present context. In the above equations n is the electron density, m the effective electron mass, T_e the electron temperature, T the lattice temperature, n_i the impurity concentration, $u(\mathbf{q})$ the impurity potential, and $M(\mathbf{q},\lambda)$ is the electron-phonon coupling matrix element for a phonon of wavenumber $\mathbf{q}$, frequency $\Omega_{\mathbf{q}\lambda}$, in branch λ. (For isotropic phonon dispersion, $\mathbf{f}(\mathbf{v})$ is always antiparallel to $\mathbf{v}$). Similar remarks pertain to the time dependent energy balance equation of the last chapter(Lei and Horing, 1991),

$$\frac{dE_e}{dt} = -\mathbf{v}\cdot\mathbf{f} - w = C_e\frac{dT_e}{dt}, \tag{4}$$

where

$$w(\mathbf{v}) = 2\sum_{\mathbf{q},\lambda}|M(\mathbf{q},\lambda)|^2\Omega_{\mathbf{q}\lambda}\Pi_2(\mathbf{q},\Omega_{\mathbf{q}\lambda}+\mathbf{q}\cdot\mathbf{v})\left[n\left(\frac{\Omega_{\mathbf{q}\lambda}}{T}\right)-n\left(\frac{\Omega_{\mathbf{q}\lambda}+\mathbf{q}\cdot\mathbf{v}}{T_e}\right)\right] \tag{5}$$

Here $C_e = dE_e/dT_e$ is the rate of change of relative electron internal energy, w is the average rate of change of phonon energy, and $\mathbf{f} \equiv \mathbf{f}(\mathbf{v}) = \mathbf{f}_i(\mathbf{v}) + \mathbf{f}_p(\mathbf{v})$. Focusing on steady state, the time dependencies of current $\mathbf{J}$, $\mathbf{v} = \mathbf{J}/ne$, T_e and phonon occupation number are null, and the bulk dc magnetotransport energy balance equation may be written in terms of a scalar resistivity function $\rho(\mathbf{v})$ defined as

$$\rho(\mathbf{v}) = -\frac{1}{n^2e^2v}f(\mathbf{v}), \tag{6}$$

(with $f(\mathbf{v}) \equiv (\mathbf{v}/v)\cdot\mathbf{f}(\mathbf{v})$) in the form

$$\rho(\mathbf{v}) = \rho_E(\mathbf{v}), \tag{7}$$

where

$$\rho_E(\mathbf{v}) \equiv \frac{1}{n^2e^2v^2}w(\mathbf{v}) \tag{8}$$

is an energy transfer "resistivity".

All of the quantities discussed above devolve upon the relative electron density-density correlation function $\Pi(\mathbf{q},\omega)$, which is closely related to the inverse dielectric function K through the exact relation

$$K = 1 + \nu_c\Pi, \tag{9}$$

where $\nu_c(q) = 4\pi e^2/\kappa q^2$ is the Coulomb potential of interacting electrons and κ is the background dielectric constant. In the random phase approximation (RPA) the density-density correlation function is given as ($\mathbf{B} = B\hat{z}$)

$$\Pi(\mathbf{q},\omega) = \frac{\Pi_0(\mathbf{q},\omega)}{1-\nu_c(q)\Pi_0(\mathbf{q},\omega)}, \tag{10}$$

where $\Pi_0(\mathbf{q},\omega)$ is the noninteracting electron density-density correlation function which may be represented in a closed-form integral representation in terms of elementary functions by(Horing, 1965) ($\Pi_0 = \Pi_{01} + i\Pi_{02}$)

$$\Pi_{01}(\mathbf{q},\Omega) = \mathcal{P}\int\frac{d\omega}{2\pi}\int\frac{d\omega'}{2\pi}\frac{\omega'}{\Omega^2-\omega'^2}f(\omega)R(\omega,\omega',\mathbf{q}), \tag{11}$$

and

$$\Pi_{02}(\mathbf{q},\Omega) = -\frac{1}{2}\int\frac{d\omega}{2\pi}f(\omega)R(\omega,\Omega,\mathbf{q}), \tag{12}$$

where

$$\begin{aligned}R(\omega,\omega',\mathbf{q}) \;&=\; \int_{-\infty}^{\infty}dx\int_{-\infty}^{\infty}dy e^{i\omega x}e^{-\frac{1}{2}i\omega' y}2i\sin\left(\frac{\omega' x}{2}\right)\frac{\pi^{3/2}}{(2\pi)^3}\left(\frac{2m}{ix}\right)^{1/2}\frac{m\omega_c}{i\tan(\omega_c x/2)}\\ &\times\exp\left(-i\frac{q_z^2}{2m}\frac{x^2-y^2}{4x}\right)\exp\left(-i\frac{q_\parallel^2}{2m}\frac{\cos(\omega_c y/2)-\cos(\omega_c x/2)}{\omega_c\sin(\omega_c x/2)}\right).\end{aligned} \tag{13}$$

In this, $q_\parallel^2 = q_x^2 + q_y^2$, $f(\omega)$ is the Fermi distribution and ω_c is the cyclotron frequency (Landau level separation has been taken equal to spin-splitting in Eq.(13)). An alternative, equivalent, representation involves expansion in a Landau-state series(Mermin and Canel, 1964; Ting, Ying, and Quinn, 1976a, 1976b, 1977; See also Fessatidis, 1991, for a discussion of the equivalence of the Landau series representation and the integral representation of Horing.)

$$\Pi_0(\mathbf{q},\omega) = \frac{m\omega_c}{2\pi}\sum_{n,n'}C_{n,n'}\left(\frac{q_\parallel^2}{2m\omega_c}\right)\Pi_0(n,n',q_z,\omega), \tag{14}$$

where

$$C_{n,n'}(x) = \left(\frac{n_2!}{n_1!}\right)x^{n_1-n_2}e^{-x}\left[L_{n_2}^{n_1-n_2}(x)\right]^2, \tag{15}$$

with $n_1 = max(n,n')$, $n_2 = min(n,n')$, and $L_n^l(x)$ is the associated Laguerre polynomial. The quantity $\Pi_0(n,n',q_z,\omega)$ is given by (null spin-splitting in Eqs.(16) and (17))

$$\Pi_{01}(n,n',q_z,\omega) = 2\sum_{p_z}\frac{f(\varepsilon(n',p_z+q_z))-f(\varepsilon(n,p_z))}{\omega+\varepsilon(n',p_z+q_z)-\varepsilon(n,p_z)}, \tag{16}$$

and

$$\Pi_{02}(n,n',q_z,\omega) = -2\pi\sum_{p_z}[f(\varepsilon(n',p_z+q_z))-f(\varepsilon(n,p_z))]\delta(\omega+\varepsilon(n',p_z+q_z)-\varepsilon(n,p_z)), \tag{17}$$

and the Landau level energy is

$$\varepsilon(n,p_z) = \left(n+\frac{1}{2}\right)\omega_c + \frac{p_z^2}{2m}. \tag{18}$$

2 Magnetotransport in Bulk Semiconductors Revisited; Linear and Nonlinear

The balance equation analysis of dc magnetotransport yields interesting new results for both the longitudinal and transverse configurations, exhibiting temperature dependence which differs from that of standard single-particle "adiabatic" transport theory, because of the rapid thermalization about the drifted state. For the longitudinal case $\mathbf{E} = E\hat{z} \parallel \mathbf{B}$, the force balance equation

$$neE + f(v) = 0, \tag{19}$$

yields the longitudinal resistivity/conductivity element as (Fessatidis, 1991)

$$\sigma_{zz}^{-1} = \rho_{zz} = \frac{E}{J} = \frac{E}{nev} = -\frac{1}{n^2e^2v}f(v) = \rho(v). \tag{20}$$

On the other hand, the transverse case involves $\mathbf{E} = E\hat{x}$, $\mathbf{v} = v\cos\phi\hat{x} + v\sin\phi\hat{y}$, with force balnce equations (Lei, Cai, and Ting, 1985)

$$neE + neBv\sin\phi + f(v)\cos\phi = 0, \tag{21}$$

$$-\,neBv\cos\phi + f(v)\sin\phi = 0, \tag{22}$$

(with $\cot\phi = -ne\ \rho(v)/B$) which lead to the conductivity tensor elements

$$\sigma_{xx} = \frac{J_x}{E} = \frac{\rho(v)}{\rho(v)^2 + B^2/n^2e^2} = \sigma_{yy}, \tag{23}$$

and

$$\sigma_{xy} = \frac{J_y}{E} = \frac{B/ne}{\rho(v)^2 + B^2/n^2e^2} = -\sigma_{yx}, \tag{24}$$

in the crossed field configuration. The rotationally invariant resistivity tensor elements are readily determined from the balance equations, or alternatively by inverting σ_{ij} as

$$\rho_{xx} = \rho(v) = \rho_{yy}, \tag{25}$$

and

$$\rho_{yx} = B/ne = -\rho_{xy}, \tag{26}$$

with corresponding Hall coefficient $R_H = 1/ne$.

Although $\rho_{xx} = \rho_{yy} = \rho(v)$ of transverse case appears to have formal similarity to $\rho_{zz} = \rho(v)$ of the longitudinal case, it should be borne in mind that anisotropy of $\Pi(\mathbf{q}, \omega = \mathbf{q}\cdot\mathbf{v})$ in the integrand of Eqs.(2) and (3) for $\mathbf{f}_i(\mathbf{v})$ and $\mathbf{f}_p(\mathbf{v})$ yields differing results.

2.1 Longitudinal Geometry

For the longitudinal configuration,

$$\begin{aligned}\rho_{zz} = \rho(v) &= -\frac{n_i}{n^2e^2v}\sum_{\mathbf{q}}|u(\mathbf{q})|^2 q_z\Pi_2(\mathbf{q}, q_z v) \\ &\quad -\frac{2}{n^2e^2v}\sum_{\mathbf{q},\lambda}|M(\mathbf{q},\lambda)|^2 q_z\Pi_0(\mathbf{q},\Omega_{\mathbf{q}\lambda}+q_z v) \\ &\quad \left[n\left(\frac{\Omega_{\mathbf{q}\lambda}}{T}\right) - n\left(\frac{\Omega_{\mathbf{q}\lambda}+q_z v}{T_e}\right)\right]. \end{aligned} \tag{27}$$

The linearization of ρ_{zz} for low drift velocity involves expanding Π to linear order in v, and in the phonon resistivity, one must also expand the Bose distribution n to linear order in v. This yields the linear longitudinal resistivity as (note that $T_e \equiv T$ in the linear limit)

$$\begin{aligned}\rho_{zz} &= -\frac{n_i}{n^2e^2}\sum_{\mathbf{q}}\frac{|u(\mathbf{q})|^2}{[\epsilon_1(\mathbf{q},0)]^2}q_z^2\left[\frac{\partial}{\partial\omega}\Pi_0(\mathbf{q},\omega)\right]_{\omega=0} \\ &\quad -\frac{1}{2n^2e^2T}\sum_{\mathbf{q},\lambda}|M(\mathbf{q},\lambda)|^2 q_z^2\Pi_0(\mathbf{q},\Omega_{\mathbf{q}\lambda})\operatorname{csch}^2\left(\frac{\Omega_{\mathbf{q}\lambda}}{2T}\right),\end{aligned} \tag{28}$$

where $\epsilon(\mathbf{q},\omega)$ is the RPA dielectric function. In the first, impurity-limited term of the above equation, $u(\mathbf{q})/\epsilon(\mathbf{q},0) \rightarrow u$ will be taken to be constant for a short range shielded scattering potential, with the result

$$\begin{aligned}\rho_{zz}^i &= -\frac{n_i m^4\omega_c^2u^2}{\pi^3n^2e^2}\sum_{n,n'}\int_0^\infty d\varepsilon f'(\varepsilon)\left[\frac{\varepsilon-(n+1/2)\omega_c}{\varepsilon-(n'+1/2)\omega_c}\right]^{1/2} \\ &\quad \times\theta(\varepsilon-(n+1/2)\omega_c)\theta(\varepsilon-(n'+1/2)\omega_c),\end{aligned} \tag{29}$$

where $\theta(x)$ is the Heaviside unit step function. The second, phonon-limited term, takes exactly the same form for acoustic phonons with deformation potential coupling with the replacement $n_iu^2 \longrightarrow \Theta^2T/dv_{sl}^2$. Here Θ is the deformation potential, d is the mass density of the crystal, and v_{sl} is the longitudinal sound velocity. Note that unless otherwise noticed, we adopt a system of units such that $k_B = \hbar = 1$ throughout this paper.

In all regimes of magnetic field strength (low field, intermediate Shubnikov-de Haas oscillatory regime, quantum strong field limit), these results are in precise agreement with the adiabatic magnetotransport results at zero temperature (Argyres, 1956, 1958; Roth and Argyres, 1966). However, the isothermal character of the Lei-Ting balance equations yields finite temperature results which are different, particularly in the nondegenerate limit. The balance equation result for linear resistivity in the nondegenerate regime for arbitrary magnetic field strength is ($x = \omega/T$)

$$\frac{\rho_{zz}^i}{\rho_0^i} = \frac{3}{8}(1-e^{-x})\sum_{n,n'}\int_0^\infty ds e^{-s}\left[\frac{s-nx}{s-n'x}\right]^{1/2}\theta(s-nx)\theta(s-n'x), \tag{30}$$

(ρ_0^i is the zero field limit), whereas Argyres' corresponding result for adiabatic linear longitudinal nondegenerate conductivity is

$$\left(\frac{\sigma_{zz}^i}{\sigma_0^i}\right)_{Ad} = 3\frac{(1-e^{-x})}{x}\int_0^\infty dse^{-s}\frac{\sum_n(s-nx)^{1/2}}{\sum_n(s-nx)^{-1/2}}. \tag{31}$$

In the quantum strong field limit where only the lowest Landau eigenstate is occupied ($\omega \gg T$), we obtain $\rho_{zz}^i/\rho_0^i = 3\omega_c/8T$ in contrast to the adiabatic result $(\rho_{zz}^i/\rho_0^i)_{Ad} = \omega_c/3T$. Taking account of the fact that ρ_0^i differs in the two theories, we find that $\rho_{zz}^i/(\rho_{zz}^i)_{Ad} = 4/\pi$, for the quantum strong field limit. Furthermore, for the zero field limit we obtain $\rho_{zz}^i/(\rho_{zz}^i)_{Ad} = 32/9\pi$, in agreement with known results(Huberman and Chester, 1975).

The nonpolar optic phonon (dispersionless) linear resistivity (with electron-phonon coupling matrix element given as $|M(\mathbf{q},\lambda)|^2 = D^2/2\Omega_0 d,)$ ρ_{zz}^p arising from Eq.(28) exhibits interesting nondegenerate magnetophonon resonance structure, analytically represented in the absence of screening by

$$\begin{aligned}\rho_{zz}^p(v\to 0) &= \frac{D^2m^{5/2}\omega_c}{2^{1/2}\pi^{3/2}ne^2d\Omega_0T^{3/2}}n\left(\frac{\Omega}{T}\right)\sum_{n=0}^{\infty}\sum_{\pm}\exp\left(\frac{\Omega_0-n\omega_c}{2T}\right)|\Omega_0-(\pm)n\omega_c| \\ &\quad \times K_1\left(\frac{(\Omega_0-(\pm)n\omega_c)}{2T}\right),\end{aligned} \tag{32}$$

where Ω_0 is the phonon frequency, and D is the shift of the band edge per unit relative displacement of the two sublattices relative to the optical mode. This structure corresponds to adiabatic results predicted earlier (Gurevich and Firsov, 1961, 1965; Kharus and Tsidilkovskii, 1971; Peterson 1972a, 1972c, 1975). Its graphic representation in terms of the second-field-derivative $\partial^2\rho_{zz}^p/\partial\omega_c^2$ is shown in Fig.1 as a function of ω_c.

Extension to the nonlinear regime yields nondegenerate unshielded nonpolar optic phonon resistivity functions as

$$\begin{aligned}\rho(v) &= \frac{D^2m^{5/2}\omega_c}{2^{9/2}\pi^{3/2}ne^2vd\Omega_0T_e^{3/2}}n\left(\frac{\Omega_0}{T}\right)\sum_{n=0}^{\infty}\sum_{\pm}\exp\left(\frac{n\omega_c}{2T_e}\right)\int_0^\infty dq_z\exp\left(-\frac{q_z^2}{8mT_e}\right) \\ &\quad\times\left\{\exp\left(-\frac{m(\Omega_0+q_zv-(\pm)n\omega_c)^2}{2T_eq_z^2}\right)\right. \\ &\quad\times\left[\exp\left(\frac{(\Omega_0+q_zv)}{2T_e}\right)-\exp\left(\frac{\Omega_0}{T}\right)\exp\left(-\frac{(\Omega_0+q_zv)}{2T_e}\right)\right] \\ &\quad-\exp\left(-\frac{m(\Omega_0-q_zv-(\pm)n\omega_c)^2}{2T_eq_z^2}\right) \\ &\quad\left.\times\left[\exp\left(\frac{(\Omega_0-q_zv)}{2T_e}\right)-\exp\left(\frac{\Omega_0}{T}\right)\exp\left(-\frac{(\Omega_0-q_zv)}{2T_e}\right)\right]\right\},\end{aligned} \tag{33}$$

and

$$\rho_E(v) = \frac{D^2m^{3/2}\omega_c}{2^{9/2}\pi^{3/2}ne^2v^2dT_e^{3/2}}n\left(\frac{\Omega_0}{T}\right)\sum_{n=0}^{\infty}\sum_{\pm}\exp\left(\frac{n\omega_c}{2T_e}\right)\int_0^\infty dq_z\frac{1}{q_z}\exp\left(-\frac{q_z^2}{8mT_e}\right)$$

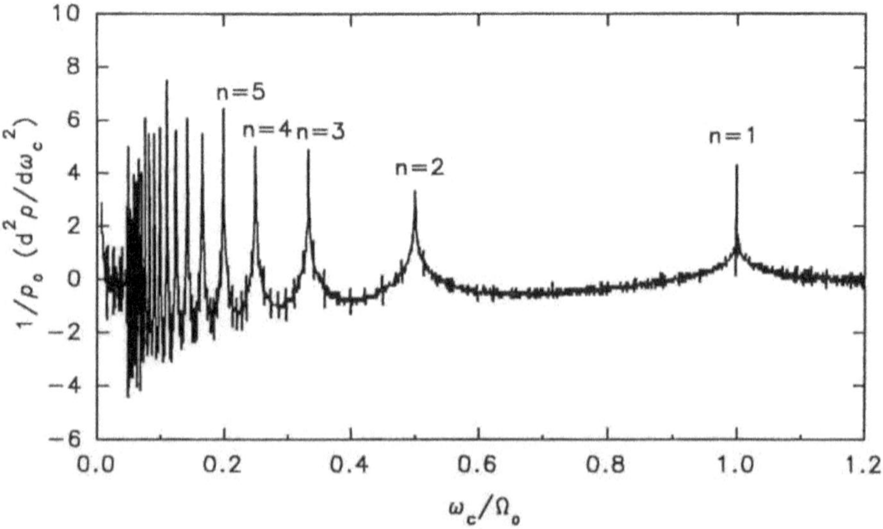

Figure 1: Plot of $\frac{1}{\rho_0}\frac{\partial^2\rho}{\partial\omega_c^2}$ as a function of ω_c/Ω_0, at a temperature $k_BT = 0.5\hbar\Omega_0$. (Fessatidis, 1991).

$$\times\left\{\exp\left(-\frac{m(\Omega_0+q_zv-(\pm)n\omega_c)^2}{2T_eq_z^2}\right)\right.$$
$$\times\left[\exp\left(\frac{(\Omega_0+q_zv)}{2T_e}\right)-\exp\left(\frac{\Omega_0}{T}\right)\exp\left(-\frac{(\Omega_0+q_zv)}{2T_e}\right)\right]$$
$$+\exp\left(-\frac{m(\Omega_0-q_zv-(\pm)n\omega_c)^2}{2T_eq_z^2}\right)$$
$$\left.\times\left[\exp\left(\frac{(\Omega_0-q_zv)}{2T_e}\right)-\exp\left(\frac{\Omega_0}{T}\right)\exp\left(-\frac{(\Omega_0-q_zv)}{2T_e}\right)\right]\right\}. \quad (34)$$

For polar optic phonon (dispersionless) scattering, it is clear that predominant contributions in the nondegenerate case come from $q_{\|}^2 \ll m\omega_c^2/T$, but such low values for q_z^2 are insignificant because of the factor $\exp(-m(\Omega_0+q_zv-(\pm)n\omega_c)^2/2T_eq_z^2)$, and higher values of $q_z^2 \sim m(\Omega_0+q_zv-(\pm)n\omega_c)^2/2T_e$ are important in the approach to resonance $\Omega_0 \sim n\omega_c$. In consequence, we may obtain an analytic upper bound estimate in the absence of screening by approximating the polar optic phonon interaction matrix element by $(q_z^2 > \bar{q}^2)$

$$|M(\mathbf{q},\lambda)|^2 = \frac{M^2}{q_{\|}^2+q_z^2} \longrightarrow \frac{M^2}{q_z^2}. \quad (35)$$

This yields an explicit analytic representation of magnetophonon resonance structure in the linear limit as

$$\rho_{zz}^p(v\to 0) = \frac{M^2m^{3/2}\omega_c}{2^{1/2}\pi^{3/2}ne^2T^{3/2}}n\left(\frac{\Omega_0}{T}\right)\sum_{n=0}^{\infty}\sum_{\pm}\exp\left(\frac{\Omega_0-n\omega_c}{2T}\right)K_0\left(\frac{\Omega_0-(\pm)n\omega_c}{2T}\right). \quad (36)$$

The log-divergent behavior of $K_0(x)$ as $\Omega_0 \longrightarrow n\omega_c$ is due to overestimating $|M(\mathbf{q},\lambda)|^2$ in Eq.(35) and would be rendered finite by refinement of our bound evaluation approximation, and further reduced by shielding. The extension of this examination of polar optic phonon resonance structure for nondegenerate magnetotransport yields the nonlinear unscreened results

$$\begin{aligned}\rho(v) &= \frac{M^2 m^{3/2}\omega_c}{2^{9/2}\pi^{3/2}ne^2vT_e^{1/2}} n\left(\frac{\Omega_0}{T}\right)\sum_{n=0}^{\infty}\sum_{\pm}\exp\left(\frac{n\omega_c}{2T_e}\right)\int_0^{\infty}\frac{dq_z}{q_z}\exp\left(-\frac{q_z^2}{8mT_e}\right)\\ &\quad\times\left\{\exp\left(-\frac{m(\Omega_0+q_zv-(\pm)n\omega_c)^2}{2T_eq_z^2}\right)\right.\\ &\quad\times\left[\exp\left(\frac{(\Omega_0+q_zv)}{2T_e}\right)-\exp\left(\frac{\Omega_0}{T}\right)\exp\left(-\frac{(\Omega_0+q_zv)}{2T_e}\right)\right]\\ &\quad-\exp\left(-\frac{m(\Omega_0-q_zv-(\pm)n\omega_c)^2}{2T_eq_z^2}\right)\\ &\quad\times\left.\left[\exp\left(\frac{(\Omega_0-q_zv)}{2T_e}\right)-\exp\left(\frac{\Omega_0}{T}\right)\exp\left(-\frac{(\Omega_0-q_zv)}{2T_e}\right)\right]\right\}, \end{aligned} \tag{37}$$

and

$$\begin{aligned}\rho_E(v) &= \frac{M^2 m^{3/2}\omega_c\Omega_0}{2^{5/2}\pi^{3/2}ne^2v^2T_e^{1/2}} n\left(\frac{\Omega_0}{T}\right)\sum_{n=0}^{\infty}\sum_{\pm}\exp\left(\frac{n\omega_c}{2T_e}\right)\int_0^{\infty}dq_z\frac{1}{q_z^3}\exp\left(-\frac{q_z^2}{8mT_e}\right)\\ &\quad\times\left\{\exp\left(-\frac{m(\Omega_0+q_zv-(\pm)n\omega_c)^2}{2T_eq_z^2}\right)\right.\\ &\quad\times\left[\exp\left(\frac{(\Omega_0+q_zv)}{2T_e}\right)-\exp\left(\frac{\Omega_0}{T}\right)\exp\left(-\frac{(\Omega_0+q_zv)}{2T_e}\right)\right]\\ &\quad+\exp\left(-\frac{m(\Omega_0-q_zv-(\pm)n\omega_c)^2}{2T_eq_z^2}\right)\\ &\quad\times\left.\left[\exp\left(\frac{(\Omega_0-q_zv)}{2T_e}\right)-\exp\left(\frac{\Omega_0}{T}\right)\exp\left(-\frac{(\Omega_0-q_zv)}{2T_e}\right)\right]\right\} \end{aligned} \tag{38}$$

Low-q_z divergencies at resonance again originate from overestimation of $|M(\mathbf{q},\lambda)|^2$, and may be cured by the indicated refinements.

2.2 Transverse Geometry

Focusing attention on the transverse configuration, we have ($\mathbf{E} = E\hat{x}$, $\mathbf{B} = B\hat{z}$) (Cui, Lei, and Horing, 1988)

$$\rho_{xx} = \rho_{xx}^i + \rho_{xx}^p, \tag{39}$$

where the impurity scattering term is

$$\rho_{xx}^i(v) = -\left(\frac{n_i}{n^2e^2v}\right)\sum_{\mathbf{q}} q_xq^2K_2(\mathbf{q},\mathbf{q}\cdot\mathbf{v})|u(\mathbf{q})|^2, \tag{40}$$

and the phonon scattering term is

$$\rho_{xx}^p(v) = -\frac{2}{n^2e^2v}\sum_{\mathbf{q},\lambda}|M(\mathbf{q},\lambda)|^2\Pi_2(\mathbf{q},\Omega_{\mathbf{q}\lambda}+\mathbf{q}\cdot\mathbf{v})\left[n\left(\frac{\Omega_{\mathbf{q}\lambda}}{T}\right)-n\left(\frac{\Omega_{\mathbf{q}\lambda}+\mathbf{q}\cdot\mathbf{v}}{T_e}\right)\right]. \tag{41}$$

The linearized impurity resistivity for short range scattering $u(\mathbf{q})/\epsilon_1(\mathbf{q},0) \longrightarrow u =$ constant,

$$\rho^i_{xx}(v \to 0) = -\frac{n_i u^2 m^4 \omega_c^2}{n^2 e^2 \pi \kappa} \sum_{n_1,n_2} \int_0^\infty dE \frac{f'(E)}{[E-(n_1+\frac{1}{2})\omega_c]^{1/2}[E-(n_2+\frac{1}{2})\omega_c]^{1/2}}, \quad (42)$$

exhibits a characteristic log-divergence for the $n_1 = n_2$ terms where the inverse-square-root state-density factors coincide for identical Landau levels (Davydov and Pomeranchuk, 1958). An alternative linear formula for the quantum strong field limit (all electrons in lowest Landau state) which exhibits the role of shielding is readily found as

$$\rho^i_{xx}(v \to 0) = \frac{n_i m^2 \omega_c}{4\pi^3 n^2 \kappa} \int d^2 q_\parallel e^{-q_\parallel^2/2m\omega_c} \int dq_z \frac{q_z^2}{|q_z|} \frac{|u(\mathbf{q})|^2}{\epsilon_1^2(\mathbf{q},0)} f'\left(\frac{\omega_c}{2} + \frac{q_z^2}{8m}\right). \quad (43)$$

It is apparent that the role of screening in the linear limit is confined to static shielding (zero-frequency) and is devoid of dynamic effects. In their analysis of the zero frequency limit of the linear dynamic conductivity, Gerlach and coworkers (Gerlach, 1974a, 1974b; Gerlach and Rautenberg, 1974; Gerlach and Harbecke, 1982) made the observation that this yields a convergent result at zero temperature for a Coulombic scattering potential. This is clearly true in view of the fact that δ-function nature of $f'(\omega_c/2 + q_z^2/8m)$ at zero temperature excludes any contribution from the neighborhood of $q_z = 0$. However, for finite temperature the singular contribution from $q_z \sim 0$ leads to the well-known log-divergence, irrespective of whether shielding, $\epsilon_1(\mathbf{q},0)$, is included or not. It is perhaps of interest to note below the convergent zero temperature linear balance equation impurity resistivities of the transverse configuration for (I) Coulombic and (II) point δ-function impurities in the (A) shielded and (B) unshielded cases (Z is the impurity charge number, $k_0 = [(2k_F)^2 - 4m\omega_c]^{1/2} = (2^{1/2}\pi^2 n)/m^{3/2}\omega_c$); k_F is Fermi wavenumber)

(I) Coulombic impurity

(A) Shielded

$$\rho^i_{xx}(v \to 0) = \frac{16 n_i m^3 \omega_c Z^2 e^4}{n^2 \kappa k_0^2} \int_0^\infty dx \frac{x e^{-x/2m\omega_c}}{(x^2 + k_0^2)\epsilon_1^2(\sqrt{x}, k_0; 0)}, \quad (44)$$

(B) Unshielded

$$\rho^i_{xx}(v \to 0) = \frac{16 n_i m^3 \omega_c Z^2 e^4}{n^2 \kappa k_0^2} \left[-1 - \left(1 + \frac{k_0^2}{2m\omega_c}\right) e^{k_0^2/2m\omega_c} \mathrm{Ei}\left(\frac{-k_0^2}{2m\omega_c}\right)\right], \quad (45)$$

where $\mathrm{Ei}(x)$ is an exponential integral.

(II) δ-function impurity

(A) Shielded

$$\rho^i_{xx}(v \to 0) = \frac{n_i m^3 \omega_c u^2}{\pi^2 n^2 \kappa k_0^2} \int_0^\infty dx \frac{x e^{-x/2m\omega_c}}{\epsilon_1^2(\sqrt{x}, k_0; 0)}, \quad (46)$$

(B) Unshielded

$$\rho^i_{xx}(v \to 0) = \frac{4 n_i m^5 \omega_c^3 u^2}{\pi^2 n^2 \kappa k_0^2}. \quad (47)$$

Several cures of the log-divergence of the transverse linear impurity resistivity (for non-zero temperature) have been cited in the review of Roth and Argyres (1966). These include (1) collision broadening, (2) non-Born scattering, (3) inelastic collisions, and (4) classical cutoff. To these, one may add yet another convergence mechanism, namely nonlinearity. Its role may be assessed from the isothermal result

$$\rho^i_{xx}(v) = -\frac{n_i}{n^2e^2v}\sum_{\mathbf{q}} q_x q^2|u(\mathbf{q})|^2 \frac{\epsilon_2(\mathbf{q},\mathbf{q}\cdot\mathbf{v})}{\epsilon_1^2(\mathbf{q},\mathbf{q}\cdot\mathbf{v})+\epsilon_2^2(\mathbf{q},\mathbf{q}\cdot\mathbf{v})}, \tag{48}$$

and one may verify convergence even in the absence of screening (point δ-function impurity)

$$\begin{aligned}\rho^i_{xx}(v) &= \frac{n_i m^2\omega_c u^2}{2\pi^2 n^2 e^2 \kappa v}\sum_{n_1,n_2}\sum_{\bar{q}} q_x C_{n_1n_2}\left(\frac{\bar{q}^2}{2m\omega_c}\right)\\ &\quad \int_0^\infty dE \frac{f(E)-f(E+q_xv)}{[E-(n_2+\frac{1}{2})\omega_c]^{1/2}[q_xv+E-(n_2+\frac{1}{2})\omega_c]^{1/2}},\end{aligned} \tag{49}$$

since the two inverse-square-root state-density factors have their arguments skewed by the nonlinear frequency shift $\mathbf{q}\cdot\mathbf{v} = q_xv$. This convergent unscreened result exhibits strong oscillations as a function of v and ω_c^{-1}, in correspondence with unscreened nonlinear adiabatic results obtained earlier (Elesin, 1969; Magarill and Sarvinykh, 1970; Agaeva, Askerov, and Gashimzade, 1976), but differing in the detailed temperature dependence. The inclusion of dynamic screening [Eq.(48)] moderates and smooths the oscillatory behavior of ρ^i_{xx} which is plotted against v in Fig.2 for GaAs at 4.2 K with carrier density $2.5\times10^{18}cm^{-3}$, effective mass $m = 0.07m_e$, background dielectric constant $\kappa = 12.9$. In this plot ρ^i_{xx} is scaled with $\rho_0 = 4n_iZ^2r_s^2k_F^2/n^2e^2\pi^2$, with $r_s = (3/4\pi n)^{1/3}a_0$ (a_0 is the Bohr radius); v is scaled with half of the Fermi velocity, and ω_c is normalized with the Fermi energy at zero magnetic field and zero temperature.

In considering nonlinear (and linear) phonon-limited magnetoresistivity, it is of interest to discuss a general low-magnetic-field approximation $\omega_c/\omega_p \ll 1$; ω_c/T_e (or ω_c/ζ) $\ll 1$; $m\omega_c^2/q^2T_e$ (or $m\omega_c^2/q^2\zeta$) $\ll 1$ for unscreened carriers scattered by nonpolar optic phonons. (Here ω_p is the plasma frequency, and ζ is the chemical potential.) In this situation, we have $\Pi_{02}(\mathbf{q},\omega) \longrightarrow \Pi_{02}(\mathbf{q},\omega) = -(m^2T_e/2\pi q)P_2(\mathbf{q},\omega)$, where(Horing, 1965; Cui and Horing, 1989)

$$\begin{aligned}P_2(\mathbf{q},\omega) &= \ln\left(\frac{1-f(y)}{1-f(x)}\right) + \left(H_1+\frac{H_2}{4}+\frac{H_3}{24}\right)[f(x)-f(y)]\\ &\quad -\left(\frac{H_2}{4}+\frac{H_3}{8}\right)[f(x)^2-f(y)^2] - \frac{H_3}{12}[f(x)^3-f(y)^3],\end{aligned} \tag{50}$$

where

$$x = \frac{m\omega^2}{2q^2}+\frac{q^2}{8m}-\frac{\omega}{2},$$

$$y = \frac{m\omega^2}{2q^2}+\frac{q^2}{8m}+\frac{\omega}{2},$$

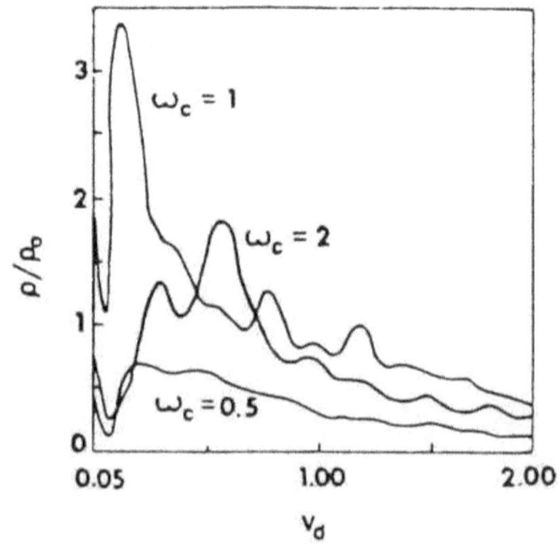

Figure 2: Plot of ρ/ρ_0 ($\equiv \rho_{xx}/\rho_0$) as a function of drift velocity (in units of $v_F/2$) at $T = 4.2K$ for several magnetic field strengths $\omega_c = 2,\ 1,\ 0.5$ (in units of $E_F/\hbar$). (Horing, Cui, and Lei, 1987).

and ($\cos\theta = q_z^2/q^2$)

$$H_1 = \frac{m\omega_c^2}{16T_e}\cos^2\theta, \tag{51}$$

$$H_2 = \frac{\omega_c^2}{3T_e^2}\left(1 + \frac{\cos^2\theta}{4}\right) - \left(\frac{m\omega_c\omega}{T_e q^2}\right)^2 \cos^2\theta, \tag{52}$$

$$H_3 = \left[\frac{\omega_c^2}{16T_e^2}\frac{q^2}{mT_e} - \left(\frac{\omega_c}{T_e}\right)^2 \frac{m\omega^2}{T_e q^2} + \frac{m\omega_c^2}{T_e q^2}\left(\frac{m\omega^2}{T_e q^2}\right)^2\right]\cos^2\theta. \tag{53}$$

Moreover, considering scattering by nondispersive nonpolar optic phonons of frequency Ω_0, one has

$$\rho(v) = \frac{m^2 T_e}{\pi n^2 e^2 v}\frac{D^2}{2\Omega_0 d}\sum_{\mathbf{q}}\frac{q_x}{|\mathbf{q}|^2}P_2(\mathbf{q}, \Omega_0 + q_x v)\left[n\left(\frac{\Omega_0}{T}\right) - n\left(\frac{\Omega_0 + q_x v}{T_e}\right)\right], \tag{54}$$

and

$$\rho_E(v) = -\frac{m^2 T_e}{\pi n^2 e^2 v^2}\frac{D^2}{2\Omega_0 d}\sum_{\mathbf{q}}\frac{\Omega_0}{|\mathbf{q}|}P_2(\mathbf{q}, \Omega_0 + q_x v)\left[n\left(\frac{\Omega_0}{T}\right) - n\left(\frac{\Omega_0 + q_x v}{T_e}\right)\right]. \tag{55}$$

As usual, the chemical potential for electron temperature T_e, magnetic field B and electron density n is determined by

$$n = \int_0^\infty dE f(E) D(E), \tag{56}$$

where the Landau-quantized density of states is (no spin)

$$D(E) = \sqrt{2m}\frac{eB}{2\pi^2}\sum_{n=0}^{\infty}[E - \left(n + \frac{1}{2}\right)\omega_c]^{-1/2}. \tag{57}$$

At high magnetic fields the magnetophonon effect(Peterson, 1970, 1972a, 1972b, 1972c, 1975; Barker, 1978a; Nicholas 1985) becomes prominent in the transverse magnetoresistivity. For polar optic phonon scattering (dispersionless, frequency Ω_0) in the absence of screening, the linear limit has $T_e = T$, and in the nondegenerate case

$$\begin{aligned}\rho(v \to 0) &= \frac{\alpha m \omega_c}{\pi^{1/2} n e^2}\left(\frac{\Omega_0}{T}\right)^{3/2}\sinh\left(\frac{\omega_c}{2T}\right)\frac{1}{e^{\Omega_0/T}-1}\sum_{n_1,n_2}\exp\left(-\frac{(n_1+n_2)\omega_c - \Omega_0}{2T}\right) \\ &\quad \times K_0\left(\frac{|(n_2-n_1)\omega_c - \Omega_0|}{2T}\right),\end{aligned} \tag{58}$$

where α is the Fröhlich coupling constant and $K_0(x)$ is the modified Bessel function of the second kind. This linear result, in precise agreement with that of Barker (1972) at high field (such that $\rho \cong \sigma_{xx}/\sigma_{yx}^2$; $\sigma_{yx} = ne^2/m\omega_c$), exhibits magnetophonon resonances through the log-divergence of $K_0(x)$ when $(n_2 - n_1)\omega_c = \Omega_0$.

Hot electron magnetophonon resonances may also be analyzed using Eqs.(52) and (53) for $\rho(v)$ and $\rho_E(v)$. In the absence of screening in the nondegenerate case, dispersionless nonpolar optic phonon scattering yields

$$\begin{aligned}\rho(v) &= \rho_0^{*}\frac{\omega_c^3}{8T_e}e^{\zeta/T_e}\frac{\sinh(\omega_c/T_e)}{\sinh(1/2T)}\int_0^{\infty}dx\frac{\exp(\xi(1-x^2))}{\sqrt{1+x^2}}\frac{1}{(\cosh(\omega_c/2T_e)-\cos(\omega_c x/2T_e))^2} \\ &\quad \times\left\{\exp\left(\frac{1}{2T}-\frac{1}{2T_e}\right)\left[\cos\left(\frac{x}{2T_e}-2\xi x\right)-x\sin\left(\frac{x}{2T_e}-2\xi x\right)\right]\right. \\ &\quad \left.+\exp\left(-\frac{1}{2T}+\frac{1}{2T_e}\right)\left[\cos\left(\frac{x}{2T_e}+2\xi x\right)-x\sin\left(\frac{x}{2T_e}+2\xi x\right)\right]\right\},\end{aligned} \tag{59}$$

and

$$\begin{aligned}\rho_E(v) &= -\rho_0^{*}\frac{\omega_c^3}{2v^2}e^{\zeta/T_e}\frac{\cosh(\omega_c/T_e)}{\sinh(1/2T)}\int_0^{\infty}dx\frac{\exp(\xi(1-x^2))}{\sqrt{1+x^2}}\frac{1}{\cosh(\omega_c/2T_e)-\cos(\omega_c x/2T_e)} \\ &\quad \times\left\{\exp\left(-\frac{1}{2T}+\frac{1}{2T_e}\right)\cos\left(\frac{x}{2T_e}+2\xi x\right)\right. \\ &\quad \left.-\exp\left(\frac{1}{2T}-\frac{1}{2T_e}\right)\cos\left(\frac{x}{2T_e}-2\xi x\right)\right\},\end{aligned} \tag{60}$$

where

$$\xi = \frac{\omega_c v^2}{8T_e^2}\frac{\sinh(\omega_c/2T_e)}{\cosh(\omega_c/2T_e)-\cos(\omega_c x/2T_e)}.$$

In this, dimensionless quantities are understood with $T/\Omega_0 \to T$, $T_e/\Omega_0 \to T_e$, $\omega_c/\Omega_0 \to \omega_c$, $\zeta/\Omega_0 \to \zeta$, and $v/v_0 \to v$, with v_0 as defined above.

It is well known that for optical phonon scattering in the absence of a magnetic field, the current saturates at high electric field(Thornber and Feynman, 1970; Seeger, 1982; Lei and Ting, 1984). This is also true in the presence of magnetic fields, as one can see

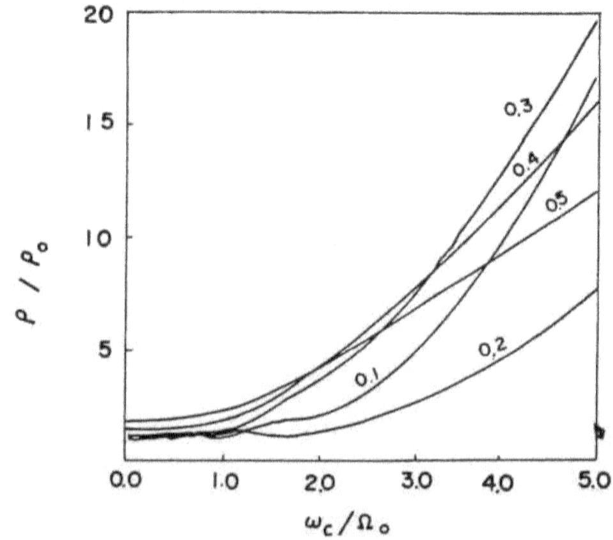

Figure 3: Transverse magnetoresistivity normalized by the linear resistivity at $B = 0$ *vs.* cyclotron frequency in units of the phonon frequency, at a lattice temperature $k_B T = 0.4\Omega_0$ and for different values of v/v_0. (Cui, Lei, and Horing, 1988a).

from an asymptotic examination of the energy balance equation, in which $\rho(v)$ and $\rho_E(v)$ are given by Eqs.(59) and (60) in the limit $T_e \to \infty$. To leading order in $1/T_e$, the energy balance equation

$$\rho(v) - \rho_E(v) = 0 \tag{61}$$

becomes

$$v_s^2 = \frac{3}{4} \tanh\left(\frac{1}{2T}\right), \tag{62}$$

where v_s denotes the saturation value of the drift velocity. This result is independent of the magnetic field, and is identical to the saturation value for $B = 0$.

The momentum balance equation (59) and the energy balance equation (61) have been examined numerically for wide ranges of electric field strength and magnetic field strength. The results are depicted graphically in Fig.3 to Fig.8 for lattice temperature of $T = 0.4\Omega_0$. Fig.3 shows the nonlinear transverse magnetoresistivity normalized to the linear resistivity at $B = 0$ as a function of the cyclotron frequency normalized to the phonon frequency, ω_c/Ω_0, for a number of drift velocities ($v/v_0 = 0.1, 0.2, 0.3, 0.4, 0.5$). The magnetophonon structures are evident for $\omega_c \cong \Omega_0$ and for $v \cong 0.1v_0$. At higher magnetic fields, a monotonic dependence on ω_c dominates, with the asymptotic behavior of $\rho(v)$ strongly affected by the drift velocity. Fig.4 shows the ratio of the electron temperature to the lattice temperature, T_e/T, as a function of the cyclotron frequency, ω_c/Ω_0, for various electron drift velocities ($v/v_0 = 0.1, 0.2, 0.3, 0.4, 0.5$). Here we observe that (1) electron resonant cooling is quite distinctive for relatively low drift velocities

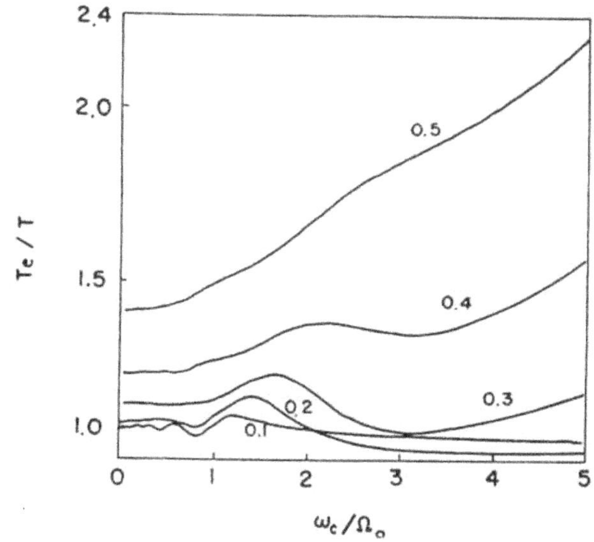

Figure 4: Ratio of the electron temperature to the lattice temperature as a function of the cyclotron frequency in units of the phonon frequency and for different values of v/v_0. (Cui, Lei, and Horing, 1988a).

around the fundamental field $\omega_c/\Omega_0 = 1$ and also around $\omega_c/\Omega_0 = 0.5$; (2) the $v/v_0 = 0.2$ curve dips significantly lower than the $v/v_0 = 0.1$ curve beyond a magnetic field of about $\omega_c/\Omega_0 = 2$; (3) resonant heating maxima are in evidence, and these peaks are broadened and shifted to the higher magnetic field region with increasing drift velocity (higher electric field). Because of the monotonic dependencies of the nonlinear magnetoresistivity and the electron temperature on the magnetic field, the magnetophonon oscillations in $\rho(v)$ and in T_e tend to be obscured at large magnetic field strengths. However, for moderate magnetic field strengths, they are readily discernable as shown in Fig.5 and Fig.6, where the low-field regions of Fig.3 and Fig.4 are shown. Fig.7 shows the normalized resistivity as a function of drift velocity, and Fig.8 gives the corresponding electron temperature scaled to the lattice temperature for several different cyclotron frequencies. Both sets of curves go to infinity at $v/v_0 = 0.79$, the saturation value of drift velocity as predicted by Eq.(62) for $T/\Omega_0 = 0.4$.

Magnetophonon oscillations in thermal noise have been examined using a balance equation approach for the current carrying state with a velocity fluctuation, and the correlation of current fluctuations about a nonzero dc bias leads to a noise conductivity (which differs from the small signal conductivity) given by

$$\sigma_n(\omega) = \frac{ine^2/m}{\omega + M(\omega)}, \tag{63}$$

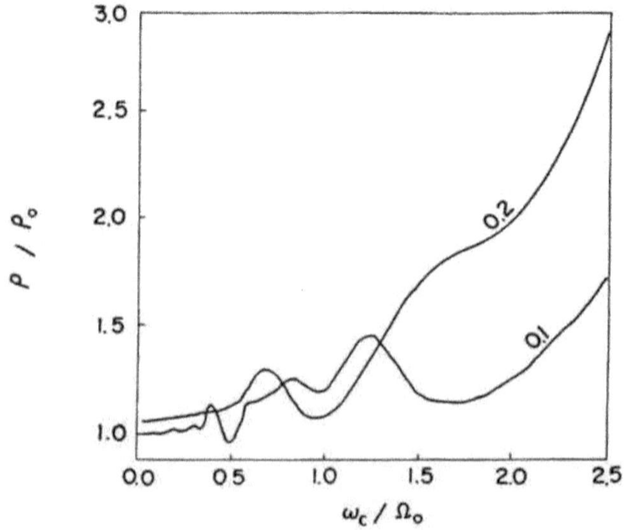

Figure 5: The two lower curves of Fig.3 ($v/v_0 = 0.1,\ 0.2$) are replotted in the region $\omega_c/\Omega_0 \leq 2.5$. (Cui, Lei, and Horing, 1988a).

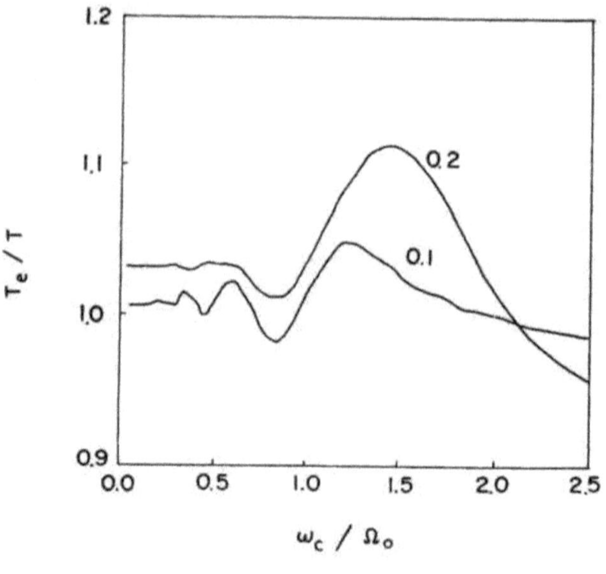

Figure 6: The two lower curves of Fig.4 ($v/v_0 = 0.1,\ 0.2$) are replotted in the region $\omega_c/\Omega_0 \leq 2.5$. (Cui, Lei, and Horing, 1988a).

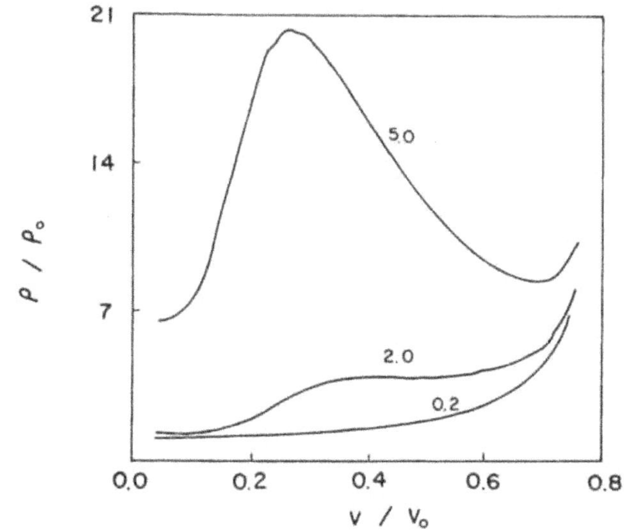

Figure 7: Normalized resistivity *vs.* electron drift velocity for various values of the cyclotron frequency. (Cui, Lei, and Horing, 1988a).

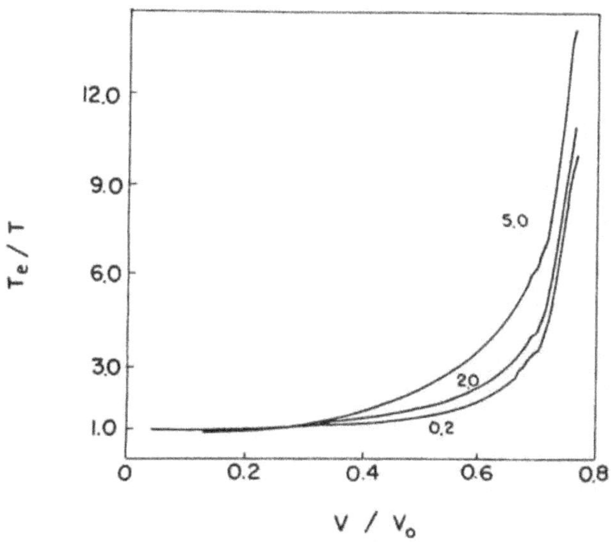

Figure 8: Ratio of electron temperature to lattice temperature as a function of the electron drift velocity for various values of the cyclotron frequency. (Cui, Lei, and Horing, 1988a).

with a noise temperature $T_n(\omega)$ defined by the maximum available noise power as

$$T_n(\omega) = \frac{S(\omega)}{2mM_2(\omega)}, \tag{64}$$

where $M_2(\omega)$ is the imaginary part of the memory function $M(\omega)$ at nonzero dc bias, given for phonon scattering as (we ignore hot-phonon phenomena due to finite phonon relaxation time in this discussion)

$$\begin{aligned} M_2(\omega) &= \frac{1}{nm\omega}\sum_{\mathbf{q},\lambda} q_\alpha^2|M(\mathbf{q},\lambda)|^2 \\ &\left\{\Pi_2(\mathbf{q},\Omega_0+\mathbf{q}\cdot\mathbf{v}+\omega)\left[n\left(\frac{\Omega_0+\mathbf{q}\cdot\mathbf{v}+\omega}{T_e}\right)-n\left(\frac{\Omega_0}{T}\right)\right]-(\omega\to-\omega)\right\} \end{aligned} \tag{65}$$

and $S(\omega)$ is the fluctuation force correlation function,

$$\begin{aligned} S(\omega) &= -\frac{1}{2n}\sum_{\mathbf{q},\lambda} q_\alpha^2|M(\mathbf{q},\lambda)|^2 \\ &\times\left\{\Pi_2(\mathbf{q},\Omega_0+\mathbf{q}\cdot\mathbf{v}+\omega)\left[\left[2n\left(\frac{\Omega_0+\mathbf{q}\cdot\mathbf{v}+\omega}{T_e}\right)+1\right]\left(2n\left(\frac{\Omega_0}{T}\right)+1\right)-1\right]\right. \\ &\left.-(\omega\to-\omega)\right\}, \end{aligned} \tag{66}$$

(α designates a Cartesian component x, y). These equations are applicable both with and without a magnetic field, which is embedded in the structure of $\Pi(\mathbf{q},\omega)$ of the unperturbed quasi-equilibrium system at the electron temperature. Based on this formulation, thermal noise for the system described above in crossed electric and magnetic fields is shown in Fig.9 for the longitudinal and transverse noise temperatures T_{nl} and T_{nt} at $\omega \cong 0$, as well as the electron temperature T_e, as functions of ω_c/Ω_0 at a bias dc drift velocity $v/v_0 = 0.2$ and lattice temperature $T/\Omega_0 = 0.4$. As above, the scattering mechanism is that of nonpolar optic phonons (frequency Ω_0). The noise temperatures show significant oscillatory behavior with magnetic field.

Finally, the subject of bulk cyclotron resonance has provided an interesting forum for comparison of the balance equation approach to quantum magnetotransport with the more traditional adiabatic theory. The former, as an isothermal theory embodying strong electron-electron interactions which create a tendency toward rapid thermalization about the drifted state, yields frequency dependent linear response conforming to the memory function approach of Götze and Wölfle (1972). Applied to cyclotron resonance, such results are at variance with the adiabatic determination of cyclotron resonance line shape, which excludes both electron-electron interactions and the concomitant drifted thermalization. While the literature has abundant discussion of adiabatic analyses of cyclotron resonance, there are some little known magnetotransport works which are particularly meritorious for thoroughness in treating impurity scattering for Landau quantized transport(Han, 1976) and, indeed, for the treatment of strong impurity scattering in the study of the quantum magnetoconductivity tensor(Malwal, 1974). Moreover, adiabatic bulk cyclotron resonance with phonon scattering has been definitively examined by Badjou and Argyres (1987, and references cited therein) and Shin *et al* (1973).

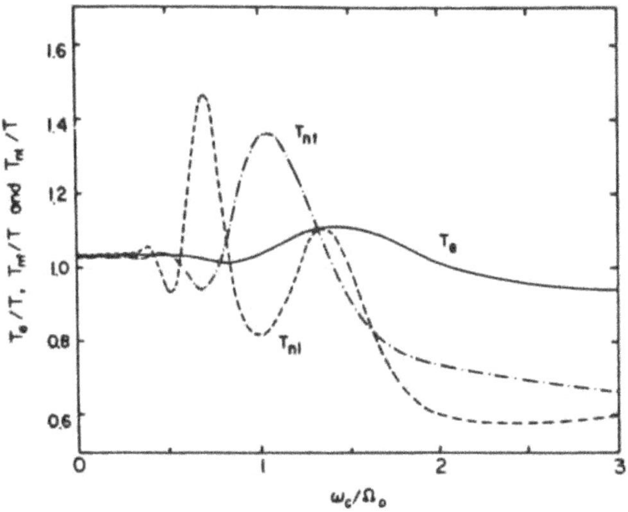

Figure 9: Longitudinal and transverse thermal noise temperatures and the electron temperature (all normalized to lattice temperature) as functions of the cyclotron frequency normalized to the phonon frequency. (Cui, Lei, and Horing, 1988a).

3 Magnetotransport in Microstructures

3.1 Type-I Superlattice

Transport in 2D systems has been extensively reviewed by Ando, Fowler, and Stern (1982). Nevertheless, there are some interesting features of 2D magnetotransport that have arisen since that review article was published, particularly in superlattices of interacting planar 2D quantum-well microstructures. The formulation of balance equation transport theory for a Type-I superlattice, consisting of an infinite array of periodically placed electron quantum wells, was described in the accompanying review article(Lei and Horing, 1991) in the absence of a magnetic field. Considering only the lowest transverse subband envelope function $\zeta_m(z) \longrightarrow \zeta(z)$ for interaction energies incapable of exciting higher subbands, a normal magnetic field induces Landau quantization of the 2D electron states in the lth well in accordance with the wavefunction

$$\psi_{lnk_y}(\mathrm{r}) = \phi_{nk_y}(x,y)\zeta(z-ld);\ l = 0, \pm 1, \pm 2, \ldots, \pm\infty, \tag{67}$$

where, in the absence of tunneling, we take

$$\zeta(z) = \begin{cases} \sqrt{(2/a)}\cos(\pi z/a), & \text{for } -a/2 \leq z < a/2 \\ 0, & \text{otherwise} \end{cases}$$

and the quantized magnetic field states for electron motion parallel to the quantum well planes are given by (Landau gauge)

$$\phi_{nk_y}(x,y) = e^{-ik_y y}(\sqrt{\pi}2^n n!\alpha)^{-1/2} H_n(\frac{x}{\alpha}+\alpha k_y)\exp\left(-\frac{1}{2}\left(\frac{x}{\alpha}+\alpha k_y\right)^2\right), \tag{68}$$

with $H_n(x)$ as the nth order Hermite polynomial and $\alpha = 1/\sqrt{(eB)}$. The associated Landau quantized electron energy levels are

$$\varepsilon_n = \left(n+\frac{1}{2}\right)\omega_c + \varepsilon_0 \tag{69}$$

in the absence of spin. ε_0 is the bottom of the lowest conduction subband, and will henceforth be taken as zero. Notwithstanding the fact that our attention is mainly focused on the lowest transverse subband here, it is straightforward to extend these considerations to higher subband excitations(Lei and Horing, 1991; Tso, 1990).

The superlattice balance equations for "parallel" transport are developed on a per-layer basis in full analogy to the corresponding zero-field analysis(Lei and Horing, 1991), with the inclusion of a Lorentz-force term due to the magnetic field, which has the same form as that of the bulk case as given by Eq.(1). However, in the present situation n is the areal density of electrons, and the 2D superlattice frictional force densities associated with impurity and phonon scatterings are given by ($\mathbf{q} = (\mathbf{q}_{\parallel}, q_z)$)

$$\mathbf{f}_i(\mathbf{v}) = \left(\frac{2\pi e^2}{\kappa}\right)^2 \sum_{\mathbf{q}} \frac{\mathbf{q}_{\parallel}}{q_{\parallel}^2}\tilde{N}(\mathbf{q})g(q_z)\Pi_2(\mathbf{q}, \mathbf{q}_{\parallel}\cdot\mathbf{v}), \tag{70}$$

$$\mathbf{f}_p(\mathbf{v}) = 2\sum_{\mathbf{q},\lambda}|M(\mathbf{q},\lambda)|^2|I(iq_z)|^2\mathbf{q}_{\parallel}\,\Pi_2(\mathbf{q},\Omega_{\mathbf{q}\lambda}+\mathbf{q}_{\parallel}\cdot\mathbf{v})\left[n\left(\frac{\Omega_{\mathbf{q}\lambda}}{T}\right)-n\left(\frac{\Omega_{\mathbf{q}\lambda}+\mathbf{q}_{\parallel}\cdot\mathbf{v}}{T_e}\right)\right], \tag{71}$$

and the superlattice energy transfer function $w(\mathbf{v})$ is

$$w(\mathbf{v}) = 2\sum_{\mathbf{q},\lambda}|M(\mathbf{q},\lambda)|^2|I(iq_z)|^2\Omega_{\mathbf{q}\lambda}\Pi_2(\mathbf{q},\Omega_{\mathbf{q}\lambda}+\mathbf{q}_{\parallel}\cdot\mathbf{v})\left[n\left(\frac{\Omega_{\mathbf{q}\lambda}}{T}\right)-n\left(\frac{\Omega_{\mathbf{q}\lambda}+\mathbf{q}_{\parallel}\cdot\mathbf{v}}{T_e}\right)\right]. \tag{72}$$

Again, the impurities considered include remote impurity sheets of ionized dopants (at a distance s from the center of each layer) having areal density N_r, and equivalent charge number Z_r, as well as randomized background impurities within the well regions having areal density N_b, and equivalent charge number Z_b. The various factors appearing in the above equations are all the same as those of the null magnetic field case (with the exception of the density-density correlation function), and are already given in the preceding chapter(Lei and Horing, 1991). We refer the reader to that article for the definitions and relevant discussion of the form-factors.

The 2D relative electron density-density correlation function $\Pi_2(\mathbf{q},\omega)$ which appears here is subject to Landau quantization, having the random phase approximation

$$\Pi(\mathbf{q},\omega) = \frac{\Pi_0(\mathbf{q}_{\parallel},\omega)}{1-V(\mathbf{q})\Pi_0(\mathbf{q}_{\parallel},\omega)} \tag{73}$$

with the Coulomb potential $V(\mathbf{q})$ representing both inter-layer and intra-layer electron-electron interactions again given in Lei and Horing (1991). The role of Landau quantization in $\Pi_0(\mathbf{q}_\parallel, \omega)$ may be represented as (lowest electric subband only)

$$\Pi_0(q_\parallel, \omega) = \sum_{n,n'} C_{n,n'}(x)\Pi_0(n, n', \omega), \tag{74}$$

where $x = \alpha^2 q_\parallel^2/2$ and the coefficient $C_{n,n'}(x)$ has been given in Eq.(15). The function $\Pi_0(n, n', \omega)$ may be constructed from the single Landau level retarded Green's function(Ting, Ying, and Quinn, 1976a, 1976b, 1977; Cai, Lei, and Ting, 1985)

$$G_n(\omega) = \frac{1}{\omega - \varepsilon_n + i\delta}, \tag{75}$$

such that

$$\Re\Pi_0(n, n', \omega) = -\frac{1}{\pi}\int_{-\infty}^{\infty} dz f(z)[\Re G_n(z+\omega)\Im G_{n'}(z) + \Re G_{n'}(z-\omega)\Im G_n(z)], \tag{76}$$

$$\Im\Pi_0(n, n', \omega) = -\frac{1}{\pi}\int_{-\infty}^{\infty} dz[f(z) - f(z+\omega)]\Im G_n(z+\omega)\Im G_{n'}(z). \tag{77}$$

The chemical potential of the quantum well in a normal magnetic field is determined, for total electron number n_s, by the relation

$$-\frac{1}{\pi}\sum_{n=0}^{\infty}\int_{-\infty}^{\infty} dz f(z)\Im G_n(z) = \frac{n_s}{p}, \tag{78}$$

where $p = 2/\pi\alpha^2$ is the degeneracy factor.

Analysis involving $\Pi_0(n, n', \omega)$ is complicated by its divergent behavior for frequencies at integral multiples of the cyclotron frequency. In an effort to remediate this, various Landau level broadening mechanisms were investigated(Ando, Fowler, and Stern, 1982; Cai and Ting 1986). Ando and Uemura (1973) and Ando (1974) employed a self-consistent Born approximation to introduce a line-width into the determination of $G_n(\omega)$ for short-range impurity scatterers, leading to an elliptic density of states and

$$G_n(\omega) = \frac{2}{\Gamma^2}\{\omega - \varepsilon_n - [(\omega - \varepsilon_n)^2 - \Gamma^2]^{1/2}\}, \tag{79}$$

which provides excellent agreement between the calculated peak value of the transverse conductivity and experimental data for Si(100) inversion layers. Here, $\Gamma^2 = 2\omega_c/\pi\tau_0$, τ_0 being the electron transport lifetime in the absence of a magnetic field at zero temperature, related to the mobility μ_0 under the same conditions by $\tau_0 = m\mu_0/e$. More recent work(Gornik *et al*, 1985) advanced a Gaussian line shape, admixed with a uniform background for GaAs/GaAlAs multilayers, in preference to a simple Gaussian or Lorentzian density of states: This brought about good agreement between theory and experimental data for electronic specific heat capacity measurements. The associated density of states $D_n(\omega)$ for this model lineshape of Gornik *et al* is given by

$$D_n(\omega) = (1 - X_{BG})\frac{1}{\Gamma}\sqrt{\frac{2}{\pi}}\exp\left\{-\frac{2}{\Gamma^2}[\omega - \varepsilon_n]^2\right\} + \frac{X_{BG}}{\omega_c}\theta(\omega)\theta(\omega - \varepsilon_n)\theta(\varepsilon_{n+1} - \omega), \tag{80}$$

where X_{BG} is the fraction of flat background states, and $\Gamma = (2\omega_c\Gamma_0/\pi)^{1/2}$ where Γ_0 is a parameter, taken to be the same for all Landau levels. Recognizing that the imaginary part of the Green's function is related to the density of states, $\Im G_n(z) = -\pi D_n(z)$, one can use the above expression for the density of states to construct the density-density correlation function. In the case of a nondegenerate 2DEG one obtains

$$\begin{aligned}
\Im\Pi_0(n_1,n_2,\omega) = & -2\pi\left[e^{\zeta/T} - e^{(\zeta-\omega)/T}\right] \\
& \times\left\{\frac{1}{\sqrt{\pi}\Gamma}\exp\left[-\frac{1}{\Gamma^2}(\omega-\varepsilon_{n_1}+\varepsilon_{n_2})^2+\frac{1}{2T}(\omega-\varepsilon_{n_1}-\varepsilon_{n_2})+\frac{\Gamma^2}{16T^2}\right]\right. \\
& +\frac{X_{BG}}{2\omega_c}(1-X_{BG})\exp\left[\frac{1}{T}\left(\omega-\varepsilon_{n_1}+\frac{\Gamma^2}{4T}\right)\right] \\
& \times\left[\mathrm{erf}\left(\frac{\sqrt{2}}{\Gamma}\left(\varepsilon_{n_2+1}+\omega-\varepsilon_{n_1}+\frac{\Gamma^2}{4T}\right)\right)\right. \\
& \left.-\mathrm{erf}\left(\frac{\sqrt{2}}{\Gamma}\left(\varepsilon_{n_2}+\omega-\varepsilon_{n_1}+\frac{\Gamma^2}{4T}\right)\right)\right] \\
& +\frac{X_{BG}}{2\omega_c}(1-X_{BG})\exp\left[\frac{1}{T}\left(-\varepsilon_{n_2}+\frac{\Gamma^2}{8T}\right)\right] \\
& \times\left[\mathrm{erf}\left(\frac{\sqrt{2}}{\Gamma}\left(\varepsilon_{n_1+1}-\omega-\varepsilon_{n_2}+\frac{\Gamma^2}{4T}\right)\right)\right. \\
& \left.-\mathrm{erf}\left(\frac{\sqrt{2}}{\Gamma}\left(\varepsilon_{n_1}-\omega-\varepsilon_{n_2}+\frac{\Gamma^2}{4T}\right)\right)\right] \\
& \frac{T}{\omega_c^2}X_{BG}^2\left[\theta(\varepsilon_{n_2}-\varepsilon_{n_1}+\omega)\theta(\varepsilon_{n_1+1}-\varepsilon_{n_2}-\omega)\right. \\
& \times\left(e^{-\varepsilon_{n_2}/T}-e^{-(\varepsilon_{n_1+1}-\omega)/T}\right) \\
& +\theta(\varepsilon_{n_1}-\varepsilon_{n_2}-\omega)\theta(\varepsilon_{n_2+1}-\varepsilon_{n_1}+\omega) \\
& \left.\left.\times\left(e^{-(\varepsilon_{n_1}-\omega)/T}-e^{-\varepsilon_{n_2+1}/T}\right)\right]\right\}, \qquad (81)
\end{aligned}$$

where $\mathrm{erf}(x)$ is an error function. The real part of the density-density correlation function may be obtained invoking the Kramers-Kronig relation

$$\Re\Pi_0(n,n',\omega) = \frac{2}{\pi}\mathcal{P}\int_0^\infty ds\frac{s\Im\Pi_0(n,n',s)}{s^2-\omega^2}. \qquad (82)$$

Similarly, in the extreme degenerate case ($T=0$), one obtains

$$\Im\Pi_0(n,n',\omega) = -\frac{2}{\pi}\int_0^\omega dz\Im G_{n_1}(\omega+\zeta-z)\Im G_{n_2}(\zeta-z), \qquad (83)$$

which can again be expressed in terms of error functions.

Both linear and nonlinear steady state, time oscillatory and transient magnetotransport in a type-I superlattice may be examined on the basis of the equations above in this section. Anticipating interesting variation of the superlattice magnetoplasmon spectrum as a function of superlattice period, we direct attention to the current response to time-oscillatory electric fields, $\mathbf{E} = \mathbf{E}_0\exp(i\omega t)$ in the linear limit, which may be employed to examine superlattice magnetoplasmon resonance as well as superlattice cyclotron resonance. In this situation, one may expect a memory function description of the dynamic a.c. conductivity $\sigma(\omega)$ from the superlattice force balance equation (Eqs.(70), (71)). Defining

$$\sigma_{\pm}(\omega) = \sigma_{xx}(\omega) \pm i\sigma_{yx}(\omega), \tag{84}$$

we have

$$\sigma_{\pm}(\omega) = \frac{in_s e^2/m}{\omega \mp \omega_c + M(\omega)}, \tag{85}$$

where the memory function $M(\omega)$ is given by

$$M(\omega) = \frac{1}{n_s m\omega} F(\omega), \tag{86}$$

with

$$\begin{aligned} F(\omega) &= \left(\frac{2\pi e^2}{\kappa}\right)^2 \sum_{\mathbf{q}_\parallel} \frac{q_z^2}{q_\parallel^2} \tilde{N}(\mathbf{q})[\Pi(\mathbf{q}_\parallel, q_z = 0, 0) - \Pi(\mathbf{q}_\parallel, q_z = 0, \omega)] \\ &+ \sum_{\mathbf{q},\lambda} |M(\mathbf{q},\lambda)|^2 |I(iq_z)|^2 q_x^2 [\Theta(\mathbf{q},\lambda,0) - \Theta(\mathbf{q},\lambda,\omega)], \end{aligned} \tag{87}$$

and

$$\begin{aligned} \Theta(\mathbf{q},\lambda,\omega) &= \Pi(\mathbf{q},\omega - \Omega_{\mathbf{q}\lambda}) \left[n\left(\frac{\Omega_{\mathbf{q}\lambda}}{T}\right) - n\left(\frac{\Omega_{\mathbf{q}\lambda} - \omega}{T}\right) \right] \\ &+ \Pi(\mathbf{q},\omega + \Omega_{\mathbf{q}\lambda}) \left[n\left(\frac{\Omega_{\mathbf{q}\lambda}}{T}\right) - n\left(\frac{\Omega_{\mathbf{q}\lambda} + \omega}{T}\right) \right]. \end{aligned} \tag{88}$$

In the absence of phonon scattering Horing, Tso, and Lei (1987) found a substantial magnetoplasmon resonance phenomenon in the close-pack superlattice limit at zero temperature, associated with the emergence of a bulk-like magnetoplasmon in the superlattice spectrum for small period. For a loosely packed superlattice, the plasmon reverts to quasi-2D behavior, with strong wavenumber dependence (square root) which smears out and thus dilutes its impact. Moreover, both finite temperature and phonon coupling also tend to reduce such plasmon resonance effects.

In addressing superlattice cyclotron resonance, it is convenient to introduce a frequency dependent mass and relaxation time in accordance with

$$m^*(\omega) = m[1 + M_1(\omega)/\omega], \tag{89}$$

$$\tau(\omega) = [1 + M_1(\omega)/\omega]/M_2(\omega), \tag{90}$$

where $M_1(\omega)$ and $M_2(\omega)$ are the real and imaginary parts of the memory function. In these terms, the conductivity takes the Drude form

$$\sigma_{\pm}(\omega) = \frac{in_s e^2/m^*(\omega)}{\omega \mp \omega_c^* + i/\tau(\omega)}, \tag{91}$$

where $\omega_c^* = eB/m^*(\omega)$ has the sense of a frequency dependent cyclotron frequency. In this description, cyclotron resonance occurs at the frequency satisfying $\omega = \omega_c^*$, and it is renormalized from the bare resonance by the factor $1+M_1(\omega)/\omega$. Another formulation of this type, formally extended to include higher subband excitations in a type-I superlattice was also discussed by Tso (1990).

Cyclotron resonance in quantum wells and superlattices has been investigated by Hu and O'Connell, employing a generalized quantum Langevin equation approach (Hu and O'Connell, 1988a, 1988b, 1988c, 1989). Singh and coworkers have also treated cyclotron resonance in multi-quantum-well structures and related problems (Singh 1978a, 1978c; Tanatar and Singh, 1990, 1991). Moreover, they studied electron-phonon scattering in the presence of a magnetic field(Prasad and Singh, 1984; Singh and Chaubey, 1988), magnetophonon resonances(Singh and Chaubey, 1986), and Landau level broadening(Chaubey and Singh, 1986; Singh, 1987b; Tanatar, Singh, and MacDonald, 1991). Linear and nonlinear magnetotransport in low-dimensional systems, particularly magnetophonon resonance, has been extensively examined by various techniques, such as the kinetic equation approach(Vasilopoulos, 1986; Vasilopoulos *et al*, 1989), Monte Carlo simulations(Warmenbol *et al*, 1985; Warmenbol, Peeters, and Devreese, 1987), and balance equation analysis(Warmenbol, Peeters, and Devreese, 1986, 1987, 1989a, 1989b, 1989c; Peeters, Warmenbol, and Devreese, 1990).

3.2 Type-II Superlattice

The formulation of balance equations for magnetotransport in a type-II superlattice, consisting of two distinct infinite sets of periodically arranged type-I quantum-well superlattices having the same period, one populated by electrons interspersed with the other populated by holes, closely follows the corresponding development for a type-II superlattice in the absence of a magnetic field(Lei, Cui, and Horing, 1988). Accordingly we employ the results described above(Lei and Horing, 1991), which are applicable in a magnetic field upon the introduction of Lorentz force terms for the electron and hole centers of mass, provided only that the relative electron and hole density-density correlation functions are understood to be subject to Landau quantization, as described above (we treat only lowest subbands here). The force and energy balance equations for the electrons and holes of the type-II superlattice system in a magnetic field are given as

$$n_1 m_1 \frac{d\mathbf{v}_1}{dt} = -n_1 e(\mathbf{E} + \mathbf{v}_1 \times \mathbf{B}) + \mathbf{f}_1(\mathbf{v}_1) + \mathbf{f}_{12}(\mathbf{v}_1 - \mathbf{v}_2), \tag{92}$$

$$n_2 m_2 \frac{d\mathbf{v}_2}{dt} = n_2 e(\mathbf{E} + \mathbf{v}_2 \times \mathbf{B}) + \mathbf{f}_2(\mathbf{v}_2) - \mathbf{f}_{12}(\mathbf{v}_1 - \mathbf{v}_2), \tag{93}$$

$$-\frac{dE_e}{dt} = \mathbf{v}_1 \cdot \mathbf{f}_1(\mathbf{v}_1) + w_1(\mathbf{v}_1) + w_{12}(\mathbf{v}_1 - \mathbf{v}_2), \tag{94}$$

and

$$-\frac{dE_h}{dt} = \mathbf{v}_2 \cdot \mathbf{f}_2(\mathbf{v}_2) + w_2(\mathbf{v}_2) + w_{21}(\mathbf{v}_2 - \mathbf{v}_1). \tag{95}$$

All the quantities appearing here have already been defined above(Lei and Horing, 1991).

In the linear weak field limit the conductivity tensor elements are(Cui and Horing, 1989)

$$\sigma_{xx} = n_1 e \mu^e_{xx} + n_2 e \mu^h_{xx}, \tag{96}$$

and

$$\sigma_{yx} = n_1 e \mu^e_{yx} + n_2 e \mu^h_{yx}, \tag{97}$$

where the electron mobility μ^e_{ij} and the hole mobility μ^h_{ij} components are given for weak magnetic field strengths, respectively, as

$$\mu^e_{xx} = \frac{\mu^e_0}{D_1}\left[(\omega_{c2}\tau_2)^2 + \left(1 + \frac{n_1 m_1 \tau_2}{n_2 m_2 \tau_{12}} - \frac{m_1\tau_2}{m_2\tau_{12}}\right)\left(1 + \frac{\tau_1}{\tau_{12}} + \frac{n_1 m_1 \tau_2}{n_2 m_2 \tau_{12}}\right)\right], \tag{98}$$

$$\begin{aligned}\mu^e_{yx} &= \frac{\mu^e_0}{D_1}(\omega_{c2}\tau_2)\left[(\omega_{c1}\tau_1)(\omega_{c2}\tau_2) + \frac{m_1}{m_2}\left(\tau_2\tau_{12} + \frac{\tau_1\tau_2}{\tau^2_{12}}\right)\right.\\ &\left.+\frac{m_2\tau_1}{m_1\tau_2}\left(1 + \frac{n_1 m_1 \tau_2}{n_2 m_2 \tau_{12}}\right)^2 - \left(\frac{\tau_1}{\tau_{12}} + \frac{2 n_1 m_1 \tau_1 \tau_2}{n_2 m_2 \tau^2_{12}}\right)\right],\end{aligned} \tag{99}$$

$$\mu^h_{xx} = \frac{\mu^h_0}{D_2}\left[(\omega_{c1}\tau_1)^2 + \left(1 + \frac{n_2 m_2 \tau_1}{n_1 m_1 \tau_{21}} - \frac{m_2\tau_1}{m_1\tau_{21}}\right)\left(1 + \frac{\tau_2}{\tau_{21}} + \frac{n_2 m_2 \tau_1}{n_1 m_1 \tau_{21}}\right)\right], \tag{100}$$

and

$$\begin{aligned}\mu^h_{yx} &= -\frac{\mu^h_0}{D_2}(\omega_{c1}\tau_1)\left[(\omega_{c1}\tau_1)(\omega_{c2}\tau_2) + \frac{m_2}{m_1}\left(\tau_1\tau_{21} + \frac{\tau_1\tau_2}{\tau^2_{21}}\right)\right.\\ &\left.+\frac{m_1\tau_2}{m_2\tau_1}\left(1 + \frac{n_2 m_2 \tau_1}{n_1 m_1 \tau_{21}}\right)^2 - \left(\frac{\tau_2}{\tau_{21}} + \frac{2 n_2 m_2 \tau_1 \tau_2}{n_1 m_1 \tau^2_{21}}\right)\right],\end{aligned} \tag{101}$$

with $\mu^e_0 = e\tau_1/m_1$ and $\mu^h_0 = e\tau_2/m_2$ being the electron and hole mobilities at $B = 0$ in the absence of electron-hole interaction; $\omega_{c1} = eB/m_1$ and $\omega_{c2} = eB/m_2$ are the cyclotron frequencies for the electrons and holes, respectively. Furthermore,

$$\begin{aligned}D_1 &= \left[(\omega_{c1}\tau_1)(\omega_{c2}\tau_2) + \left(1 + \frac{\tau_1}{\tau_{12}} + \frac{n_1 m_1 \tau_2}{n_2 m_2 \tau_{12}}\right)\right]^2\\ &+\left[(\omega_{c1}\tau_1)\left(1 + \frac{n_1 m_1 \tau_2}{n_2 m_2 \tau_{12}}\right) - (\omega_{c2}\tau_2)\left(1 + \frac{\tau_1}{\tau_{12}}\right)\right]^2,\end{aligned} \tag{102}$$

and

$$\begin{aligned}D_2 &= \left[(\omega_{c1}\tau_1)(\omega_{c2}\tau_2) + \left(1 + \frac{\tau_2}{\tau_{21}} + \frac{n_2 m_2 \tau_1}{n_1 m_1 \tau_{21}}\right)\right]^2\\ &+\left[(\omega_{c2}\tau_2)\left(1 + \frac{n_2 m_2 \tau_1}{n_1 m_1 \tau_{21}}\right) - (\omega_{c1}\tau_1)\left(1 + \frac{\tau_2}{\tau_{21}}\right)\right]^2.\end{aligned} \tag{103}$$

The various relaxation times in the above equations are defined as

$$\tau_1 = -\frac{n_1 m_1}{f_1'(0)}, \tag{104}$$

$$\tau_2 = -\frac{n_2 m_2}{f_2'(0)}, \tag{105}$$

$$\tau_{12} = -\frac{n_1 m_1}{f'_{12}(0)}, \tag{106}$$

and

$$\tau_{21} = -\frac{n_2 m_2}{f'_{12}(0)}. \tag{107}$$

$\mathbf{v}_j f'_j(0)$ (with $j = 1, 2$ corresponding to electrons and holes) are the linearized frictional forces experienced by the electrons and holes, respectively:

$$\begin{aligned} f'_j(0) &= \left(\frac{2\pi e^2}{\kappa}\right)^2 \sum_{\mathbf{q}_\parallel} \frac{q_z^2}{q_\parallel^2} \tilde{N}_j(\mathbf{q}) \left[\frac{\partial}{\partial\omega}\Pi_2^{(j)}(\mathbf{q}_\parallel, q_z = 0, \omega)\right]_{\omega=0} \\ &+ \frac{2}{T}\sum_{\mathbf{q},\lambda} |M_j(\mathbf{q},\lambda)|^2 |I_j(iq_z)|^2 q_x^2 \Pi_2^{(j)}(\mathbf{q}, \Omega_{\mathbf{q}\lambda}) \left[-n'\left(\frac{\Omega_{\mathbf{q}\lambda}}{T}\right)\right], \end{aligned} \tag{108}$$

and $(\mathbf{v}_1 - \mathbf{v}_2) f'_{12}(0)$ is the linearized Coulomb force between the electron system and the hole system:

$$\begin{aligned} f'_{12}(0) &= \frac{1}{T}\frac{d}{\pi}\int_0^{\pi/d} dq_z \sum_{\mathbf{q}_\parallel} q_x^2 \left(\frac{e^2}{2\epsilon_0 \kappa q_\parallel}\right)^2 [I_1(q_\parallel)]^2 [I_2(q_\parallel)]^2 S_{12}(q_\parallel, q_z) \\ &\int_{-\infty}^{\infty} \frac{d\omega}{\pi}\left[-n'\left(\frac{\omega}{T}\right)\right] \Pi_2^{(1)}(\mathbf{q},\omega)\Pi_2^{(2)}(-\mathbf{q},-\omega), \end{aligned} \tag{109}$$

with $n'(x) = dn(x)/dx$ being the derivative of the Bose factor, and d the period of the type-I superlattice. Again, only the electron(hole) density-density correlation function depends on the magnetic field. All other quantities are defined in the preceding chapter(Lei and Horing, 1991).

3.3 Single Quantum Well

A particular interesting special case of a type-II superlattice is that of a single quantum well populated by both electrons and holes (infinite period, electron and hole wells coincident). Originally predicted for an electron-hole plasma in a bulk semiconductor(McLean and Paige, 1960), the phenomenon of negative absolute minority electron mobility was observed to occur in such a 2D electron-hole plasma in a GaAs/AlGaAs quantum well for temperatures below 100 K(Höpfel *et al*, 1986). As described above(Lei and Horing, 1991), the steady-state balance equation analysis of this two component system demonstrates that a large concentration of majority holes drags minority electrons against their natural direction of motion in the electric field because of strong electron-hole Coulomb attraction, resulting in negative absolute electron mobility (Cui, Lei, and Horing, 1988b; Cai, Zheng, and Lax, 1988). Moreover, the transient time development of negative absolute minority electron mobility in a quantum well was recently analyzed using coupled electron-hole Wigner function equations derived from microscopic nonequilibrium Green's function equations for the interacting electrons and holes in an electric field, complete with both electron-electron and hole-hole Coulomb interactions, and phonon and impurity scatterings for both the electrons and holes. As such, this approach incorporates rather interesting phenomenology such as a phonon mediated electron-hole interaction (as well as Coulombic electron-hole attraction) which induces a small irregularity-oscillation into the knee of the dc negative minority electron mobility curve at temperature about 60 K, which nicely parallels the experimental data(Tso, 1990; Tso and Horing, 1991). This transient analysis traces the electron and hole mobilities through the ballistic regime, followed by their overshoot phenomena and finally saturation. The details of this study show that there are discernable overshoots at temperatures corresponding to positive absolute minority electron dc mobility, but at temperatures where it becomes negative, the overshoots cease to exist. This general study of such transients provides interesting insight into the time scale over which electron-hole drag becomes effective, and its ultimate competitive role in relation to phonon scattering in generating negative absolute dc minority electron mobility in the steady state.

The introduction of a magnetic field normal to the quantum well may be represented by specializing the type-II superlattice magnetotransport balance equations given above to coincident electron and hole quantum wells, with infinite superlattice period. Simple physical considerations lead one to expect that the magnetic field will significantly modify the carrier-drag effect: Instead of simply drifting parallel or antiparallel to the electric field direction in the plane of the quantum well, electrons and holes will execute circular motions due to the magnetic Lorentz force (with or without Landau quantization), in addition to the drifting of the orbit centers. In this crossed-field geometry the electron(hole) drift velocity divides into two components, one parallel to the electric field and one perpendicular to it. Only the former is directly related to the negative electron mobility, if any. This parallel component is invariably smaller in magnitude than its counterpart in the field-free ($B = 0$) case for a given electric field. Hence, considered alone, the magnetic field leads to a reduced electron mobility (positive or negative). This is analogous to the positive change of magnetoresistivity ($\rho(B) > \rho(0)$) for a two-component bulk system, neglecting interactions between the two components (Ziman, 1972). This geometric tendency toward reduced mobility is opposed by the enhancement of the effectiveness of the electron-hole interaction due to the circularity of magnetic field orbits, increasing the likelihood of electrons and holes being in proximity of each other. As it does not substantially alter carrier-lattice scattering, the magnetic field in this respect tends to render the minority carriers more susceptible to being dragged along by the majority carriers. In our analysis of the competition of these opposing influences, we have considered the role of the magnetic field in carrier screening, and more importantly, its effect on the attractive interaction between electrons and holes.

In the steady state, our analysis of the role of the magnetic field in linear quantum well transport based on these balance equations demonstrates that the field tends to reduce the velocity of electrons relative to holes, and thus drastically reduces the effectiveness of electron-hole scattering, leading to a breakdown of carrier drag in favor of Hall drift(Cui and Horing, 1989). In this, the negative 2D minority electron mobility at low temperature

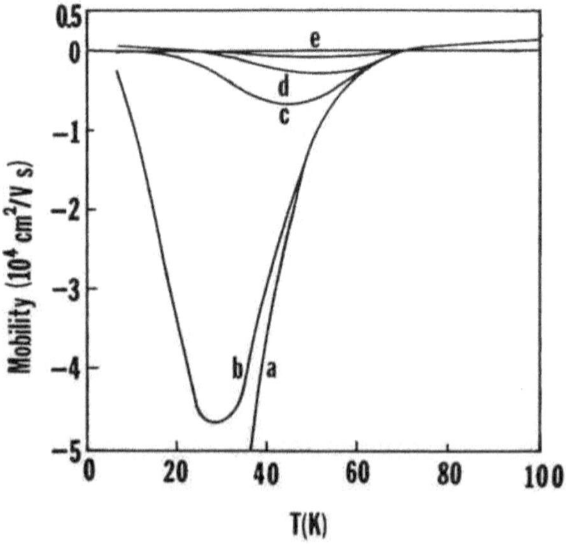

Figure 10: μ^e_{xx} *vs.* temperature for various magnetic field strengths. (a) $B = 0.0$, (b) $B = 0.1$, (c) $B = 0.5$, (d) $B = 1.0$, (e) $B = 5.0T$. (Cui and Horing, 1989).

is dramatically forced to very much smaller negative values as the magnetic field is increased from 0 to 5T, as shown in Fig.10 to Fig.13.

One of the most novel 2D systems examined in recent years involves the application of a weak periodic density-potential modulation, with concomitant resonance effects due to the interplay of the length scale of the cyclotron radius R_c (at the Fermi energy) with the modulation period a (Weiss *et al*, 1989a). The governing resonance condition is $2R_c = (n + \phi)a$, where ϕ is a constant phase shift and n is an integer. A host of phenomena attending the commensurability of the length scales in this interplay emerge in the thermodynamic, interaction and transport properties of such a modulated 2DEG(Weiss *et al*, 1989a; Gerhardts, Weiss, and Klitzing, 1989; Winkler, Kotthaus, and Ploog, 1989; Weiss *et al*, 1989b; Vasilopoulos and Peeters, 1989; Beenakker, 1989; Cui, Fessatidis, and Horing, 1989; Alves *et al*, 1989; Beton *et al*, 1990a, 1990b; Fang and Stiles, 1990; Středa and MacDonald, 1990; Peeters and Vasilopoulos, 1990; Zhang and Gerhardts, 1990; Fessatidis and Cui, 1991). They are characterized by the period $\Delta(1/B)_C = ea/2\hbar k_F$ as opposed to the Shubnikov-de Haas-van Alphen period $\Delta(1/B)_{SdH} = e/\hbar k_F^2$ (Kahn and Frederikse, 1959; Hajdu, 1991).

In essence, the periodic density/potential modulation broadens the Landau levels into minibands, whose widths oscillate as functions of the magnetic field. The associated spectrum of states is substantially altered, and with this the density of states is correspondingly modulated, as shown in magneto-capacitance measurements. Indeed, the 2D plasmon spectrum is modified by the density modulation, with the emergence of a new low-frequency "commensurability plasmon" branch, as well as modulation corrections to

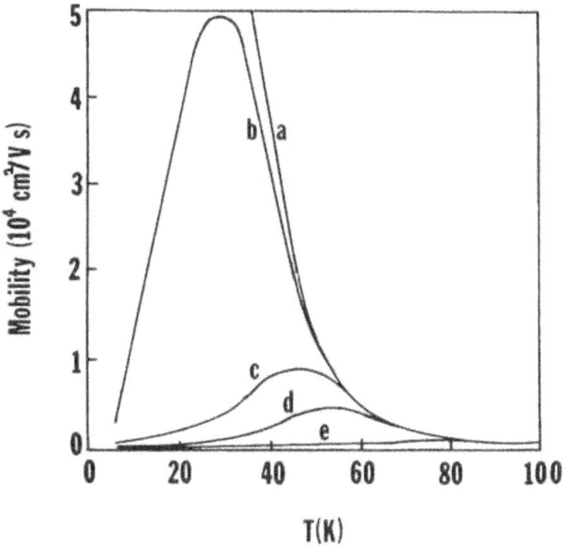

Figure 11: μ^h_{xx} *vs.* temperature for various magnetic field strengths. (a) $B = 0.0$, (b) $B = 0.1$, (c) $B = 0.5$, (d) $B = 1.0$, (e) $B = 5.0T$. (Cui and Horing, 1989).

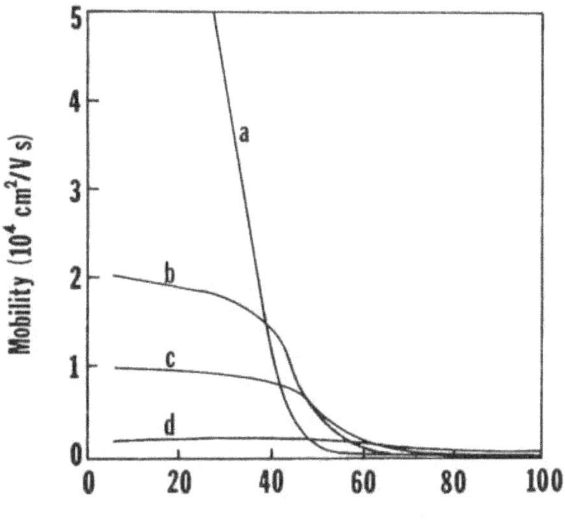

Figure 12: μ^e_{yx} *vs.* temperature for various magnetic field strengths. (a) $B = 0.1$, (b) $B = 0.5$, (c) $B = 1.0$, (d) $B = 5.0$. (Cui and Horing, 1989).

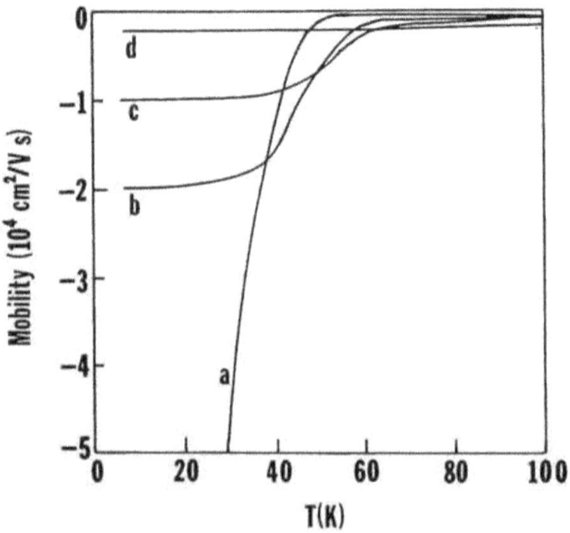

Figure 13: μ^h_{yx} *vs.* temperature for various magnetic field strengths. (a) $B = 0.1$, (b) $B = 0.5$, (c) $B = 1.0$, (d) $B = 5.0$. (Cui and Horing, 1989).

the local 2D hybrid magnetoplasmon(Cui, Fessatidis, and Horing, 1989; Fessatidis and Cui, 1991). In interaction with a passing charged particle, the density modulated 2DEG produces a significant commensurability oscillation in the low velocity energy loss for the particle passing parallel to the 2DEG or perpendicular through it. At high velocity, the commensurability effects are relatively small corrections to the energy loss by excitation of the local 2D hybrid magnetoplasmon(Fessatidis, 1991; Horing, Fessatidis, and Cui, 1991a, 1991b).

Magnetotransport measurements(Weiss *et al*, 1989a; Gerhardts, Weiss, and Klitzing, 1989; Winkler, Kotthaus, and Ploog, 1989; Alves *et al*, 1989; Beton *et al*, 1990a, 1990b; Fang and Stiles, 1990) reveal similar significant commensurability oscillations superposed as an envelope upon Shubnikov-de Haas oscillations of the linear conductivity of a periodically modulated 2DEG. From the balance equation standpoint, the weak periodic potential induces a first order shift into 2D Landau level energies, spreading them into minibands in accordance with

$$\varepsilon(n, x_0) = \left(n + \frac{1}{2}\right)\omega_c + V_n \cos\frac{2\pi x_0}{a}. \tag{110}$$

(Wavefunction corrections are of second order in the modulation potential, and may be ignored.) In this, $x_0 = -k_y/m\omega_c$, and the Landau bandwidth is $V_n = V_0 \exp(-X/2)L_n(X)$, with $X = (2\pi/a)^2/2m\omega_c$. The Landau bandwidth oscillates as a function of the Landau level index n (since $L_n(x)$ oscillates as a function of its index). These Landau minibands are involved in forming the RPA density-density correlation function, inducing commensurability oscillations in all phenomena in which it plays a role. In the absence of electron-electron interactions, it is given by

$$\Pi(\mathbf{q}_{\parallel},\omega)=\frac{m\omega_c}{\pi a}\sum_{n,n'}C_{nn'}\left(\frac{q_{\parallel}^2}{2m\omega_c}\right)\int_0^a dx_0\frac{f(\varepsilon(n,x_0+x_0'))-f(\varepsilon(n',x_0))}{\varepsilon(n,x_0+x_0')-\varepsilon(n',x_0)+\omega+i\delta},\tag{111}$$

where $x_0'=-q_y/m\omega_c$, and the coefficient $C_{nn'}(x)$ has been given in Eq.(15). It should be noted that the determination of the Fermi energy at zero-temperature is given in terms of

$$n_{2D}=\int_0^\infty d\varepsilon f(\varepsilon)D(\varepsilon),\tag{112}$$

where $D(\varepsilon)$ is the electronic density of states given by

$$D(\varepsilon)=\frac{m\omega_c}{\pi^2}\sum_n\frac{\theta(|V_n|-|\varepsilon-\varepsilon_n|)}{\sqrt{V_n^2-(\varepsilon-\varepsilon_n)^2}}.\tag{113}$$

In passing, it is important to comment that the commensurability oscillatory phenomena associated with a periodic density/potential modulation in a magnetic field are not specifically quantum phenomena (unlike Shubnikov-de Haas-van Alphen oscillations). As they reflect the interplay of the modulation period and the cyclotron radius, a semiclassical description is adequate, and has been developed(Beenakker, 1989). In the same vein, a hydrodynamic theory of density modulated 2DEG commensurability oscillations for both one and two dimensional modulations has been advanced(Fessatidis and Cui, 1991; Fessatidis, Cui, and Horing, 1991). Of course, such semiclassical theories are devoid of Shubnikov-de Haas-van Alphen oscillations, and differ from quantum theories in some other details, but they all faithfully reproduce the commensurability oscillations.

Needless to say, there are now other interesting semiconductor microstructures available, including quantum wire and quantum dot superlattices. The magnetotransport properties of such structures are already under investigation(Vasilopoulos and Tso, 1991; Tso and Vasilopoulos, 1991), and one may expect that the magnetic field will be a valuable probe in revealing their unique properties.

4 Concluding Remarks and Summary

Although much of our attention has been focused on the balance equation description of normal, dissipative magnetotransport characterized by rapid thermalization about the drifted state due to strong electron-electron interactions, this review would not be complete without mention of the pioneering work of P.N. Argyres. Within the context of his elegant and penetrating superoperator analysis of quantum magnetotransport, he examined the role of the electric field in electron acceleration during the collision process, discovering the "intra-collisional-field effect", also known as the "interference effect" (Argyres, 1960, 1963a Argyres, 1963b; Horing and Argyres, 1962; Argyres and Kelley, 1964; Argyres and Kirkpatrick, 1967; Sigel and Argyres, 1969, 1970, 1972) and showing it to be vitally important in high magnetic fields even for weak electric fields, particularly with respect to indirect power absorption. Moreover, Argyres demonstrated that the consistency of quantum kinetic equations for transport devolves upon the proper inclusion of the intra-collisional field effect as an essential ingredient. This effect was also later examined by other authors(Levinson, 1970; Barker, 1973; Barker, 1973, 1978b; Barker and Ferry, 1979). In fact, much attention has been given to it in both linear and nonlinear transport theories(Thornber, 1978a, 1978b; Calecki and Pottier, 1981; Pottier and Calecki, 1982; Seminozhenko, 1982; Arora, 1983) and it is still an active subject of interest in the literature(Sarkar, 1986; Khan, Davies, and Wilkins, 1987; Krieger and Iafrate, 1987; Cai *et al*, 1989; Abdolsalami and Khan, 1990; Lipavsky *et al*, 1991; Rossi and Jacoboni, 1991). The appearance of the intra-collisional field effect in balance equation transport theory was pointed out by Xing and Ting (1978). Moreover, carrier acceleration by the electric field during a collision is naturally incorporated when one treats the electric field in the unperturbed Hamiltonian(Lei and Horing, 1989). Particularly interesting aspects of the intra-collisional field effect in the context of nonlinear transport in a strong high frequency electric field(Cai *et al*, 1989) are also discussed in this volume(Cai and Lax, 1991).

In order to clarify the distinctive features of the isothermal balance equation approach to magnetotransport, we revisited bulk magnetotransport, both linear and nonlinear. In this, we explicitly exhibited its differences from adiabatic magnetotransport theory at finite temperature, even in the linear limit, examining both impurity and phonon scatterings and the longitudinal and transverse configurations. Transverse linear impurity conductivity divergencies arising in conjunction with Landau quantization were seen to be cured by an additional convergence mechanism—nonlinearity. Magnetophonon resonances in the linear transport regime and for hot electrons were also examined within the framework of the balance equation theory.

The relative simplicity of the balance equation approach to magnetotransport is of advantage in exploring the current response of semiconductor microstructures. In this area, much has transpired (by various methods) since the important review of Ando, Fowler, and Stern (1982). We have focused on "parallel" transport for Landau quantized type-I and type-II quantum-well superlattices, treating aspects of linear and nonlinear d.c. steady state conduction, as well as a.c. frequency response and transient behavior. In the special case of a single quantum well, attention was directed at the particularly interesting phenomena of negative absolute minority electron mobility and commensurability oscillations in the case of a density modulated two-dimensional electron gas. Nevertheless, much remains to be done in the analysis of "vertical" magnetotransport across superlattice planes, and quantum wire and quantum dot arrays. In this, the balance equation method may be expected to be informative in the examination of the role of the magnetic field in negative differential conductance and Bloch oscillations, as well as other important features.

Acknowledgements

The authors are pleased to acknowledge the active participation of Dr. V. Fessatidis in exploring various aspects of balance equation magnetotransport theory in conjunction with his Ph.D. thesis research at Stevens Institute of Technology. We also wish to thank Dr. W. Cai for useful discussions. One of us (NJMH) wishes to thank Professor P.N. Argyres for helpful advice generously proferred.

Dedication

I dedicate this work to the memory of Mildred Gendell with deep love and profound sorrow.

> "In jedem Kleide werd' ich wohl die Pein
> Des engen Erdenlebens fühlen.
> Ich bin zu alt, um nur zu spielen,
> Zu jung, um ohne Wunsch zu sein.
> Was kann die Welt mir wohl gewähren?
> Entbenhren sollst du! sollst entbehren!
> Das ist der ewige Gesang,
> Der jedem an die Ohren Klingt,
> Den unser ganzes Leben lang
> Uns heiser jede Stunde singt."
>
> (Goethe, J.W.v., 1878)

May Millie's graceful presence among us and her untimely and unmarked passing be herewith remembered.

וְיָשֹׁב הֶעָפָר עַל הָאָרֶץ כְּשֶׁהָיָה, וְהָרוּחַ תָּשׁוּב אֶל
הָאֱלֹהִים אֲשֶׁר נְתָנָהּ.
(Ecclesiastes)

Norman Horing

REFERENCES

Agaeva, R.G., B.M. Askerov, and F.M. Gashimzade, 1976, Sov. Phys. Semicond. **10**, 1268.

Alves, E.S. *et al*, 1989, J. Phys. Condens. Matter **1**, 8257.

Ando, T and Y. Uemura, 1973, J. Phys. Soc. Jpn, **36**, 959.

Ando, T., 1974, J. Phys. Soc. Jpn, **36**, 1521.

Ando, T., A.B. Fowler, and F. Stern, 1982, Rev. Mod. Phys. **54**, 437.

Argyres, P.N. and E.N. Adams, 1956, Phys. Rev. **104**, 900.

Argyres, P.N., 1958, J. Phys. Chem. Solids **4**, 19.

Argyres, P.N., 1960, Phys. Rev. **117**, 315.

Argyres, P.N., 1963, in *Proc. Eindhoven Conf. on Magnetic and Electric Resonance and Relaxation*, pg. 555 Ed. J. Smidt, North Holland.

Argyres, P.N., 1963, Phys. Rev. **132**, 1527.

Argyres, P.N. and P.L. Kelley, 1964, Phys. Rev. **134**, A98.

Argyres, P.N., 1966, in *Lectures in Theoretical Physics–Boulder, Colorado*, Vol. VIII-A, pg. 183, Ed. A. Barut.

Argyres, P.N. and E.S. Kirkpatrick, 1967, Ann. Phys. (N.Y.) **42**, 513.

Argyres, P.N. and J.L. Sigel, 1974, Phys. Rev. B**10**, 1139.

Badjou, S. and P.N. Argyres, 1987, Phys. Rev. B**35**, 5964.

Barker, J.R., 1972, J. Phys. C: Solid State Phys. **5**, 1657.

Barker, J.R., 1973, J. Phys. C: Solid State Phys. **6**, 2663.

Barker, J.R., 1978a, Solid State Electronics **21**, 197.

Barker, J.R., 1978b, Solid State Electronics **21**, 267.

Barker, J.R. and D.K. Ferry, 1979, Phys. Rev. Lett. **42**, 1779.

Beenakker, C.W.J., 1989, Phys. Rev. Lett. **62**, 2020.

Beton, P.H., E.S. Alves, P.C. Main, L. Eaves, M.W. Dellow, M. Henini, O.H. Hughs, S.P. Beaumont, and C.D.W. Wilkinson, 1990a, Phys. Rev. B**42**, 9229.

Beton, P.H., P.C. Main, M. Davidson, M.W. Dellow, R.P. Taylor, E.S. Alves, L. Eaves, S.P. Beaumont, and C.D.W. Wilkinson, 1990b, Phys. Rev. B**42**, 9689.

Cai, W., X.L. Lei, and C.S. Ting, 1985, Phys. Rev. B**31**, 4070.

Cai, W. and C.S. Ting, 1986, Phys. Rev. B**33**, 3967.

Cai, W., T.F. Zheng, and M. Lax, 1988, Phys. Rev. B**37**, 8205.

Chaubey, M.P. and M. Singh, 1986, Phys. Rev. B**34**, 2385.

Cui, H.L., X.L. Lei, and N.J.M. Horing, 1988a, Phys. Stat. Sol. (b) **146**, 189.

Cui, H.L., X.L. Lei, and N.J.M. Horing, 1988b, Phys. Phys. Rev. B**37**, 8223.

Cui, H.L. and N.J.M. Horing, 1989, Phys. Rev. B**40**, 2901.

Cui, H.L., V. Fessatidis, and N.J.M. Horing, 1989, Phys. Rev. Lett. **63**, 2598.

Davydov, B. and I. Pomeranchuk, 1958, J. Phys. USSR (Moscow) **2**, 279.

Doran, N.J. and E. Gerlach, 1976a, Phys. Status Solidi (b) **73**, 203.

Doran, N.J. and E. Gerlach, 1976b, Phys. Status Solidi (b) **75**, K15.

Ecclesiastes 12:7, *Old Testament*

"The dust returns to the earth as it was,
But the spirit returns to God who gave it."

Elesin, V.F., 1969, Sov. Phys. JETP **28**, 410.

Fang, H. and P.J. Stiles, 1990, Phys. Rev. B**41**, 10171.

Fessatidis, V. and H.L. Cui, 1991, Phys. Rev. B**43**, 11725.

Fessatidis, V., 1991, Ph.D. Thesis, Stevens Institute of Technology.

Fessatidis, V., H.L. Cui, and N.J.M. Horing, 1991, in *Proc. 20th Int. Conf. Phys. Semicond., Salonica, Greece.*

Gerhardts, R.R., D. Weiss, and K.v. Klitzing, 1989, Phys. Rev. Lett. **62**, 1173.

Gerlach, E., 1974a, Phys. Status Solidi (b) **61**, K97.

Gerlach, E., 1974b, Phys. Status Solidi (b) **62**, K43.

Gerlach, E. and M. Rautenberg, 1974, Phys. Status Solidi (b) **65**, K13.

Gerlach, E. and B. Harbecke, 1982, Phys. Status Solidi (b) **112**, 143.

Goethe, J.W.v., 1878, *Faust*, line 1544 ff., (Putnam, New York)

"In any garb would I surely feel the pain of narrow earthly life.
I am too old to merely play, too young to be without wish.
What can the world grant me? Forego, you shall forego!
That is the eternal song that rings in the ears of each of us,
that sings to us hoarsely every hour our whole life long."

Gornik, E., R. Lassnig, G. Strasser, H.L. Strömer, A.C. Gossard, and W. Wiegmann, 1985, Phys. Rev. Lett. **54**, 1820.

Götze, W. and P. Wölfle, 1972, Phys. Rev. B**6**, 122.

Gurevich, V.L. and Yu. A. Firsov, 1961, Sov. Phys. JETP **13**, 137.

Gurevich, V.L. and Yu. A. Firsov, 1965, Sov. Phys. JETP **20**, 489.

Hajdu, J., 1991, in *Landau Level Spectroscopy*, edited by L. Landwehr and E.I. Rashba, Elsevier Science Publishers.

Han, S.C., 1976, Ph.D. Thesis, Northeastern University.

Höpfel, R.A., J. Shah, P.A. Wolff, and A.C. Gossard, 1986, Phys. Rev. Lett. **25**, 2736.

Horing, N.J.M. and P.N. Argyres, 1962, in *Report Int. Conf. Phys. Semicond. at Exeter*, pg. 58, Institute of Physics, London.

Horing, N.J.M., 1965, Ann. Phys. (N.Y.) **31**, 1.

Horing, N.J.M., X.L. Lei, and H.L. Cui, 1986, Phys. Rev. B**33**, 6929.

Horing, N.J.M., H.L. Cui, and X.L. Lei, 1987, Phys. Rev. B**35**, 6438.

Horing, N.J.M., H.C. Tso, and X.L. Lei, 1987, Phys. Rev. B**35**, 851.

Horing, N,J.M., V. Fessatidis, and H.L. Cui, 1991a, to appear in Superlattices and Microstructures.

Horing, N,J.M., V. Fessatidis, and H.L. Cui, 1991b, unpublished.

Hu, G.Y. and R.F. O'Connell, 1988a, Physica A **151**, 33.

Hu, G.Y. and R.F. O'Connell, 1988b, Phys. Rev. B**37**, 10391.

Hu, G.Y. and R.F. O'Connell, 1988c, Solid State Comm. **63**, 33.

Hu, G.Y. and R.F. O'Connell, 1989, Phys. Rev. B**40**, 11701.

Huberman, M. and G.V. Chester, 1975, Adv. Phys. **24**, 489.

Kahn, A.H. and H.P.R. Frederikse, 1959, in *Solid State Physics*, edited by F. Seitz and D. Turnbull, Vol. 9, pg. 257, (Academic, New York).

Kharus, G.I. and I.M. Tsidilkovskii, 1971, Sov. Phys. Phys. Semicond. **5**, 534.

Kubo, R., N. Hashitsume, and S.J. Miyake, 1965, in *Solid State Physics*, edited by F. Seitz and D. Turnbull, Vol. 17, pg. 269, (Academic, New York).
Lei, X.L. and C.S. Ting, 1984, Phys. Rev. B **30**, 4809.
Lei, X.L., W. Cai, and C.S. Ting, 1985, J. Phys. C: Solid State Phys. **18**, 4315.
Lei, X.L., H.L. Cui, and N.J.M. Horing, 1988, Phys. Phys. Rev. B**38**, 8230.
Lei, X.L. and N.J.M. Horing, 1991, preceding article, this volume.
Levinson, I.B., 1970, Sov. Phys. JETP **30**, 362.
Magarill, L.I. and S.K. Sarvinykh, 1970, Sov. Phys. JETP **30**, 1128.
Malwal, K., 1974, Ph.D. Thesis, Northeastern University.
McLean, T.P. and E.G.S. Paige, 1960, J. Phys. Chem. Solids **16**, 220.
Mermin, N.D. and E. Canel, 1964, Ann. Phys. (N.Y.) **26**, 247.
Nicholas, R.J., 1985, Progr. Quantum Electronics, **10**, 1.
Peeters, F.M., P. Warmenbol, and J.T. Devreese, 1990, Europhys. Lett. **12**, 435.
Peeters, F.M. and P. Vasilopoulos, 1990, Phys. Rev. B**42**, 5899.
Peterson, R.L., 1970, Phys. Rev. B**2**, 4135.
Peterson, R.L., 1972a, Phys. Rev. Lett. **28**, 431.
Peterson, R.L., 1972b, Phys. Rev. B**5**, 3994.
Peterson, R.L., 1972c, Phys. Rev. B**6**, 3756.
Peterson, R.L., 1975, in *Semiconductors and Semimetals*, edited by R.K. Willardson and A.C. Beer, Vol. 10, pg. 221, (Academic, New York).
Prasad, M. and M. Singh, 1984, Phys. Rev. B**29**, 4803.
Roth, L.M. and P.N. Argyres, 1966, in *Semiconductors and Semimetals*, edited by R.K. Willardson and A.C. Beer, Vol. 1, pg. 159, (Academic, New York).
Seeger, K., 1982, *Semiconductor Physics*, (Springer-Verlag, Berlin).
Shin, E.E.H., P.N. Argyres, N.J.M. Horing, and B. Lax, 1972, Phys. Rev. B**7**, 5408.
Sigel, J.L. and P.N. Argyres, 1969, Phys. Rev. **178**, 1016.
Sigel, J.L. and P.N. Argyres, 1970, Phys. Rev. B**1**, 1845.
Sigel, J.L. and P.N. Argyres, 1972, Ann. Phys. (N.Y.) **74**, 352.
Singh, M. and M.P. Chaubey, 1986, Phys. Rev. B**34**, 4026.
Singh, 1987a, Phys. Rev. B**35**, 9301.
Singh, M, 1987b, Phys. Rev. B**36**, 1178.
Singh, M., 1987c, Phys. Rev. B**36**, 5062.
Singh, M. and M.P. Chaubey, 1988, Phys. Rev. B**38**, 10069.
Středa, P. and A.H. MacDonald, 1990, Phys. Rev. B**41**, 11892.
Tanatar, B. and M. Singh, 1990, Phys. Rev. B**42**, 3077.
Tanatar, B., M. Singh, and A.H. MacDonald, 1991, Phys. Rev. B**43**, 4308.
Tanatar, B. and M. Singh, 1991, Phys. Rev. B**43**, xxxx.
Thornber, K.K. and R.P. Feynman, 1970, Phys. Rev. B**1**, 4099.
Ting, C.S., S.C. Ying, and J.J. Quinn, 1976a, Phys. Rev. B**14** 4439.
Ting, C.S., S.C. Ying, and J.J. Quinn, 1976b, Phys. Rev. Lett. **37** 215.
Ting, C.S., S.C. Ying, and J.J. Quinn, 1977, Phys. Rev. B**16** 5394.
Tso, H.C., 1990, Ph.D. Thesis, Stevens Institute of Technology.
Tso, H.C. and N.J.M. Horing, 1991, Phys. Lett. A**152**, 498.

Tso, H.C. and P. Vasilopoulos, 1991, to appear in Phys. Rev. B**44**.

Vasilopoulos, P., 1986, Phys. Rev. B**33**, 8587.

Vasilopoulos, P., P. Warmenbol, F.M. Peeters, and J.T. Devreese, 1989, Phys. Rev. B**40**, 1810.

Vasilopoulos, P. and F.M. Peeters, 1989, Phys. Rev. Lett. **63**, 2120.

Vasilopoulos, P. and H.C. Tso, 1991, *Proc. 9th Int. Conf. Electronic Properties of 2D Systems, Nara, Japan.*

Warmenbol, P., F.M. Peeters, J.T. Devreese, G.E. Alberga, and R.G. van Welzenis, 1985, Phys. Rev. B**31**, 5285.

Warmenbol, P., F.M. Peeters, and J.T. Devreese, 1986, Phys. Rev. B**33**, 1213.

Warmenbol, P., F.M. Peeters, and J.T. Devreese, 1987, Phys. Rev. B**36**, 8094.

Warmenbol, P., F.M. Peeters, and J.T. Devreese, 1988, Phys. Rev. B**37**, 4694.

Warmenbol, P., F.M. Peeters, and J.T. Devreese, 1989a, Phys. Rev. B**39**, 7821.

Warmenbol, P., F.M. Peeters, and J.T. Devreese, 1989b, Phys. Rev. B**40**, 6258.

Warmenbol, P., F.M. Peeters, and J.T. Devreese, 1989c, Solid State Electron. **32**, 1545.

Weiss, D., K.v. Klitzing, K. Ploog, and G. Weimann, 1989, Europhys. Lett. **8**, 179.

Weiss, D., C. Zhang, R.R. Gerhardts, K.v. Klitzing, and G. Weimann, 1989, Phys. Rev. B**39**, 13020.

Winkler, R.W., J.P Kotthaus, and K. Ploog, 1989, Phys. Rev. Lett. **62**, 1177.

Zhang, C. and R.R. Gerhardts, 1990, Phys. Rev. B**41**, 12805.

Ziman, J.M., 1972, *Principles of the Theory of Solids*, (Cambridge University Press, Cambridge).

International Journal of Modern Physics B, Vol. 6, No. 7 (1992) 975–1006

EFFECT OF NONEQUILIBRIUM PHONONS ON THE ELECTRON RELAXATION AND TRANSPORT

M. Lax

Department of Physics, City College of New York, New York, N.Y. 10031 and AT&T Bell Laboratories, Murray Hill, New Jersey 07974.

W. Cai,

Department of Physics, City College of New York, New York, New York 10031

We review the recent theoretical study of the effect of nonequilibrium phonons on hot-carrier relaxation and transport. In a quantum well, the proper treatment of the electron-phonon coupling between electrons confined to two dimensions (2-D) by phonons traveling freely in three dimensions (3-D) requires special care because phonon heating produces a bottleneck in the rate of transfer of energy from the carriers to the phonons. Because the carriers interact with phonons primarily when the latter are close to the quantum well, the latter should be described, not by plane waves, but by packets adapted to the shape of the carrier confinement. A quasi-equilibrium technique that retains off-diagonal elements in the phonon wave-vector permits an unrestricted treatment of the density operator equation. That in turn leads to a choice of wave packet that comes from solving the integrodifferential equations rather than by imposition. Moreover, if the carrier distribution is assumed in quasi-equilibrium with a given drift and temperature, the coupled partial differential equations are reduced to coupled ordinary differential equations that can be solved with modest computer power. Comparison with experimental results for steady flow of energy from carriers to phonons, and for time-dependent relaxation yields quantitative agreement.

1. THE PROBLEM

The production of microelectronic devices by molecular beam epitaxy, and the high mobility of carriers particularly in modulated heterostructures created the importance of this area of study. Moreover, for transport in small high mobility systems, moderate voltages can lead to strong fields and non-linear effects. An excellent review of two-dimensional transport, with an extensive list of references has been given by Ando, Fowler, and Stern[1]

There already was experimental evidence by Shah *et al*[2,3] and Ryan *et al*[4] that the rate of energy transfer from electrons to phonons was an order of magnitude less than perturbation theory would yield.[5–7] On the other hand, controversial experimental results were also reported by Yang *et al.*[8] Several causes that may contribute to this reduction of the energy-loss rate were considered. The first one is the effect of reduced dimensionality. Theory[5–7] showed merely insignificant dependence of the energy-loss rate on dimensionality, when the predominant loss is by emission of longitudinal-optical (LO) phonons. The second is screening of electron-phonon interaction by free carriers. Screening has been appraised by Rühle *et al*[9] to play only a minor role up to electron densities of $n = 10^{17}$ cm^{-3}

It was clearly perceived by the experimentalists, and a qualitative theory was developed by Price,[10] that the inability of the longitudinal optical (LO) phonon system to dissipate heat fast enough was creating a bottleneck. A reabsorption of phonon energy by electrons is presumed to decelerate carrier cooling. In fact, this bottleneck effect has been discussed by Pötz and Kocevar[11] in the 3D case. In the 2-D case, Price's theory is qualitative, because he was forced to introduce an ad hoc parameter - the number of phonons that interact with an electron: Price recognized the need, and called for a more rigorous treatment.

2. OUR RESOLUTION

During that time, I (M. Lax) perceived a need for understanding electron (and hole) transport in quantum wells and heterostructures and hired two research associates, W. Cai and M. C. Marchetti, to work in this area. Cai was already an expert in semiconductor physics and Marchetti an expert on transport in liquids. My role was advice and criticism.

We concluded[12,13] that the problem was a general one: how should electrons, whose transport is confined to two dimensions, interact with phonons that can propagate freely in three dimensions. More specifically, since the electrons only interact with phonons when they are in the vicinity of the quantum well, a plane wave description for the phonons is inconvenient. One possibility is to retain the plane wave description, but quantize the phonons (in the z direction) over a thickness L comparable to the well width. But such a treatment would be equivalent to Price's with L as the arbitrary parameter.

Cai proposed a resolution of this problem by using a basis set for the z direction (normal to the well walls) that consists in a Gaussian times a set of Hermite polynomials. But it was not known how many terms were needed. The use of a single term, the Gaussian, has as an arbitrary parameter, the width of the Gaussian.

2.1. Quasi-Equilibrium

After reviewing the work, I suggested that the shape of the phonon "packet" should come out of the problem, not be imposed. Marchetti then suggested the use of a quasi-equilibrium procedure of the sort introduced by Bogolyubov in dealing with classical liquid transport and by Zwanzig[14,15] in a variety of problems. See also Zubarev[16] and Peletminskii and Yatsenko[17] The basic idea is the quasi-equilibrium assumption that the variables of a problem can be divided into slow variables and fast ones. The fast variables are assumed to be in equilibrium with the current values of the slow variables. When the fast variables are inserted into the equations for the slow ones, we get an effective set of equations for the slow variables.

2.2. The Choice of Slow Variables

The success of such a procedure clearly depends on the appropriate choice of slow variables. The electronic variables $a^{\dagger}_{n\mathbf{k}}a_{n\mathbf{k}}$ describe the occupancy of a state of transverse momentum $\mathbf{k}$ in the nth subband. The average of this set of variables

$$f_{n\mathbf{k}}(t) = \langle a^{\dagger}_{n\mathbf{k}}a_{n\mathbf{k}} \rangle \tag{2.1}$$

is the familiar distribution function for these carriers. These variables must clearly be included in the slow set.

The phonons are described by the three-dimensional wave vector

$$\mathbf{Q} = (\mathbf{q}, q_z) \tag{2.2}$$

associated with a plane wave representation $\exp(i\mathbf{Q}\cdot\mathbf{R})$. The variables to be used in this case are $b^{\dagger}_{\mathbf{q},q_z}b_{\mathbf{q}q'_z}$ with average value.

$$n_{\mathbf{q}}(q_z, q'_z) = \langle b^{\dagger}_{\mathbf{q},q_z}b_{\mathbf{q}q'_z} \rangle \tag{2.3}$$

Marchetti made the crucial proposal to retain the off-diagonal elements in Eq. (2.3). Although the work starts in the plane-wave representation, by allowing off-diagonal elements with respect to q_z, we have prepared the way for an eventual transformation to packets whose shape is as yet unknown.

2.3. Description of the Hamiltonian

The Hamiltonian $\hat{H}$ consists in an electron part $\hat{H}_e$, a phonon part $\hat{H}_p$, an electron-phonon interaction $\hat{V}_{ep}$ and a phonon-phonon interaction $\hat{V}_{pp}$ with

$$\hat{H}_e = \sum_{n,\mathbf{k}} E_{n\mathbf{k}} \, \hat{a}^{\dagger}_{n\mathbf{k}} \, \hat{a}_{n\mathbf{k}} \tag{2.4}$$

where $E_{n\mathbf{k}}$ is the energy associated with the state

$$\Psi_{n,\mathbf{k}}(\mathbf{r}, z) = A^{-1/2}\zeta_n(z)\exp(i\mathbf{k}\cdot\mathbf{r}) \, , \tag{2.5}$$

associated with the transverse wave-vector $\mathbf{k}$, and quantum well state n. Here $\zeta_n(z)$ is the n-th quantum state in the well. The unperturbed phonon part is

$$\hat{H}_p = \sum_{\mathbf{Q}} \hbar\omega_L \, \hat{b}^{\dagger}_{\mathbf{Q}} \hat{b}_{\mathbf{Q}} \tag{2.6}$$

$\mathbf{Q} = (\mathbf{q}, q_z)$

The electron-phonon interaction is given by the Fröhlich interaction [18]

$$\hat{V}_{ep} = \int d\mathbf{R} \int d\mathbf{R}' \, e \, \hat{n}(\mathbf{R}) \frac{1}{|\mathbf{R} - \mathbf{R}'|} [-\nabla'\cdot\hat{\mathbf{P}}(\mathbf{R}')] \, , \tag{2.7}$$

namely, the Coulomb interaction between the electron charge density $e\hat{n}$ and the charge density $-\nabla\cdot\mathbf{P}$, where $\mathbf{P}$ is the phonon induced polarization, so that

$$\hat{V}_{ep} = \frac{i\alpha}{\sqrt{A}} \sum_{n,\mathbf{k}} \sum_{n',\mathbf{k}'} \sum_{\mathbf{Q}} \left\{ \hat{b}_{\mathbf{Q}} \, \delta_{\mathbf{k}',\mathbf{k}+\mathbf{q}} \, G_{n'n}(\mathbf{q}, q_z) - \hat{b}^{\dagger}_{\mathbf{Q}} \, \delta_{\mathbf{k}',\mathbf{k}-\mathbf{q}} \, G^{*}_{n'n}(\mathbf{q}, q_z) \right\} \hat{a}^{\dagger}_{n'\mathbf{k}'} \hat{a}_{n\mathbf{k}} \, , \tag{2.8}$$

where the matrix element G takes the form

$$G_{n'n}(\mathbf{q}, q_z) = \frac{1}{Q\sqrt{L}} \int_{-\infty}^{+\infty} dz \, \zeta^{*}_{n'}(z) \exp(\, iq_z z) \, \zeta_n(z). \tag{2.9}$$

In Eq. (2.8), α is the Fröhlich interaction constant[18]

$$\alpha = [2\pi e^2 \hbar\omega_L (1/\varepsilon_\infty - 1/\varepsilon_0\,]^{1/2} \, . \tag{2.10}$$

2.4. Form of the Equations of Motion

One can derive the kinetic equations for a set of macroscopic observables, $\{\gamma_i(t)\}$ from the quantum-mechanical Liouville equation. Here we choose the following observables: $\{\gamma_i(t)\} = \{\, f_{n\mathbf{k}}(t), \; n_{\mathbf{q}}(q_z, q_z', t)\}$ (see Eqs. (2.1) and (2.3)). If the Liouville equation is solved to the lowest order, the rate of change of any function of $\gamma(t)$, denoted by $F(\gamma(t))$, is given by

$$\frac{\partial F(\gamma(t))}{\partial t} = \frac{i}{\hbar}\mathrm{Tr}[\, \hat{H}_0(t), F(\hat{\gamma}(t))\,]\hat{\rho}_0(t) + \left(\frac{i}{\hbar}\right)^2 \lim_{\varepsilon\to 0^+} \int_{-\infty}^{0} d\tau \, e^{\varepsilon\tau} \, \mathrm{Tr}\{[\, \hat{V}(\tau), [\hat{V}, F(\hat{\gamma})]\,]\, \hat{\rho}_0(t)\,\} \, , \tag{2.11}$$

where $\hat{V}(\tau)$ is the coupling perturbation in the interaction representation:

$$\hat{V}(\tau) = \exp[\ (i/\hbar)(\hat{H}_e + \hat{H}_P)\tau]\ \hat{V}\ \exp[\ (-i/\hbar)(\hat{H}_e + \hat{H}_P)\tau]\ \hat{V} \tag{2.12}$$

and $\hat{\rho}_0(t)$ the unperturbed density matrix. Here, neither the electron nor the phonon system is assumed to be even close to equilibrium. From Eq. (2.11), we obtain coupled integrodifferential equations of the form

$$\frac{\partial f_{n\mathbf{k}}}{\partial t} = \sum_{q}\sum_{q_z}\sum_{q_z'}\sum_{n'\mathbf{k}'} M[f_{n\mathbf{k}}, f_{n'\mathbf{k}'}, n_{\mathbf{q}}(q_z, q_z', t)] \tag{2.13}$$

$$\left[\frac{\partial n_{\mathbf{q}}(q_z, q_z', t)}{\partial t}\right]_{ep} = \sum_{n\mathbf{k}}\sum_{n'\mathbf{k}'}\sum_{q_z''} f_{n\mathbf{k}}\ [1 - f_{n'\mathbf{k}'}] \times [\text{ Four terms }] \tag{2.14}$$

where a typical term is given by

$$\delta_{\mathbf{k}',\ \mathbf{k}-\mathbf{q}}\ \frac{G_{n'n}(\mathbf{q}, q_z)\ G^*_{n'n}(\mathbf{q}, q_z'')}{\varepsilon - i[E_{n\mathbf{k}} - E_{n'\mathbf{k}'} - \hbar\omega_L]}[\delta_{q_z'', q_z'} + n_{\mathbf{q}}(q_z'', q_z', t)] \tag{2.15}$$

The detailed equations will be presented in the Appendix. Here we emphasize their form. In particular, let us regard $\mathbf{q}$ and q_z' as parameters. Then Eq. (2.14) with Eq. (2.15) has the form (for the term shown in Eq. (2.15))

$$\frac{\partial n(q_z, t)}{\partial t} = \phi(q_z, t) + K(q_z)M(t) \tag{2.16}$$

where

$$M(t) = \int H(q_z'')n(q_z'', t)dq_z'' \tag{2.17}$$

after all parameters such as $\mathbf{q}$, q_z' are suppressed. Multiplication of Eq. (2.15) by $H(q_z)$ and integration leads to the reduced *ordinary differential* equation

$$\frac{dM(t)}{dt} = I(t) + AM(t) \tag{2.18}$$

where

$$I(t) = \int H(q_z)\phi(q_z, t)dq_z\ ;\quad A = \int H(q_z)K(q_z)\,dq_z \tag{2.19}$$

2.5. The Shape of the Packet

Not only is the remaining computational task greatly simplified, Eq. (2.17) already informs us that (aside from a choice of normalization) the phonon wave-packet operators are

$$\hat{c}^{\dagger}_{n'n} \sim \sum_{q_z} G^*_{n'n}(\mathbf{q}, q_z)\hat{b}^{\dagger}_{\mathbf{q}, q_z} \tag{2.20}$$

If there are S subbands of importance, there is one packet operator for each choice of (n', n) or $S(S+1)/2$ differently shaped packets. For the important case in which only the

$n = 0$ states participate there is one such packet (for each $\mathbf{q}$):

$$G_{00}^{*}(\mathbf{q}, q_z) = \frac{1}{\sqrt{L}} \int_{-\infty}^{\infty} |\zeta_0(z')|^2 dz' \frac{\exp(-iq_z z')}{\sqrt{q^2 + q_z^2}} \tag{2.21}$$

The shape of this packet in ordinary share may be obtained by multiplication by $\exp(iq_z z)$ and integrating over q_z:

$$\frac{1}{\sqrt{L}} \int R(z - z') dz' \, |\zeta_0(z')|^2 \tag{2.22}$$

where

$$R(z - z') = \int \frac{\exp[iq_z(z - z')]}{\sqrt{q^2 + q_z^2}} dq_z = 2K_0(q\,|z - z'|) \tag{2.23}$$

has the form of a modified Bessel function. This form arises from the Coulomb nature of the interaction. If we had used a point interaction $(q^2 + q_z^2)^{-1/2}$ would have been replaced by unity and $R(z - z')$ by $\delta(z - z')$ so that the packet shape would simply be $|\zeta_0(z)|^2$. More generally, the packet shape is a convolution of the electron density $|\zeta_0(z)|^2$ with $K_0(q\,|z - z'|)$ for each q.

2.6. Further Simplifications

We have not written the explicit form of the phonon-phonon interaction $\hat{V}_{pp}$ by means of which the relevant longitudinal optical phonons decay into acoustic phonons because we have replaced that process by a relaxation process of the form

$$\left[\frac{\partial n_{\mathbf{q}}(q_z, q_z', t)}{\partial t}\right]_{pp} = -\frac{n_{\mathbf{q}}(q_z, q_z', t) - \delta_{q_z, q_z'} n_L}{\tau_{\rm op}}, \tag{2.24}$$

where the decay time for optical phonons $\tau_{\rm op}$ has been estimated in the experimental papers[19] to be 7 psec.

For times larger than a picosecond it has been found by Monte Carlo calculations[20–23] and by our own quasi-analytical procedure[24] that the electron distribution has equilibrated relative to two macroscopic parameters: an electron temperature $T_e(t)$ and a drift velocity $\mathbf{v}_e(t)$, both of which may be time dependent. By introducing these parameters, the equation for the electron distribution is replaced by ordinary differential equations for these parameters.

Finally, since we have been dealing with variables such as $n_q(q_z, q'_z)$ it was convenient to use a wave-packet construction on both left and right wave-vectors. Thus our reduced phonon variables are

$$N_{n'n,\, m'm}(q,\, t) = \frac{\sum\limits_{q_z q'_z}\sum G^*_{n'n}(\mathbf{q},\, q_z)\, n_{\mathbf{q}}(q_z,\, q'_z,\, t)\, G_{m'm}(\mathbf{q},\, q_z')}{F_{n'n,\, m'm}(q)/2q}\,, \tag{2.25}$$

where

$$F_{m'm,\, n'n}(q) = \int dz_1 \int dz_2\, \zeta^*_{m'}(z_1)\zeta_m(z_1)\zeta^*_{n'}(z_2)\zeta_n(z_2) e^{-q\,|z_1 - z_2|} \tag{2.26}$$

Thus, in general, we will get coupled ordinary differential equations for $T_e(t)$, $\mathbf{v}_e(t)$, and $N_{n'n,\, m'm}(t)$.

2.7. The reduced Equations

For simplicity, we shall write here only the special case in which only the lowest subband contributes. Thus we shall set

$$N_0(\mathbf{q}, t) = N_{00,\,00}(\mathbf{q}, t) \tag{2.27}$$

The phonon equation can then be written

$$\frac{\partial N_0(q,\, t)}{\partial t} = \frac{N_0(q,\, t) + 1}{\tau_e(q,\, T_e(t))} - \frac{N_0(q,\, t)}{\tau_a(q,\, T_e(t))} - \frac{N_0(q,\, t) - N_L(T_L)}{\tau_{\rm op}} \tag{2.28}$$

where the rate of phonon emission $1/\tau_e$ and the rate of phonon absorption $1/\tau_a$ are given by

$$\frac{1}{\tau_e(q,\, T_e)} = \frac{\pi\alpha^2}{\hbar}\, \frac{1}{q}\, \frac{F_{00,\,00}(q)\, I^{(+)}_{00}(q)}{|\varepsilon_{00,00}(q, \omega_L)|^2}\,, \tag{2.29}$$

$$\frac{1}{\tau_a(q,\, T_e)} = \exp\,[\,\hbar\omega_L / k_B T_e(t)]\, \frac{1}{\tau_e(q,\, T_e(t))}. \tag{2.30}$$

where $F_{00,\,00}(\mathbf{q})$ was defined in Eq. (2.26), and the matrix element $I^{(+)}$ is given by

$$I^{(+)}_{00}(q,\, T_e(t)) = \frac{m}{\pi^2\hbar^2 q} \int\limits_{\frac{q}{2}+\frac{m\omega}{\hbar q}}^{\infty} kdk\, \frac{1}{[k^2 - (\frac{q}{2} + \frac{m\omega}{\hbar q})^2]^{1/2}}$$

$$\times f\,(E_k,\, T_e(t))[1 - f\,(E_k - \hbar\omega_L,\, T_e(t))]\,, \tag{2.31}$$

Eq. (2.30) shows that the ratio of rates of absorption and emission is governed by the instantaneous electron temperature. Dynamic screening effects are contained in the dielectric response function $\varepsilon_{00,00}(q,\, \omega_L)$ used in Eq. (2.29).

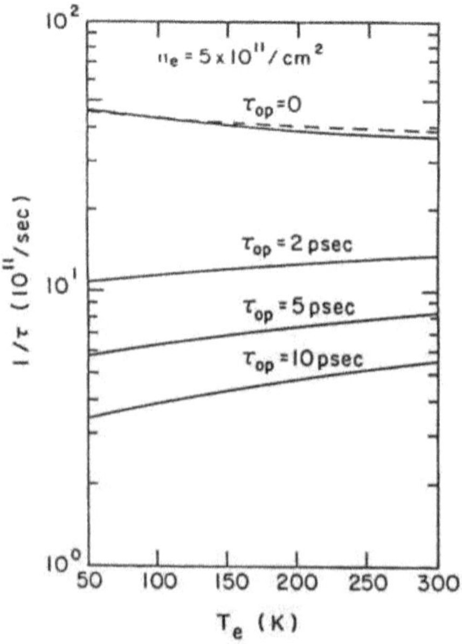

Fig. 1. $1/\tau$ as a function of the electron temperature T_e, for $\tau_{op} = 0$, 2, 5, and 10 psec and ($n = 0$) lowest subband occupation. Dynamic screening is included (solid curves). The dashed curve shows for comparison a calculation for $\tau_{op} = 0$ when static screening is used.[13] Comparison between the $\tau = 0$ (no bottleneck case) and $\tau = 10$ psec shows a reduction of one order of magnitude in the latter case.

2.8. Energy Loss Rate in Steady State Case

The first explicit calculation is made for the steady state case. In that case, the electron temperature $T_e(t) = T_e$ will be independent of time and assumed given. The problem is to calculate the rate of energy transfer from the electron gas into the LO phonons, and to express the result in the form used by the experimentalists. The latter fit their experimental data with an expression of the form:

$$P_e(t) = \frac{\hbar\omega_L}{\tau} \exp\left[-\frac{\hbar\omega_L}{k_B T_e}\right] \tag{2.32}$$

Then $1/\tau$ is plotted as a function of electron temperature.

We calculate $1/\tau$ by using the expression:

$$P_e(t) = \frac{\hbar\omega_L}{N_e}\sum_{\mathbf{q}} \frac{\partial n_{\mathbf{q}}(q_z, q_z, t)}{\partial t}\Big|_{ep} . \qquad (2.33)$$

Comparison with the previous equation yields $1/\tau$. A plot of $1/\tau$ obtained in this manner is shown in Fig. 1.

We note that in the case in which $\tau_{op} = 0$, equilibration of the LO phonon modes takes place instantaneously. Thus no bottleneck effect will occur. But the effective energy transfer rate is reduced by an order of magnitude if one takes $\tau = 10$ psec. The dashed curve demonstrates that neglect of screening would have little effect on the results.

If we apply the steady state condition:

$$\frac{dN_0(\mathbf{q}, t)}{dt} = 0 \qquad (2.34)$$

we can solve for $N_0(\mathbf{q})$, the non-equilibrium phonon occupancy associated with the transverse phonon wave-vector $\mathbf{q}$.

Using the Planck formula,

$$N_0(\mathbf{q}) = 1/\{\exp[\hbar\omega_L/k_B T(\mathbf{q})] - 1\} \qquad (2.35)$$

the results can be expressed in the form of a temperature $T(\mathbf{q})$ for the phonons of a given transverse wave-vector $\mathbf{q}$. The results displayed in Fig. 2 show that the predominant heating occurs for small wave-vector phonons.

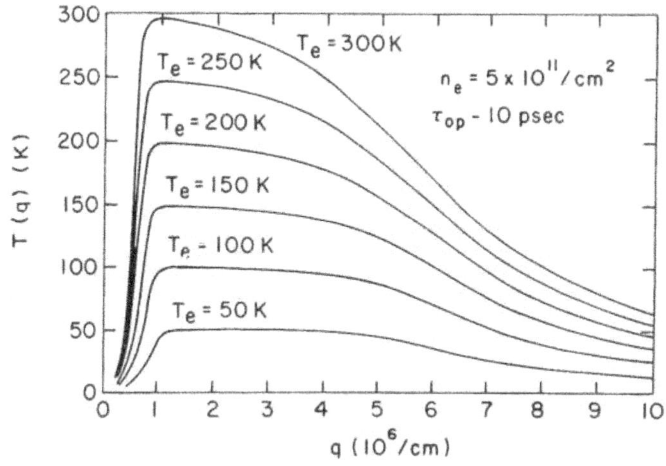

Fig.2 "Phonon temperature" $T(q)$ as a function of q for 50 K $\le T_e \le$ 300 K and lowest subband occupation.[13]

We have also made calculations when electrons and holes are simultaneously present,[25] and when more than one sub-band is occupied.[26] A comparison between experiment and theory for this case is shown in Fig. 3.

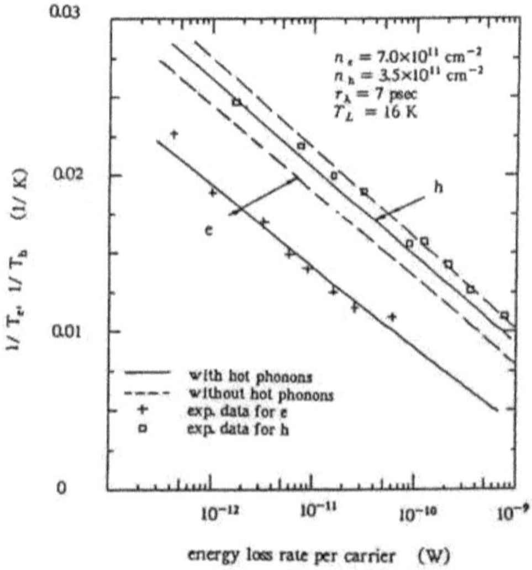

Fig. 3. The carrier temperature is plotted against the energy loss rate per carrier for the electron and hole case. For the electron case, the effect of hot phonons is found to be an order of magnitude, whereas for holes it is much less than an order of magnitude. The points are the experimental data of Shah et al.[27]

In Fig. 3 the curve for electrons is obtained by including only the polar interaction with LO phonons. The calculation for holes includes polar coupling to LO phonons and coupling to both LO and TO phonons via the deformation potential. Only heavy holes are considered. It is shown that the hot phonon effect is strong for electron relaxation, but is weak for hole relaxation. This follows because for holes the phonon emission rate $1/\tau_h(q)$ for given q, which is similar to Eq. (2.29) for $1/\tau_e(q)$, is much smaller than that for electrons. From Eqs. (2.29) and (2.31), we see that $1/\tau_e(q)$ is proportional to $f(E_k, T_e(t))$, which is proportional to $1/m$, with m the effective mass. The large effective mass of holes leads to a weaker build-up of hot phonons by the hole gas. On the other hand, heavy holes preferentially emit phonons with larger momentum $\mathbf{q}$, because of the large effective mass of hole. Therefore, the phase space of $\mathbf{q}$ that contributes to the total cooling rate for holes is much larger than that for electrons. This leads to the larger cooling rate for holes than that for electrons.

2.9. Time-dependent relaxation

Our starting equations are valid for the time-dependent case. We simply do not assume time derivatives vanish. The electron energy can be written as a sum over the transverse **k** vector:

$$<E_e(t)> = \frac{2}{A} \sum_{\mathbf{k}} E_{\mathbf{k}}\, f(E_{\mathbf{k}}, T_e(t)) \tag{2.36}$$

For simplicity, we again consider here only one sub-band. The rate of electron temperature change is then given by:

$$\frac{\partial T_e}{\partial t} = \frac{1}{C_\upsilon(T_e(t))} \frac{\partial <E(t)>}{\partial t} \tag{2.37}$$

where C_υ is the specific heat: $C_\upsilon = \partial <E(t)>/\partial T_e(t)$. Since the phonon equations depend on the instantaneous temperature $T_e(t)$, we now have coupled ordinary differential equations connecting the phonon occupancies $N_0(\mathbf{q})$ and the electron temperature. It is assumed, of course, that $T_e(t = 0)$ is given. A comparison is given in Fig. 4 of our theoretical results with experiment [8]

As the electron gas relaxes, its temperature decreases as shown by the solid curve in Fig. 5. The associated rise in the phonon temperature is shown by the dashed curve which merges with the solid curve as the combined system relaxes to the lattice temperature. When τ_{op} is set equal to zero, the phonon bottleneck effect disappears, and the electron temperature falls more quickly as shown by the dot-dash curve.

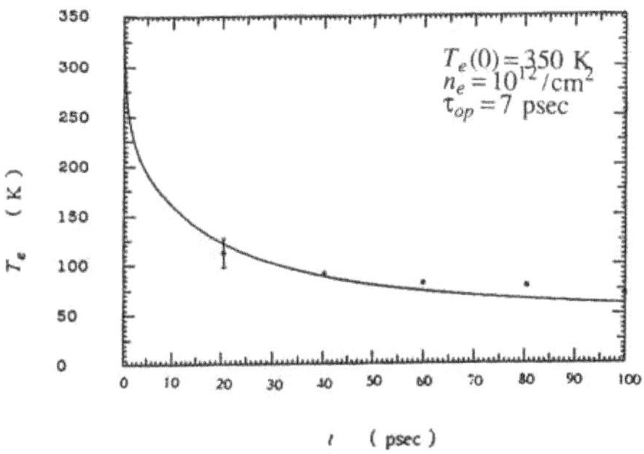

Fig. 4. Time dependent relaxation of an electron gas of density $n_e = 10^{12}/\text{cm}^2$ starting at an initial temperature of $T_e(0) = 350\text{K}$. The optical phonon decay rate has been given the accepted value of $\tau_{op} = 7$ psec. The experimental data are from Ryan *et al.*[4] corresponding to a 3D electron density $n = 5\times 10^{18}\, cm^{-3}$ and to a maximum power absorbed by the sample of ~ 50 mW.

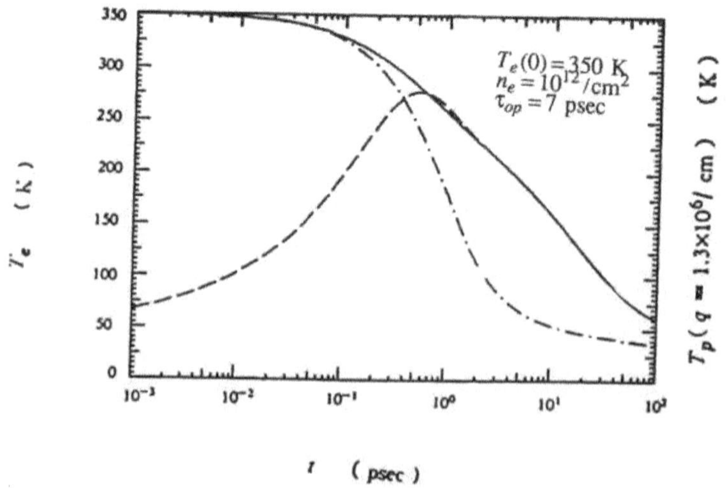

Fig. 5. A plot of electron temperature T_e against time (solid curve for $\tau_{op}=7$ psec and dot-dash curve for $\tau_{op}=0$) and of the phonon temperature $T_p(q)$ at $q = 1.3\times10^6 cm^{-1}$

3. FURTHER APPLICATIONS OF THE THEORY

3.1. Hot electron transport

When a strong dc field is applied, the electron gas acquires an elevated temperature as well as a drift velocity. Moreover, the drift mobility is reduced more in the presence of a phonon bottleneck than in its absence. This issue was studied by Lei and Horing[28] and by us[29] The methods described earlier involving the introduction of a phonon packet remain valid. It is convenient, in addition to separate the center of mass motion of the electron gas from its relative motion. The density of electrons will again be assumed sufficiently high that the electron gas can be assumed in quasi-equilibrium with a given drift velocity and a given temperature. A term, which describes the center-of-mass motion under an applied field should be added upon to the Hamiltonian described in Sec. 2.3:

$$\hat{H}_c = \hat{\mathbf{P}}^2/2M - Ne\mathbf{E}\cdot\hat{\mathbf{R}} , \tag{3.1}$$

with $\hat{\mathbf{P}}$ and $\hat{\mathbf{R}}$ the center-of-mass momentum and coordinate operators, respectively, N the total number of electrons, and $M = Nm$, with m the electron effective mass. The electronic states are labeled by the 2D wave vector in the relative coordinate system, $\mathbf{k}$, and the discrete subband index, n; $\hat{a}^{\dagger}_{n\mathbf{k}}$ and $\hat{a}_{n\mathbf{k}}$ (Eq. (2.1)) now are the corresponding electron creation and annihilation operators in the relative coordinates and $E'_{n\mathbf{k}}$ is the energy of an electron in the

$(n, \mathbf{k})$ state. Therefore, the energy exchange of electron in the relative coordinates obeys the conservation condition:

$$E'_{n'\mathbf{k}'} - E'_{n\mathbf{k}} = \hbar(\omega_L - \mathbf{q}\cdot\mathbf{v}_e)\,, \tag{3.2}$$

with $\mathbf{v}_e$ the drift velocity of electrons. The intracollisional effect[30,31], namely the effect of the electric field during the course of a collision can be neglected for the fields considered in our work[29]. If only the lowest subband is considered, we obtain a pair of time evolution equations for the drift velocity $\mathbf{v}_e$ and the total energy of electrons in the relative coordinates:

$$\frac{\partial M\mathbf{v}_e(t)}{\partial t} = -Ne\mathbf{E} - \sum_{\mathbf{q}} \hbar\mathbf{q} L_{\text{e-LO}}(\mathbf{q}, t)\,, \tag{3.3}$$

$$\left(\frac{\partial E'_e}{\partial t}\right)_{\text{e-LO}} = -\sum_{\mathbf{q}} (\hbar\omega_L - \mathbf{q}\cdot\mathbf{v}_e) L_{\text{e-LO}}(\mathbf{q}, t) \tag{3.4}$$

Here

$$L_{\text{e-LO}}(\mathbf{q}, t) = \frac{2\pi\alpha^2}{\hbar}\left(\sum_{q_z} G^*_{00}(\mathbf{q}, q_z) G_{00}(\mathbf{q}, q_z)\right)$$

$$\times\{I^{(+)}_{00}(\mathbf{q}, \omega_{\text{LO}})[1 + N_0(\mathbf{q}, t)] - I^{(-)}_{00}(\mathbf{q}, \omega_{\text{LO}}) N_0(\mathbf{q}, t)\}\,, \tag{3.5}$$

where $N_0(\mathbf{q}, t)$ is defined by Eq. (2.27) and

$$I^{(\pm)}_{00}(\mathbf{q}, \omega_L) = \frac{2}{A}\sum_{\mathbf{k}}\sum_{\mathbf{k}'} f_{0\mathbf{k}}(t)\,[\,1 - f_{0'\mathbf{k}'}(t)]\,\delta_{\mathbf{k}',\mathbf{k}+\mathbf{q}}\,\delta(E'_{0\mathbf{k}} - E'_{0'\mathbf{k}'} \pm \hbar\mathbf{q}\cdot\mathbf{v}_e \mp \hbar\omega_L)\,, \tag{3.6}$$

where the electron distribution function in the relative coordinates $f_{0\mathbf{k}}(t)$ is assumed to be a Fermi-Dirac distribution with temperature $T_e(t)$. With the above approximations, the problem is again reduced to coupled ordinary nonlinear differential equations. Solving Eqs. (3.3) and (3.4) and corresponding equation for N_0, which is similar to Eq. (2.28), we can determine $\mathbf{v}_e$, T_e as well as $N_0(\mathbf{q})$

Results for the mobility reduction and temperature increase in a steady applied field are shown in Fig. 6.

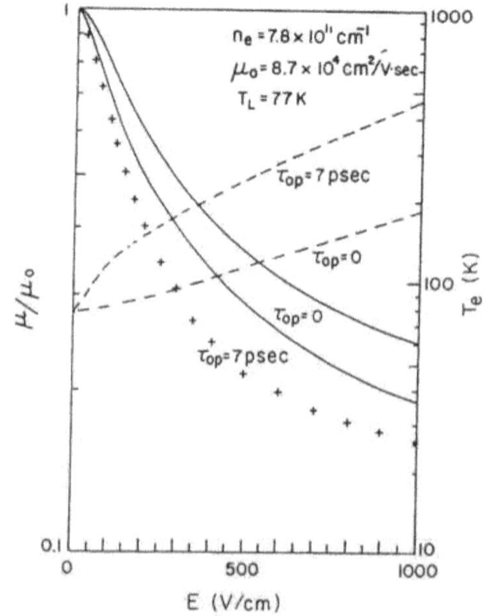

Fig. 6. The normalized mobility of electrons, $\mu(\mathbf{E})/\mu(\mathbf{E} \to \mathbf{0})$ (solid curves), and the electron temperature, T_e, (dashed curves) as functions of the external electric field $|\mathbf{E}|$ at $\tau_{op} = 0$ and $\tau_{op} = 7$ psec. The crosses represent the experimental data from Fig. 2(c) of Keever *et al*[32] at $T_L = 77$ K.

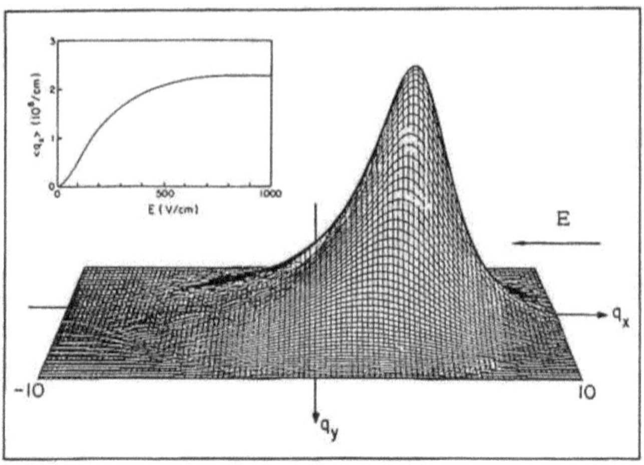

Fig. 7. "Optical phonon temperature", $T_{op}(\mathbf{q})$, as a function of $\mathbf{q}$ (unit: 10^6 cm^{-1}) at $|\mathbf{E}| = 500$ V/cm. The q_x-q_y plane represents $T_{op} = 77$ K, the peak value of T_{op} is 696 K. Inset: The average wave vector of nonequilibrium LO phonons, $< q_x >$, as a function of the electric field.

Phonon heating produced by the strong electric field is not isotropic, but is more effective for phonons whose propagation direction is in the direction of motion of the electrons. The phonons also acquire a mean "momentum" $<q_x>$. A contour plot of the phonon temperature rise, and an inset of the mean momentum are shown in Fig. 7.

We also studied the time evolution of electrons beginning from "switch on" of a applied electric field until arrival at a steady state. Fig. 8 displays the results. The presence of hot phonons leads to time delay in arriving at a steady state.

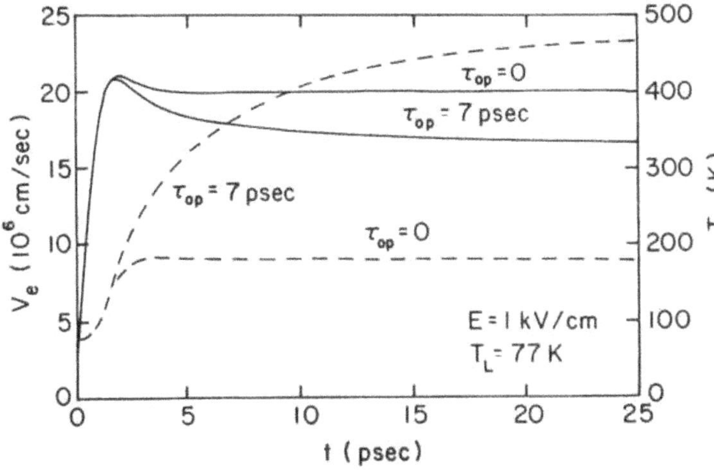

Fig. 8. The drift velocity of electrons, $|v_e|$ (solid curves) and the electron temperature, T_e (dashed curves) as functions of time t. The parameters are the same as those of Fig. 6.

3.2. Relaxation of the electron-hole plasma

When an undoped quantum well device is illuminated by a high-power laser, photo-excited electrons and holes are produced simultaneously. Therefore, it is necessary to study the cooling process of a photo-generated electron-hole plasma. This process involves, not only the carrier-phonon coupling, but also the electron-hole interaction. This issue was studied by Pötz[33] in bulk GaAs, and then by Marchetti and Pötz[34] in a GaAs-GaAlAs quantum well using the approach described above, since hot-phonon reabsorption can also slow down considerably the carrier cooling in the electron-hole plasma.

The energy of photo-excited electrons is assumed below that of the L valley, so inter-valley scattering is not included. The carrier distributions are modeled as a time-dependent Fermi-Dirac distribution functions. The temperatures in the distribution functions for electrons and holes are allowed to be different to account for noninstantaneous energy transfer between two carrier systems. The equations for the time evolution of total electron energy (E_e) and hole energy (E_h) is derived as

$$\frac{\partial E_e}{\partial t} = -R_{e\text{-}h} - R_{e\text{-LO}} + R_L^e \,, \tag{3.7}$$

and

$$\frac{\partial E_h}{\partial t} = +R_{e\text{-}h} - R_{h\text{-LO}} - R_{h\text{-TO}} + R_L^h \,. \tag{3.8}$$

Here $R_{e\text{-}h}$ is the power loss by electrons to holes via the screened Coulomb interaction. $R_{e\text{-LO}}$ and $R_{h\text{-LO}}$ are the rate of energy loss by carriers to LO phonons via the polar Fröhlich coupling, $R_{h\text{-TO}}$ is the rate of energy loss by hole to TO phonons via the deformation potential coupling. R_L^e and R_L^h are the power input from the laser to electrons and holes, respectively. Only the lowest subband in the quantum well is considered.

The corresponding time evolution equations for non-equilibrium LO and TO phonons (similar to Eq. (2.28)) are coupled to Eqs. (3.7) and (3.8). Therefore, the temperatures T_e, T_h and the phonon distributions, $N_0^{\text{LO}}(q)$ and $N_0^{\text{TO}}(q)$, can be determined as function of time t.

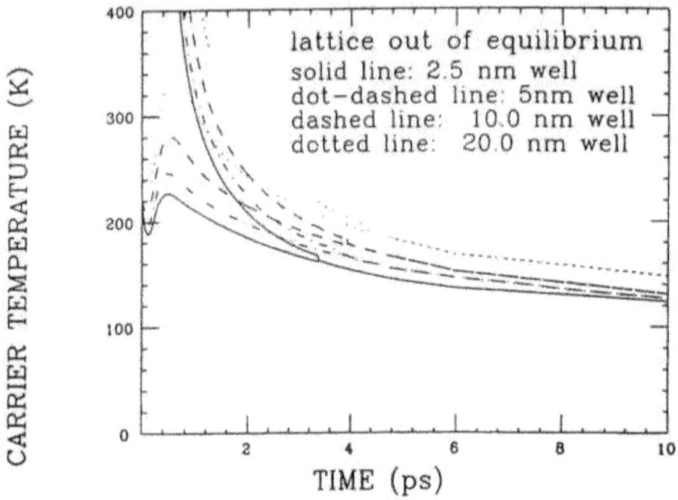

Fig. 9. Electron and hole temperatures as functions of time for $n_{2\text{D}} = 0.5 \times 10^{12}\ \text{cm}^{-2}$ and four values of well thickness: $L = 25$ Å, solid line; $L = 50$ Å, dot-dashed line; $L = 100$ Å, dashed line; $L = 200$ Å, dotted line. The four curves starting at $T = 200$ K represent hole temperatures, the others represent electron temperatures. This figure is Fig. 1 from Marchetti and Pötz.[34]

In Fig. 9, the electron and hole temperatures are displayed as functions of time for different layer thickness L. The exchange of energy between electrons and holes plays an important role in the initial stage of the relaxation. The main portion of the photon excess energy is given to the photogenerated electrons. Initially, the kinetic energy of excited holes is below the threshold for optical phonon emission. However, the $e-h$ coupling rapidly transfers energy from electrons to holes and thus ensures significant participation of the holes in the cooling process. This energy loss channel is also important after the buildup of LO-phonon modes which couple to electrons slows the cooling of electrons. Reabsorption of phonons becomes important within less than a picosecond after the onset of the laser pulse and leads to a considerable reduction of the carrier cooling rate in the later stage of the relaxation.

One can see, from Fig. 9, that the carrier cooling rate is rather insensitive to variations in the well thickness, if the sheet density is kept constant. This result is in agreement with experiment.[35] On the other hand, their calculation[34] indicates that for given layer thickness and times beyond 0.5 ps, cooling occurs at a slow rate at higher values of sheet density, because of a strong buildup of nonequilibrium optical phonons at higher sheet density.

3.3. Electron-hole Transport and Negative Mobility

The luminescence measurements of the photoexcited electron-hole (e-h) plasma in quasi-2D quantum wells and the relative theoretical study, which was discussed in Sec. 3.2, are important to derive information about relaxation of this system. On the other hand, transport measurements of a photoexcited electron-hole system under an applied electric field have also provided some interesting results.[36,37] Recently, Höpfel, Shah, Wolff, and Gossard[37] found that in a such system the minority electrons, which are injected by laser pumps on the p-modulation-doped quantum wells, can move in the direction of the external electric field. This negative absolute mobility of electrons occurs because of strong electron-hole drag. This subject was theoretically studied by Cui, Lei, and Horing[38] and by us[39].

We first briefly discuss the condition for negative absolute mobility of electrons in the region of weak electric field where the conductivity is linear. In the steady state we have the force balance equation for carriers:

$$n_\mu e_\mu \mathbf{E} - \mathbf{F}^{\mu-\nu} - \mathbf{F}^{\mu-L} = 0 \,, \tag{3.9}$$

where $\mathbf{E}$ is the external electric field, $\mathbf{F}^{\mu-L}$ represents the frictional force due to the carrier-lattice interaction and $\mathbf{F}^{\mu-\nu}$ represents the frictional force upon the μ-type carriers due to the carrier-carrier interaction with the ν-type carriers. It is obvious that $\mathbf{F}^{\mu-\nu} = -\mathbf{F}^{\nu-\mu}$. In the region of linear conductivity we have

$$\mathbf{F}^{\mu\text{-L}} = n_\mu A^{\mu\text{-L}} \mathbf{v}_\mu \,, \tag{3.10}$$

$$\mathbf{F}^{\mu\text{-}\nu} = n_\mu n_\nu A^{\mu\text{-}\nu}(\mathbf{v}_\mu - \mathbf{v}_\nu) \,, \tag{3.11}$$

where $\mathbf{v}_\mu$ is the drift velocity of μ type of carriers, $A^{\mu\text{-L}}$ represents the contribution to the resistivity (per carrier) from μ–L scattering, and $A^{\mu\text{-}\nu}$ relates to the contribution to the resistivity (per carrier μ) from μ–ν scattering normalized to per carrier ν. From Eqs. (3.9) - (3.11) we immediately obtain the mobility for electrons, μ_e:

$$\mu_e = \frac{(n_e - n_h)A^{e\text{-}h} + A^{h\text{-}L}}{A^{e\text{-}L}A^{h\text{-}L} + n_e A^{e\text{-}h}A^{e\text{-}L} + n_h A^{e\text{-}h}A^{h\text{-}L}} \tag{3.12}$$

Equation (3.12) indicates that for electrons a negative absolute mobility is only possible when $n_e < n_h$. (The corresponding statment for hole mobility could require $n_h < n_e$.) It is determined by the difference between the density of electrons and the density of holes, and the competition between electron-hole drag and hole-lattice scattering. At low temperature and under a weak electric field, the former dominates, so the mobility of electrons is negative. When the lattice temperature or the electric field increase, the latter tends to dominate and the mobility of minority electrons becomes positive.

The carrier dynamics is derived in a similar way as discussed in Sec. 3.1 and Sec. 3.2. We separate the center-of-mass motion of each type of carrier from its relative motion. The 2D momentum for μ-type carriers in the relative coordinates is defined as $\hbar\bar{\mathbf{k}} = \hbar\mathbf{k} - m_\mu \mathbf{v}_\mu$ and a Fermi-Dirac distribution function is assumed at temperature T_μ, $\bar{f}^{\mu}_{i\bar{\mathbf{k}}}(T_\mu)$, in the relative coordinates. The exchange of momentum and energy in the relative coordinates can then be written as

$$\bar{\mathbf{k}} - \bar{\mathbf{k}}' = \mathbf{q} \,, \quad \bar{E}^{\mu}_{i\bar{\mathbf{k}}} - \bar{E}^{\mu}_{i'\bar{\mathbf{k}}'} = \hbar(\,\omega - \mathbf{q}\cdot\mathbf{v}_\mu\,) \tag{3.13}$$

A set of coupled equations for the time evolution of the drift velocities of the center-of-mass and energies in the relative coordinates for each type of carrier can be derived. We obtain

$$\frac{\partial n_\mu m_\mu \mathbf{v}_\mu(t)}{\partial t} = e_\mu n_\mu \mathbf{E} - \sum_{\mathrm{L}} \mathbf{F}^{\mu\text{-L}}(t) - \mathbf{F}^{\mu\text{-}\nu}(t) \,, \tag{3.14}$$

and

$$\frac{\partial \bar{E}^{\mu}(t)}{\partial t} = -\sum_{\mathrm{L}} \frac{\partial \bar{E}^{\mu\text{-L}}(t)}{\partial t} - \frac{\partial \bar{E}^{\mu\text{-}\nu}(t)}{\partial t} \,. \tag{3.15}$$

$\mathbf{F}^{\mu\text{-L}}$ in Eq. (3.14) represents the frictional force due to carrier-lattice (impurity) interaction. The expression of, for example, LO phonons is similar to Eqs. (3.3) (3.6). Here, the carrier interactions with LO, TO, acoustic phonons, and impurities are considered. Also, the hot-phonon effects for LO and TO phonons are included. $\mathbf{F}^{\mu-\nu}$, which represents the frictional force due to e-h scattering, with ν a different type of carrier than μ, is given by

$$\mathbf{F}^{\mu-\nu} = \frac{2\pi}{\hbar} \sum_{\mathbf{q}} \sum_{i',i} \sum_{j',j} \sum_{\bar{\mathbf{k}}',\bar{\mathbf{k}}} \sum_{\bar{\mathbf{p}}',\bar{\mathbf{p}}} \hbar\mathbf{q}\, \bar{f}^{\mu}_{i\bar{\mathbf{k}}}(1-\bar{f}^{\mu}_{i'\bar{\mathbf{k}}'})\bar{f}^{\nu}_{j\bar{\mathbf{p}}}(1-\bar{f}^{\nu}_{j'\bar{\mathbf{p}}'}) \,|\tilde{V}^{\mu\nu}_{i'i,\,j'j}(\mathbf{q},\,\bar{\omega})\,|^2 \delta_{\bar{\mathbf{k}}',\bar{\mathbf{k}}-\mathbf{q}}\delta_{\bar{\mathbf{p}}',\bar{\mathbf{p}}+\mathbf{q}}$$

$$\times\delta[\bar{E}^{\mu}_{i\bar{\mathbf{k}}} + \bar{E}^{\nu}_{j\bar{\mathbf{p}}} - \bar{E}^{\mu}_{i'\bar{\mathbf{k}}'} - \bar{E}^{\nu}_{j'\bar{\mathbf{p}}'} + \hbar\mathbf{q}\cdot(\mathbf{v}_{\mu}-\mathbf{v}_{\nu})]\,, \tag{3.16}$$

where the term $\hbar\mathbf{q}\cdot(\mathbf{v}_{\mu}-\mathbf{v}_{\nu})$ appears in the last δ function because the energies $\bar{E}$ for electrons and holes are defined in different coordinates, and $\bar{\omega} = (\bar{E}^{\mu}_{i\bar{\mathbf{k}}} - \bar{E}^{\mu}_{i'\bar{\mathbf{k}}'})/\hbar$. The last term in Eq. (3.15) represents the energy loss rate in the relative system due to e-h scattering. An equation for the rate of energy loss can be obtained by replacing $\mathbf{F}^{\mu-\nu}$ in Eq. (3.16) by $-\partial\bar{E}^{\mu-\nu}(t)/\partial t$ and the prefactor $\hbar\mathbf{q}$ on the right hand side of Eq. (3.16) by $\bar{E}^{\mu}_{i\bar{\mathbf{k}}} - \bar{E}^{\mu}_{i'\bar{\mathbf{k}}'}$.

The term $\tilde{V}^{\mu\nu}_{i'i,\,j'j}(\mathbf{q},\,\bar{\omega})$ in Eq. (3.16) represents the 2D screened scattering matrix between carriers. Here, one should be careful in treatment of screening effect when two kinds of carriers coexist. The unscreened electron-electron, hole-hole, and electron-hole scatterings occur via Coulomb interactions given by

$$[V^{\mu\nu}(q)]_{i'i,\,j'j} = \frac{2\pi e_{\mu}e_{\nu}}{\varepsilon_0 qA}\, F^{\mu\nu}_{i'i,\,j'j}(q)\,, \tag{3.17}$$

where the form factor $F^{\mu\nu}_{i'i,\,j'j}(q)$ is given by Eq. (2.26), and ε_0 is the static dielectric constant. Since the strength of the electron-hole interaction is of the same order as the electron-electron interaction and the hole-hole interactions, the dynamic screening effect should be expressed by a more complex form than that in the case of a single type of carrier. These interactions are shown in Fig. 10 in the random phase approximation (RPA). According to Fig. 10, the screened carrier-carrier scattering matrix, $\tilde{V}^{\mu\nu}$, satisfies the following equation:

$$\tilde{V}^{\mu\nu} = V^{\mu\nu} + \sum_{\eta} V^{\mu\eta}\Pi^{\eta}\tilde{V}^{\eta\nu}\,, \tag{3.18}$$

where Π^{η} is a diagonal matrix, its elements are the density-density correlation functions for 2D carriers. If we define

$$\tilde{V} = \begin{bmatrix} \tilde{V}^{ee} & \tilde{V}^{eh} \\ \tilde{V}^{he} & \tilde{V}^{hh} \end{bmatrix}, \qquad \Pi = \begin{bmatrix} \Pi^{e} & 0 \\ 0 & \Pi^{h} \end{bmatrix}, \qquad V = \begin{bmatrix} V^{ee} & V^{eh} \\ V^{he} & V^{hh} \end{bmatrix}, \tag{3.19}$$

The screened carrier-carrier scattering matrix can then be expressed in matrix form:

$$\tilde{V} = [\,1 - V\cdot\Pi\,]^{-1}\,V \tag{3.20}$$

When only the lowest band for electrons and holes are occupied, $\tilde{V}$ can be straightforwardly written as

$$\tilde{V} = \frac{1}{\Delta}\begin{bmatrix} (1-V^{hh}\Pi^{h})V^{ee}+V^{eh}\Pi^{h}V^{he} & V^{eh} \\ V^{he} & (1-V^{ee}\Pi^{e})V^{hh}+V^{he}\Pi^{e}V^{eh} \end{bmatrix}, \tag{3.21}$$

with

$$\Delta = (1-V^{ee}\Pi^{e})\,(1-V^{hh}\Pi^{h}) - V^{eh}\Pi^{h}V^{he}\Pi^{e}\,. \tag{3.22}$$

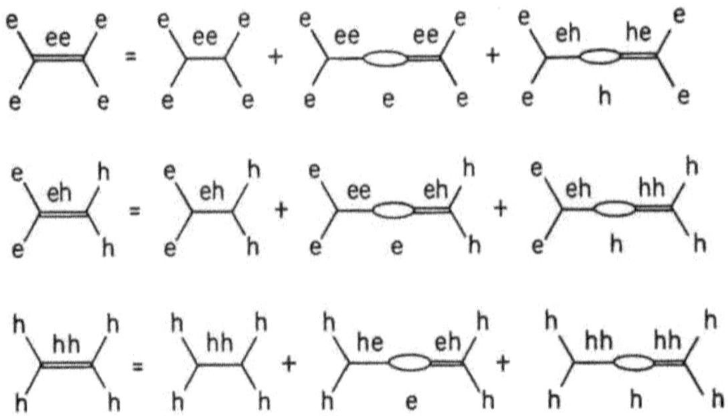

Fig. 10. Diagrams for screening effect of carrier-carrier potential in the RPA. The single and double horizontal solid lines represent, respectively, the unscreened and screened carrier-carrier interaction, $V^{\mu\nu}$ and $\tilde{V}^{\mu\nu}$ The bubble represents the density-density correlation function, Π^{μ}.

Solving a set of coupled equations, Eqs. (3.14) and (3.15) and corresponding equations for hot-phonons, we can determine the $\mathbf{v}_e$, $\mathbf{v}_h$, T_e, T_h as well as the distribution of hot-phonon wavepackets (LO, TO). The mobilities for both carriers in a weak electric field are shown in Fig. 11 as functions of lattice temperature, T_L. Fig. 12 shows the weak field mobility of electrons and holes as functions of the electron density, n_e.

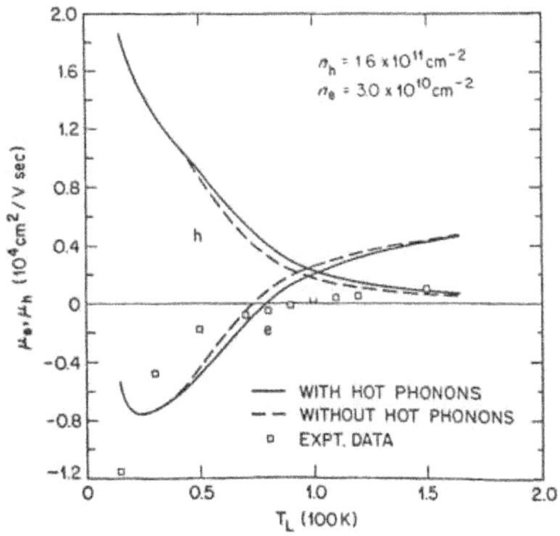

Fig. 11. Mobility of electrons, μ_e, and mobility of holes , μ_h, as functions of lattice temperature, T_L, in a weak electric field. Data for electron mobility come from Höpfel *et al*[37].

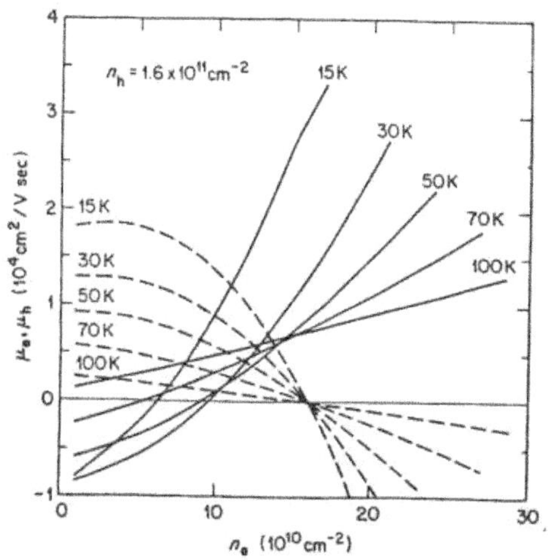

Fig. 12. Mobility of electrons, μ_e (solid curves), and mobility of holes, μ_h (dashed curves), as functions of density of electrons, n_e, at different lattice temperature, T_L (in a weak electric field).

4. OTHER THEORETICAL WORKS

4.1. Monte Carlo Simulation of Nonequilibrium Phonon effect

In the above discussion a simplifying assumption for the carrier distribution function, namely, that is a Fermi-Dirac function at the carrier temperature T, is assumed. This assumption usually is a good one for the high carrier density and for the time scale larger than pecosecond. However, recent progress in ultrafast experiments leads to processes in which the carrier distribution can not be regard as in quasi-equilibrium. Use of Monte Carlo techniques to solve the dynamical equation on the supercomputer is the way of obtaining the nonequilibrium carrier distribution. Lugli *et al* have presented a series theoretical results on nonequilibrium phonon effects based on a novel Monte Carlo algorithm.[20,40–42] They have calculated the results for both the bulk and quantum well cases. In the case of a quantum well, the effect of several subbands has been included. However, a simple model was used, that is, that the q_z, components of the phonon distribution are localized in a region of wave-vector space of extent $1/L$, with L the width of well.

The physics of the dynamical evolution of the carrier-phonon system is very similar to what we have discussed. The coupled Boltzmann equations are given by

$$\frac{\partial f_{\mathbf{k}}}{\partial t} = \left[\frac{\partial f_{\mathbf{k}}^{(i)}}{\partial t}\right]_{c-\text{ph}} + \left[\frac{\partial f_{\mathbf{k}}^{(i)}}{\partial t}\right]_{c-c} + \left[\frac{\partial f_{\mathbf{k}}^{(i)}}{\partial t}\right]_{c-\text{imp}}, \tag{4.1}$$

$$\frac{\partial N_{\mathbf{q}}^{(j)}}{\partial t} = \left[\frac{\partial N_{\mathbf{q}}^{(j)}}{\partial t}\right]_{\text{ph}-c} + \left[\frac{\partial N_{\mathbf{q}}^{(j)}}{\partial t}\right]_{\text{ph-ph}}, \tag{4.2}$$

where the superscripts i and j indicate, respectively, the type of carrier (electron or holes) and of phonon modes (LO, TO, . . .) considered. $(\partial N_{\mathbf{q}})/\partial t\ |_{\text{ph-ph}}$ is obtained by relaxation time approximation, given by Eq. (2.24). Fig. 13 shows the evolution of the electron total energy as a function of time in an n-type quantum well during and after the laser pulse. The excited electrons lose energy mainly through the interaction with the background electrons and through the emission of LO phonons. A much slower relaxation is found when nonequilibrium phonon effects are included. Fig. 14 shows the LO-phonon distribution at different times.

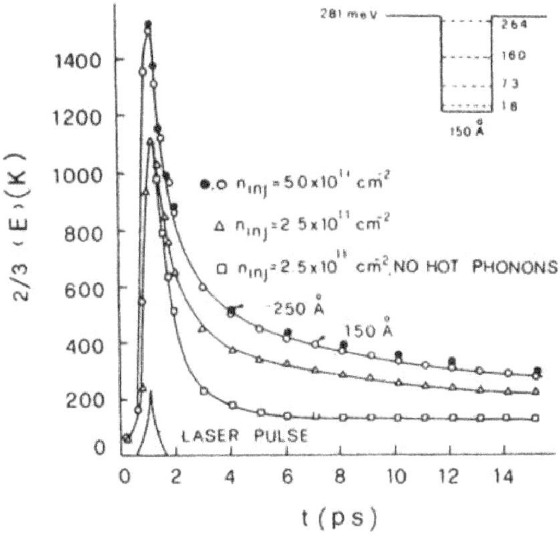

Fig. 13. Average electron energy, measured in equivalent temperature, as a function of time during and after the laser excitation for two different excitation sheet densities. The position of the energy levels in the well is shown in the insert. This figure is Fig. 12 from Lugli *et al.*[42]

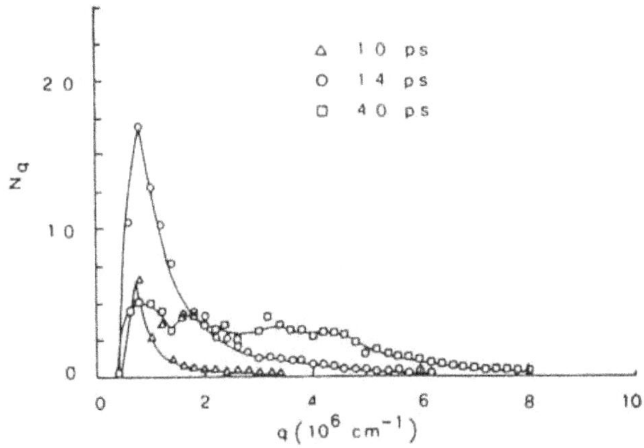

Fig. 14. LO-phonon distribution as a function of total parallel momentum (for $q_z = 0$) for times during and after the laser excitation. This figure is Fig. 13 from Lugli *et al.*[42]

When the laser photon energy is about 2 eV, photon-excitation in bulk GaAs includes transition from three hole bands. The energy distribution of photo-excited electrons is characterized by three distinct peaks, as shown in the insert of Fig. 15. In GaAs, about 60% of the photoexcited carriers transfer to the satellite valleys during the laser pulse (the average time for $\Gamma-L$ transition via phonon emission or absorption is about 80 fs). Carriers return slowly to the Γ valley, with characteristic time of 2 or 3 ps, because of smaller effective mass in the Γ valley. This leads a slow rise of luminescence in GaAs, as shown in Fig. 15.[42] In Fig. 15, a comparison to the InP case is also provided, where the L valley are located at a much high energy, and do not significantly contribute to the cooling process. By fitting data with the Monte Carlo calculation, it is determined that the Γ–L deformation potential $D_{\Gamma-L}$, is $(6.5 \pm 1.5)\times 10^8$ eV/cm. This Γ–L exchange depletes the Γ electron distribution in GaAs in the high-energy region, above 0.3 eV, as shown in Fig. 16.

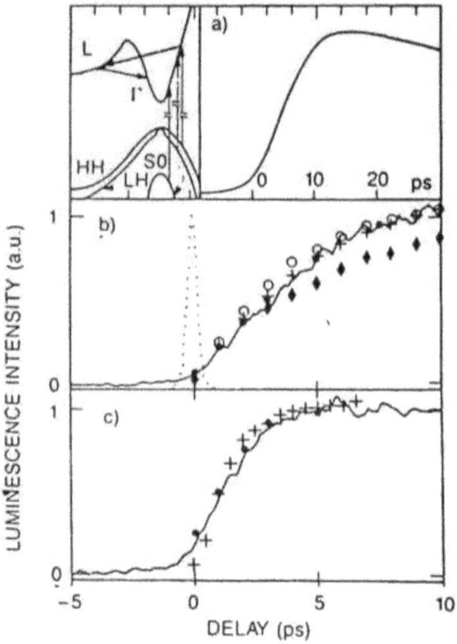

Fig. 15. Luminescence intensity at 300 K vs delay. (a) GaAs at 1.45 eV; (experiments) (b) the solid curve is the same as (a) on expanded scale; the filled circles show the spectrally integrated intensity; (c) InP, same as (b); Results of ensemble Monte Carlo calculations are also shown: (b) $D_{\Gamma-L} = (4, 6, \text{and } 8)\times 10^8$ eV/cm for GaAs (lozenges, crosses, and open circles, respectively) and (c) InP. This figure is Fig. 2 from Shah *et al.*[41]

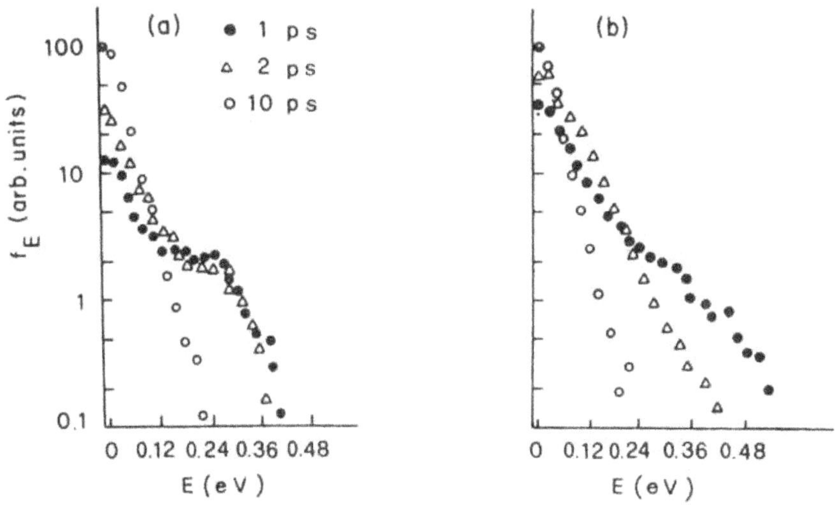

Fig. 16. The Monte Carlo results of the Energy distribution function for Γ-valley electrons in (a) GaAs and (b) InP at three different time delays after the excitation. The excitation density is $5\times10^{16}\,\text{cm}^{-1}$ and the lattice temperature is 300 K. This figure is Fig. 16 from Lugli *et al.*[42]

4.2. Effect of LO-phonon renormalization

A series of works by Das Sarma and his coworkers[43–46] studied the hot-electron relaxation by emission of LO phonons in both 2D and 3D cases. In addition to the hot-phonon effect and dynamical screening, they found that many-body renormalization of the LO phonons also plays a crucial role in the power-loss process at low electron temperatures. Due to phonon-plasmon coupling, the density states of an LO phonon has three branches: (1) the bare-phonon-like branch near the bare-phonon energy, (2) the plasmonlike branch near the plasmon energy, and (3) the low-energy quasiparticle-excitation-like branch in the quasiparticle-excitation region. Even though the oscillator strength of plasmonlike and the quasiparticle-excitation-like branchs are extremely small, they dominate the power-loss process at low enough electron temperatures that most electrons have energy below the threshold for bare LO phonon emission. This produces an enhancement of the power loss at low electron temperatures by many orders of magnitude relative to the power loss to bare LO phonons.

The phonon density of states, $A(q, \omega)$ is given by

$$A(q,\omega)=-\pi^{-1}\,\mathrm{Im}\,D(q,\omega)\_ \tag{4.3}$$

where D is the phonon propagator. For bare phonons, D is replaced by D^0,

$$D^0(q,\omega) = \frac{2\omega_L}{\omega^2 - \omega_L^2}, \tag{4.4}$$

and $\mathrm{Im}\, D^0(q,\omega) = -\pi[\delta(\omega-\omega_L) - \delta(\omega+\omega_L)]$, which has vanishing weight everywhere except at the bare energy. By introducing the phonon-plasmon coupling in the random phase approximation (RPA), the renormalized phonon propagator is obtained by

$$D(q,\omega) = \frac{2\omega_L}{\omega^2 - \omega_L^2 - 2\omega_L M_q^2 \Pi(q,\omega)}, \tag{4.5}$$

where M_q is the electron-phonon matrix element, and $\Pi(q,\omega)$ is the electron density-density correlation function. Fig. 17 displays the renormalized phonon spectral function for the 3D and 2D cases. Three branches of phonon modes are shown. Notice that the scale factor γ in Fig. 17 is different for different curves. The strength of the plasmonlike phonon is very small, the quasiparticle-excitation-like (QPE-like) mode is also weak. However, It is found that the QPE-like phonons dominate the hot-electron power-loss process at low electron temperatures.

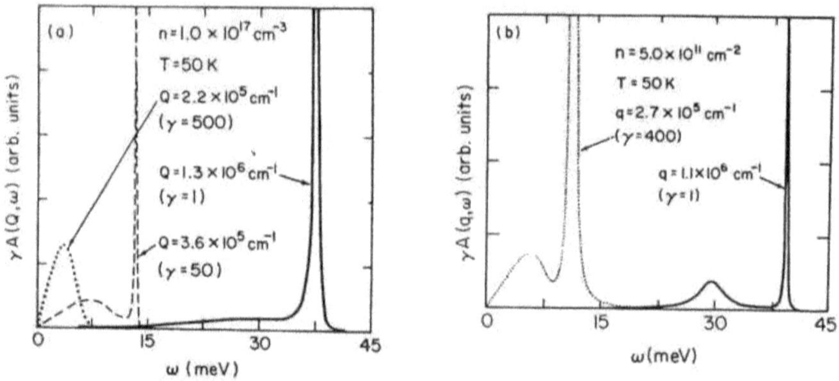

Fig. 17. (a) Phonon spectral function for three wave vectors in a 3D electron gas at an electron temperature of $T=50\,\mathrm{K}$ ($T_L=0$). For $Q=2.2\times10^5\,\mathrm{cm}^{-1}$, the QPE branch extends from 0 to 7 meV, and there are δ-function peaks near the plasmon energy ($\omega_p \approx 14$ meV) and the phonon energy (36.8 meV), which are not shown explicitly in order to avoid too much detail. For $Q=3.6\times10^5\,\mathrm{cm}^{-1}$ the QPE branch extends to ≈ 14 meV, and there is again a δ-function peak (not shown) at ω_L. For $Q=1.3\times10^6\,\mathrm{cm}^{-1}$ the plasmonlike phonon is Landau damped, and the QPE-like phonons extend all way up to ω_L. The electron density is $10^{17}\,\mathrm{cm}^{-3}$, giving a Fermi vector of $1.4\times10^6\,\mathrm{cm}^{-1}$ (b) Phonon spectral function for two different wave vector in a 2D electron gas. For $q=62.7\times10^5\,\mathrm{cm}^{-1}$ there is a δ-function peak (not shown) at ω_L. The electron density is $5\times10^{11}\,\mathrm{cm}^{-2}$ giving a Fermi vector of $1.8\times10^6\,\mathrm{cm}^{-1}$ This Figure is Fig.4 from Das Sarma.[46]

At low temperature (about 50 K) the number of electrons which can emit phonon with bare-mode phonons becomes extremely small. However, phonon-plasmon coupling creates new low-energy modes, even though its strength is rather small, they could make important contribution in energy loss at low electron temperature. In this case, Eq. (2.32) for the energy loss rate is no longer valid since it was derived by considering only bare LO phonon modes. Instead, Das Sarma *at al* derived the following expression for power loss by generalizing the Kogan formula[47] to including phonon self-energy correction:

$$P=\sum_q \int \frac{d\omega}{\pi}\omega M_q^2\,[n_{T_L(\omega)}-n_T(\omega)]\,\mathrm{Im}\Pi(q,\omega)\mathrm{Im}D(q,\omega)\ , \tag{4.6}$$

where $n_T=[\exp(\omega/k_BT)-1]^{-1}$

In order to include the hot-phonon effect, one must first rewrite Eq. (4.6) as

$$P=\sum_q \int d\omega\,\omega R(q,\omega)\,n_T(\omega)\ , \tag{4.7}$$

then replace $R(q,\omega)$ in the integrand by $R(q,\omega)/[1+\tau_{op}R(q,\omega)]$. Das Sarma *et al* suggest that only the bare phonon mode is important for hot-phonon effects.

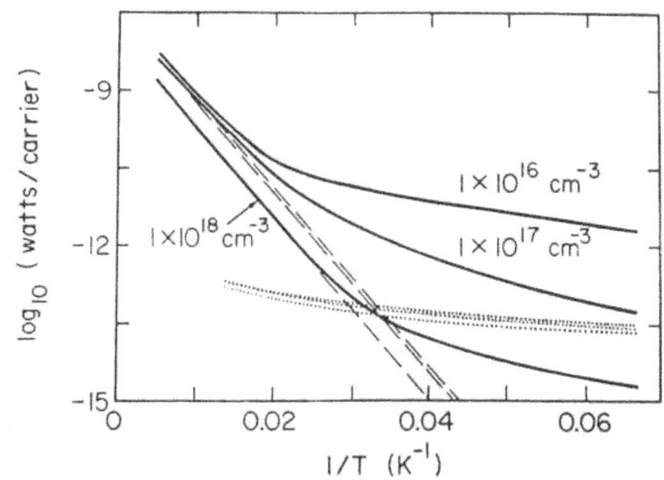

Fig. 18. Power loss as a function of the inverse electron temperature for 3D electron gas (solid lines). the dashed lines correspond to power loss to bare LO phonons only. The doted lines give the power loss to acoustic-phonon modes. For dotted curves, the uppermost curve is for $10^{18}\,\mathrm{cm}^{-3}$, the middle one for $10^{16}\,\mathrm{cm}^{-3}$, and the lowest one for $10^{17}\,\mathrm{cm}^{-3}$. $T_L=0$, and the power loss is expressed in watts per carrier. This figure is Fig. 5 from Das Sarma *et al.*[46]

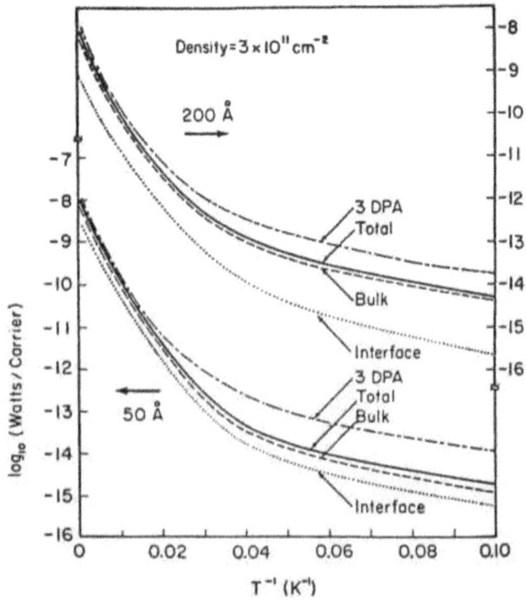

Fig. 19. Power loss as function of inverse electron temperature for two quantum wells (width 50 and 200Å) calculated in 3D phonon approximation (3DPA) and in slab-model. $T_L=0$, and the power loss is expressed in watts per carrier. This figure is Fig. 1 from Das Sarma *et al.*[45]

At high temperatures ($T \geq 50$ K) the bare LO phonons dominate the power loss, at intermediate temperatures ($20 < T \leq 50$ K) power is lost predominantly to the plasmonlike or QPE-like branch of the LO-phonon spectrum, and at still lower temperatures ($T \leq 20$ K) the acoustic phonons become important. Fig. 18 shows the power loss for 3D electron gas at lattice temperature $T_L=0$. Fig. 19 shows the results in quantum wells.

Recently, M. W. C. Dharma-wardana[48] deals with the same subject using nonequilibrium Green's function approach. He considers not only the coupled-mode spectral density, but also the coupled-distribution functions. In contrast to Das Sarma and co-workers, he found that the energy-loss rate is significantly suppressed by coupled-mode formation and quasiparticlelike modes do not contribute to the energy-loss rate. The underlying physical picture is that although the coupled-mode phonon spectral density has some weight in the quasiparticlelike region of energies, these coupled-phonon modes are effectively at electron temperature, not at lattice temperature. Hence there is no enhancement. Nevertheless, coupled-mode phonons are hotter than bare phonons and this leads to a suppression of energy-lose rate.

5. SUMMARY

We have reviewed in this paper the effect of reabsorption of nonequilibrium optical phonons on the carrier relaxation and transport. Both experiments and theoretical study confirm that the build-up of hot-phonons is mainly responsible for reducing the cooling rate of photo-excited electrons, while the dimensionality and dynamical screening effect play only a minor role. In the case of heterojunctions or quantum wells, our approach can determine the accurate shape of the phonon wave packet and produce a simple procedure for calculating the effect of nonequilibrium phonons. Our results are in quantitative agreement with experiments. On the other hand, the cooling of photo-excited carriers is a rich field, which may involve the electron-hole (three kind of holes) interaction, multi-valleys and multi-subband transitions, multi-scattering processes, from very high temperature down to very low-temperature. As examples, we refer to the effects of Γ–L transitions, which delay the build-up of hot Γ electrons in a ultrafast process, and what is the role of phonon-plasmon modes. There are many other factors, which may affect the relaxation process, such as phonon modes (interface and slab modes in quantum wells)[45] and LO phonon relaxation mechanism.[41] This subject is worthy of further study.

ACKNOWLEDGMENT

The work in the City College of City University of New York was partly supported by the U. S. Army Research Office and the U. S. Department of Energy.

Appendix

In this appendix the complete expressions for Eqs. (2.13) and (2.14) are given as follows:

$$\left[\frac{\partial f_{n\mathbf{k}}}{\partial t}\right]_{ep} = \frac{2\alpha^2}{\hbar A} 2\pi \sum_{\mathbf{q}}\sum_{q_z}\sum_{q_z'}\sum_{n'\mathbf{k}'} G^*_{nn'}(\mathbf{q}, q_z) G_{nn'}(\mathbf{q}, q_z')$$

$$(\{ f_{n'\mathbf{k}'}(t)[1-f_{n\mathbf{k}}(t)]\delta_{\mathbf{k}',\mathbf{k}+\mathbf{q}}\delta(E_{n'\mathbf{k}'}-E_{n\mathbf{k}}-\hbar\omega_L)$$

$$-f_{n\mathbf{k}}(t)[1-f_{n'\mathbf{k}'}(t)]\delta_{\mathbf{k}',\mathbf{k}-\mathbf{q}}\delta(E_{n\mathbf{k}}-E_{n'\mathbf{k}'}-\hbar\omega_L)\}[\delta_{q_z q_z'}+n_{\mathbf{q}}(q_z, q_z', t)]$$

$$+\{f_{n'\mathbf{k}'}(t)[1-f_{n\mathbf{k}}(t)]\delta_{\mathbf{k}',\mathbf{k}-\mathbf{q}}\delta(E_{n'\mathbf{k}'}-E_{n\mathbf{k}}+\hbar\omega_L)$$

$$-f_{n\mathbf{k}}(t)[1-f_{n'\mathbf{k}'}(t)]\delta_{\mathbf{k}',\mathbf{k}+\mathbf{q}}\delta(E_{n\mathbf{k}}-E_{n'\mathbf{k}'}+\hbar\omega_L)\}n_{\mathbf{q}}(q_z,q_z',t)) . \quad (A.1)$$

$$\left[\frac{\partial n_{\mathbf{q}}(q_z,q_z',t)}{\partial t}\right]_{ep} = \frac{2\alpha^2}{\hbar A}\lim_{\varepsilon\to 0^+}\sum_{n\mathbf{k}}\sum_{n'\mathbf{k}'}\sum_{q_z''} f_{n\mathbf{k}}(t)\,[1-f_{n'\mathbf{k}'}(t)]$$

$$\times\left[\delta_{\mathbf{k}',\mathbf{k}-\mathbf{q}}\frac{G_{n'n}(\mathbf{q},q_z)\,G^*_{n'n}(\mathbf{q},q_z'')}{\varepsilon-i[E_{n\mathbf{k}}-E_{n'\mathbf{k}'}-\hbar\omega_L]}[\delta_{q_z'',q_z'}+n_{\mathbf{q}}(q_z'',q_z',t)]\right.$$

$$+\delta_{\mathbf{k}',\mathbf{k}-\mathbf{q}}\frac{G^*_{n'n}(\mathbf{q},q_z')\,G_{n'n}(\mathbf{q},q_z'')}{\varepsilon+i[E_{n\mathbf{k}}-E_{n'\mathbf{k}'}-\hbar\omega_L]}[\delta_{q_z'',q_z}+n_{\mathbf{q}}(q_z,q_z'',t)]$$

$$-\delta_{\mathbf{k}',\mathbf{k}+\mathbf{q}}\frac{G_{n'n}(\mathbf{q},q_z)\,G^*_{n'n}(\mathbf{q},q_z'')}{\varepsilon+i[E_{n\mathbf{k}}-E_{n'\mathbf{k}'}+\hbar\omega_L]}n_{\mathbf{q}}(q_z'',q_z',t)]$$

$$\left.-\delta_{\mathbf{k}',\mathbf{k}+\mathbf{q}}\frac{G^*_{n'n}(\mathbf{q},q_z')\,G_{n'n}(\mathbf{q},q_z'')}{\varepsilon-i[E_{n\mathbf{k}}-E_{n'\mathbf{k}'}+\hbar\omega_L]}n_{\mathbf{q}}(q_z,q_z'',t)\right] . \quad (A.2)$$

[1]T. Ando, A. B. Fowler, and F. Stern, Rev. Mod. Phys. 54, 437 (1982)

[2]J. Shah, A. Pinczuk, A. C. Gossard, and W. Wiegmann, Phys. Rev. Lett. 54, 2045 (1985)

[3]J. Shah, A. Pinczuk, H. L. Störmer, A. C. Gossard, and W. Wiegmann, Appl. Phys. Lett. 44, 322 (1984)

[4]J. F. Ryan, R. A. Taylor, A. J. Turberfeld, A. Maciel, J. M. Worlock, A. C. Gossard and W. Wiegmann, Phys. Rev. Lett. 53, 1841, (1984).

[5]D. K. Ferry, Surf. Sci. 75, 86 (1978)

[6]K. Hess, Appl. Phys. Lett. 35, 484 (1979)

[7]B. K. Ridley, J. Phys. C 15, 5899 (1982); 16, 6971 (1983)

[8]C. H. Yang, J. M. Carlson-Swindle, S. A. Lyon, and J. A. Worlock, Phys. Rev. Lett 55, 2359 (1985).

[9]W. W. Rühle and H. J. Polland, Phys. Rev. B 36, 1683 (1987).

[10]P. Price, Ann. Phys. (NY) 313, 217, (1981), Surf. Sci. 31B, 199, (1985)

[11]W. Pötz and P. Kocevar, Phys. Rev. B 28, 7040 (1983).

[12]W. Cai, M. C. Marchetti, and M. Lax, Phys. Rev. B 334, 8573 (1986, Dec. 15)

[13]W. Cai, M. C. Marchetti, and M. Lax, Phys. Rev. B 35, 1369 (1987, Jan. 15)

[14]R. Zwanzig, J. Chem. Phys. 33, 1338 (1960)

[15]R. Zwanzig, Phys. Rev. 124, 983 (1961)

[16]D. N. Zubarev, Nonequilibrium Statistical Thermodynamics (Consultant's Bureau, NY, 1974)

[17]S. Peletminskii and A. Yatsenko, Zh. Eksp. Teor. Fiz. 53, 1327 (1967) Sov. Phys. JETP 26, 773 (1968)

[18]See for instance: M. Born and K. Huang, *Dynamical Theory of Crystal Lattices* (Oxford University, Oxford, 1985), pp. 82-100.

[19]D. Von der Linde, J. Kuhl, and H. Klingenberg, Phys. Rev. Lett. 44, 1505 (1980).

[20]P. Lugli and S. M. Goodnick, Phys. Rev. Lett. 59, 716 (1987)

[21]M. A. Osman and D. K. Ferry, Phys. Rev. B 36, 6018 (1987)

[22]D. W. Bailey, M. A. Artaki, C. J. Stanton, and K. Hess, Appl. Phys. Lett. 62, 4638 (1987)

[23]S. M. Goodnick and P. Lugli, Phys. Rev. B 38, 10135 (1988)

[24]T. F. Zheng, W. Cai, P. Hu, and M. Lax, Phys. Rev. B 40, 1271 (1989)

[25]M. Lax, W. Cai, and M. C. Marchetti, "Nonequilibrium Phonon Dynamics in a Semiconductor Quantum Well", Proceeding of ultrafast laser probe phenomena in bulk and microstructure semiconductors II SPIE symposium, 208, (1988).

[26]M. C. Marchetti and W. Cai, Phys. Rev. B 35, 7725 (1987).

[27]J. Shah, A. Pinczuk, A. C. Gossard, W. Weigmann, and K. Kash, Surf. Sci. 174, 363 (1986)

[28]X. L. Lei and N. J. M. Horing, Phys. Rev B 35, 6281 (1987).

[29]W. Cai, M. C. Marchetti and M. Lax, Phys. Rev. B 37, 2636 (1988).

[30]J. R. Barker, J. Physics C 6, 1663 (1973)

[31]S.K. Sarker, Phys. Rev. B 33, 7263 (1986)

[32]M. Keever, W. Kopp, T. J. Drummond, H. Morko, and K. Hess, Jpn. J. Appl. Phys. 21, 1489 (1982)

[33]W. Pötz, Phys. Rev. B 36, 5016 (1987).

[34]M. C. Marchetti and Pötz, J. Vac. Sci. Techonol. B 6, 1341 (1988); Phys. Rev. B 40, 12391 (1989).

[35]K. Leo, W. W. Rühle, H. J. Queisser, and K. Ploog, Phys. Rev. B 37, 7121 (1988).

[36]R. A. Höpfel, J. Shah, and A. C. Gossard, Phys. Rev. Lett. 56, 765 (1986).

[37]R. A. Höpfel, J. Shah, P. A. Wolff, and A. C. Gossard, Phys. Rev. Lett. 56, 2736 (1986).

[38]H. L. Cui, X. L. Lei, and N. J. M. Horing, Phys. Rev. B 37, 8223 (1988).

[39]W. Cai, T. F. Zheng, and M. Lax, Phys. Rev. B 37, 8205 (1988).

[40]P. Lugli, C. Jacoboni, L. Reggiani, and P. Kocevar, Appl. Phys. Lett. 50, 1251 (1987).

[41]J. Shah, B. Deveaud, J. C. Damen, W. T. Tsang, A. C. Gossard, and P. Lugli, Phys. Rev. Lett. 59, 2222 (1987).

[42]P. Lugli, P. Bordone, L. Reggiani, M. Rieger, P. Kocevar, and S. M. Goodnick, Phys. Rev. B 39, 7852 (1989).

[43]Das Sarma, J. K. Jain, and R. Jalabert, Phys. Rev. B 37, 1228 (1988); 37, 4560 (1988); 37, 6290 (1988).

[44]K. Jain, R. Jalabert, and S. Das Sarma, phys. Rev. Lett. 60, 353 (1988); 61, 2005(E) (1988).

[45]K. Jain and S. Das Sarma, Phys. Rev. Lett. 62, 2305 (1989).

[46]Das Sarma, J. K. Jain, and R. Jalabert, Phys. Rev. B 41, 3561 (1990).

[47]Sh. M. Kogan, Sov. Phys.--Solid State 4, 1813 (1963).

[48]M. W. C. Dharma-wardana, Phys. Rev. Lett. 66, 197 (1991).

International Journal of Modern Physics B, Vol. 6, No. 7 (1992) 1007–1036

NONLINEAR TRANSPORT OF ELECTRONS UNDER A STRONG HIGH FREQUENCY ELECTRIC FIELD IN SEMICONDUCTORS

W. Cai

Department of Physics, City College of the City University of New York, New York, New York 10031

M. Lax

Department of Physics and the Graduate Center of the City University of New York, New York, New York 10031, and AT&T Bell Laboratories, Murray Hill, New Jersey 07974.

The nonlinear transport of electrons in semiconductors is studied when a strong high frequency (HF) electric field is applied together with a direct current (dc) electric field. A set of equations is developed, from which the amplitude and phase of each harmonic component of the electron drift velocity and the electron temperature can be computed. In the weak HF field limit our approach reduces to the well-known memory function method. The DC conductivity decreases and definitely becomes negative with increase of the first and second harmonic components of the applied HF field. Comparison is made with experiment.

1. INTRODUCTION

The electron transport under a strong high frequency (HF) electric field is an important issue in semiconductor and other realms of physics.[1] Microwave devices often work in the region of a strong HF field. The free-carrier absorption of laser light is also related to the transport of electrons in a strong HF (possibly optical) field. There are many other applications of strong micro-waves. However, it seems that the theoretical study on this subject is still at a primitive stage.

The linear transport of electrons when a weak high frequency electric field is applied, was studied in 1960s.[2,3] It has become since then a well-understood problem. The Kubo formula or the memory function approach are typical methods to deal with this problem.[2,3] Recently, the approach of Kubo formula was applied to the carrier transport of two-dimensional electrons in a superlattice or in a quantum well.[4] The balance equation approach was also developed to deal with the HF linear response, with results the same as from the memory function approach.[5] There are also several works that study the response to a strong dc field (hot electrons) together with a weak HF field.[6–9]. We will discuss the subject of linear HF conductivity in Sec. 3.

When a strong HF electric field is applied to the carrier system, non-linear effects are produced. Namely, when a strong electric field with frequency ω is applied, the current, in addition to frequency ω, has the harmonic components 2ω, 3ω, . . ., and a dc component. Therefore, the problem becomes difficult to work out.

If a strong HF field is applied, the trajectory of electron during the collisional process is no longer straight, as shown in Fig. 1. As well known, when a strong dc electric field is applied on the carrier system, there is the so called "intracollisional effect", namely, the electron accelerates during the collisional process because of the strong electric field.[12,13] Including this effect requires transport theory beyond the usual Boltzmann approach: the motion of electrons after time t, depends not only on the the condition at time t, but also on the drift motion of electrons during a period before t (a memory effect). The intracollisional effect of a strong HF field, described by Fig. 1, is more complex. The electric force not only changes the speed of the drift motion of electron, but also changes the direction of the drift motion, so the electron oscillates during the collisional process. This complicated oscillation of the drift motion, which is superposed on the random motion of electron caused by the scattering processes, can not be completely known before the transport problem is solved.

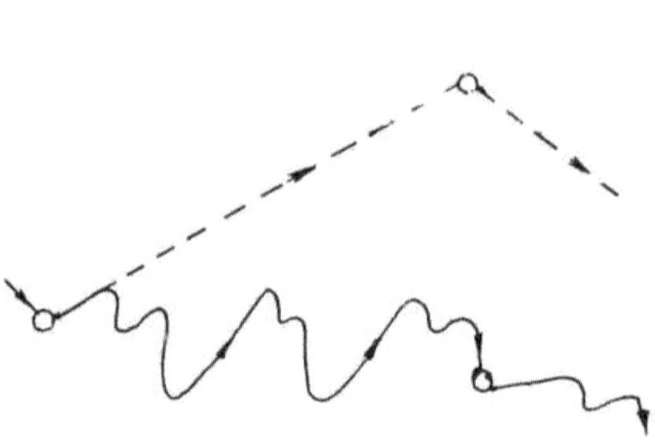

Fig. 1. The trajectory of a electron during the collisional process under a strong HF field. The solid line is for the actual case, the dashed line is an approximation by the Boltzmann approach.

1.1. Previous theories

Most previous theoretical studies of this subject are based on phenomenological models.[1] Only a few papers deal with it using a dynamic theory.[8,9] These calculations rely on Monte Carlo simulation to solve the time dependent Boltzmann equation. Lebwohl[8] in 1972 extended the Monte Carlo simulation to calculation of the periodic response of the carriers to the periodic external field, in both linear and nonlinear response regime. A brief sketch of the approach is described as follows. When an electric field is applied:

$$\mathbf{E} = [\,\mathbf{E}_0 + \mathbf{E}_1 \sin(\omega t)]\ , \tag{1.1}$$

the change in the wave vector of the electron between successive scatterings at times Δt is given by

$$\mathbf{k}(t+\Delta t) = \mathbf{k}(t) + (e\Delta t/\hbar m)\mathbf{E}_0 + (e\omega^{-1}/\hbar m)\{\cos[\omega(t+\Delta t)] - \cos(\omega t)\}\mathbf{E}_1\ , \tag{1.2}$$

with m the effective mass of electrons. Here the dc and HF intracollisional effect have been included to the lowest order. All transport coefficients can be calculated by a time average over the particular trajectory in momentum space, which, as in stationary calculations, is identical to averaging them over the distribution. For a periodic applied electric field, this average is over the periodic distribution function and gives a periodic response. In particular, the average velocity can be written as

$$\langle \mathbf{v}(t)\rangle = \mathbf{v}_0 + \sum_n [\mathbf{v}_1^{(n)} \sin(n\omega t) + \mathbf{v}_2^{(n)} \cos(n\omega t)] \tag{1.3}$$

The harmonic components can be obtained by

$$\mathbf{v}_0 = t_N^{-1} \sum_{i=1}^{N} \left(\frac{\hbar}{m(i)}\right) \int_0^{S_i} \mathbf{k}(t_{i-1}+t)\, dt\ , \tag{1.4a}$$

$$\mathbf{v}_1 = 2t_N^{-1} \sum_{i=1}^{N} \left(\frac{\hbar}{m(i)}\right) \int_0^{S_i} \mathbf{k}(t_{i-1}+t) \sin[n\omega(t_{i-1}+t)]\, dt\ , \tag{1.4b}$$

$$\mathbf{v}_2 = 2t_N^{-1} \sum_{i=1}^{N} \left(\frac{\hbar}{m(i)}\right) \int_0^{S_i} \mathbf{k}(t_{i-1}+t) \cos[n\omega(t_{i-1}+t)]\, dt\ , \tag{1.4c}$$

$t_i = \sum_{j=1}^{i} S_j$ is the time at which the ith scattering occurs, S_i is the time interval between the $(i-1)$st and ith scattering, $m(i)$ is the effective mass of the particle during this time interval, and $\mathbf{k}(t)$ is the wave vector at time t. Substituting Eq. (1.2) into Eq. (1.4) the expressions for $\mathbf{v}_0$, $\mathbf{v}_1^{(n)}$, and $\mathbf{v}_2^{(n)}$ can be numerically obtained.

The determination of the frequency response is possible because the time of an event in the simulated history can be interpreted as the true physical time of that event. This interpretation allows one to calculate the periodic response, in time, of the simulated system. To calculate the periodic response, all events that occur at the same phase of the electric field, i. e., all events that occur at times differing from one another by an integral number of periods, T, of the HF field, can be treated as if they occur simultaneously.

The net effect of this operation is to obtain an ensemble of concurrent trajectories which can be averaged over in order to derive the value of some transport property at a given time. The actual computation of the response is implemented by dividing one period of the sinusoidal electric field into N equal intervals, each of duration $\Delta t = T/N$, and then dividing the simulated trajectory of the particle in momentum space and time into segments of duration Δt such that each segment can be uniquely assigned to one of the N time intervals. The particular interval to which any segment is assigned is determined by the phase of the electric field during that segment. The response is calculated at time $(i-0.5)\Delta t$, with $i = 1, 2, \ldots, N$ by averaging over all particle trajectory segments in each interval. The periodicity of the response guarantees that this procedure, which determines the response for one complete period, determines the response for all time.

The response of GaAs to an 8 kV/cm HF field of $\omega = 1.88\times10^{11}\,\mathrm{sec}^{-1}$ superimposed on a dc field of 10 kV/cm is calculated using Monte Carlo simulation. Fig. 2(a) is a plot of the drift velocity of the carriers in the low valley L, the upper valley U, and both valleys combined, T. The total drift velocity, as well as the drift velocity of carriers in the lower valley, contain significant HF components at higher harmonics of fundamental frequency. Fig. 2(b) is a plot of the total drift velocity, of the separate contributions of the dc drift velocity and the first three harmonics, and the sum of the latter separate contributions. Except at the largest values of the drift velocity, three harmonics plus the dc component appear to be sufficient to fit the velocity curve.

Another calculations was done by Pozhela and Reklaitis[9] in 1978 for InSb at frequency up to $f = 134$GHz, with the amplitude of the HF electric field up to 700 V/cm. (see Fig. 3)

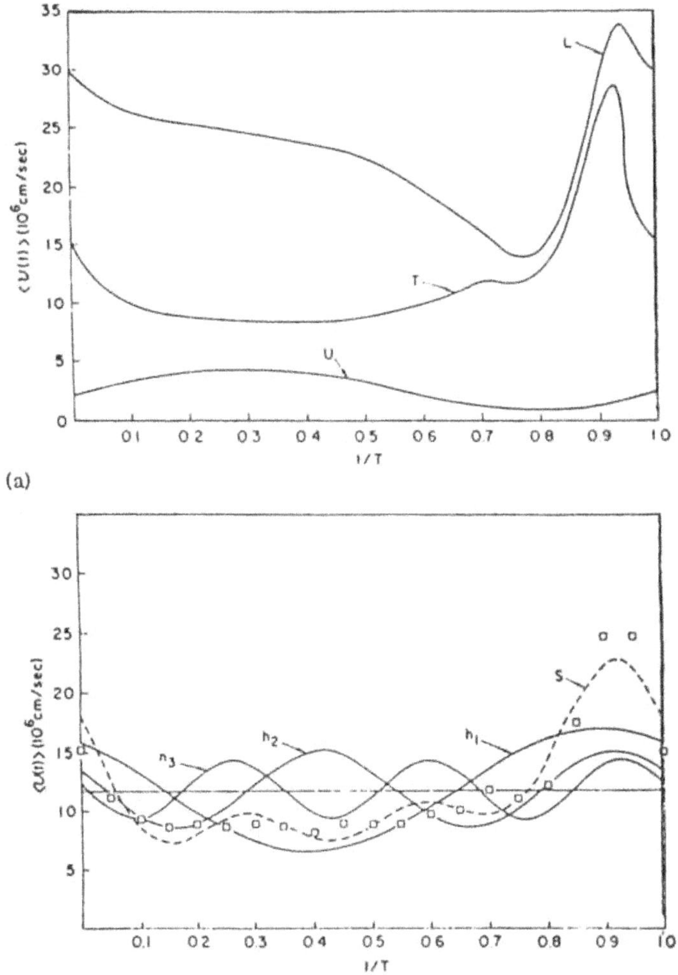

Fig. 2. The time-dependent velocity of GaAs at $T = 300$ K, $E_0 = 10$ kV/cm, $E_1 = 8$ kV/cm, and $\omega = 1.88 \times 10^{11}$ sec^{-1}. (a) is the separate contribution of the upper, U, and lower, L, valleys as well as the total, T, velocity. (b) shows the dc velocity and first three harmonics, as well as the sum of these; the computed velocity is plotted as □. This figure is Fig. 9 from Lebwohl[8]

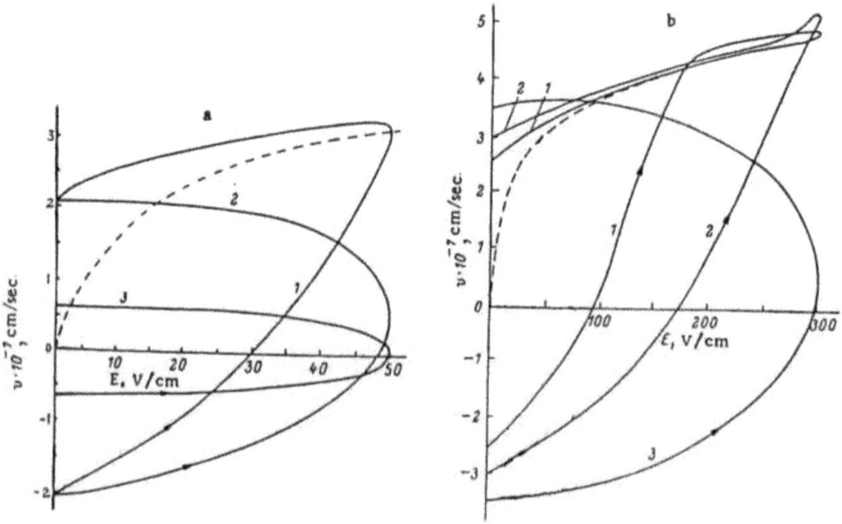

Fig. 3. Dependence of the drift velocity of electrons on the value of the microwave field during the positive half-period plotted for different amplitudes: a) 50 V/cm; b) 300 V/cm. Frequency f (GHz): 1) 10; 2) 35; 3) 134. The dashed curves are the field dependences of the static drift velocity. Impurity scattering is ignored. This figure is Fig. 2 from Pozhela[9].

1.2. Experiments

Experiments have also shown some interesting effects; for example, the direct current conductivity decreases with increase of the strength of HF field.[10,11] However, there is a difference between the results of different experimental groups. Ref. 10 reported that dc current in a n-GaAs sample placed in a strong microwave field decreases dramatically with increase of HF field, and dc conductivity definitely becomes negative (see Fig. 4). The results in Ref. 11 show a much slower decrease of dc conductivity with increasing HF field, declining only 40% from its static value even when HF electric field increases up to 3 kV/cm.

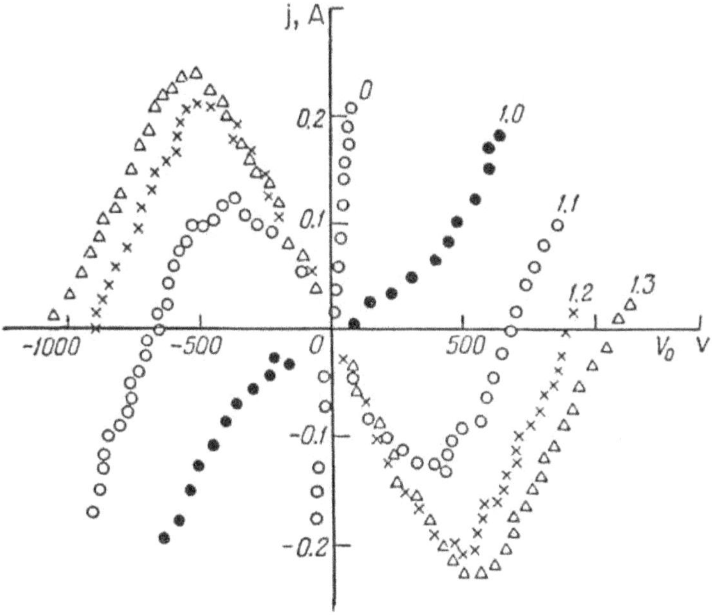

Fig. 4. The dependence of the direct current on the constant voltage applied to the sample for different values of the strong electric HF field strength with frequency 10 GHz measured experimentally (the number near the curves are in units E_c) in n-GaAs ($n_0 = 2.56\times10^{15}$ cm^{-3} $\mu = 6850\,\mathrm{cm^2/V\cdot sec}$). This figure is Fig. 6.10 from Pozhela[1].

We have proposed a new analytical approach to this issue.[14] The above mentioned memory effect is included in our approach. The dynamical equation we used, therefore, is different from Boltzmann equation. (If the memory effect for the drift motion is neglected, our formula reduce to the Boltzmann approach). Using this approach we study the HF "steady" state transport. We derive a set of evolution equations for drift velocity of electrons. Assuming the distribution function of electrons in the relative coordinates is a Maxwellian, we also obtain a set of evolution equations for energy of electrons, which is related to the electron temperature. Solving the group of equations, we can obtain dc component and each harmonic components of electron drift velocity and the electron temperature.

We review in this paper the formulation and the main results on the non-linear transport of electrons in GaAs semiconductors. The paper is organized as follows. In Sec. 2 we sketch the derivation of the kinetic equations for the harmonic components of electron drift velocity and the electron temperature. In Sec. 3 we briefly discuss the case of a weak HF electric field. In Sec. 4 we present the results of the numerical solution for nonlinear HF conductivity in a n-GaAs bulk sample. Sec. 5 is devoted to discussion. Finally, in Appendix we use the functional integral approach to derive formulas, that prove an important statement in the text [Eq. (2.12b)].

2. TRANSPORT EQUATIONS UNDER A STRONG HF ELECTRIC FIELD

2.1. Hamiltonian

For convenience in the theoretical description, we study a three-dimensional electron system. We consider N electrons, scattered by n_i randomly distributed impurities and phonons. An spatially uniform HF electric field $\mathbf{E}(t)$ is applied on this system. The following form of the applied electric field $\mathbf{E}(t)$ is assumed:

$$\mathbf{E}(t) = \mathbf{E}_0 + \sum_n \mathbf{E}_n e^{in\omega t}\,, \qquad (n = \pm 1, \pm 2, \ldots\ ) \tag{2.1}$$

here we have $\mathbf{E}_{-n} = \mathbf{E}_n^*$, with $\mathbf{E}^*$ the complex conjugate of $\mathbf{E}$, because $\mathbf{E}(t)$ is a real quantity. In order to describe the dynamical properties in the presence of a spatially uniform electric field, it is convenient to separate the center-of-mass motion from the relative motion of the electrons.[15] The Hamiltonian of the free electron system can then be written as

$$\hat{H}_e = \hat{H}_c + \hat{H}_e' + \hat{V}_{ee}'\,, \tag{2.2}$$

where

$$\hat{H}_c = \hat{\mathbf{P}}^2/2M - Ne\mathbf{E}(t)\cdot\hat{\mathbf{R}} \tag{2.3}$$

is the center-of-mass Hamiltonian, with $\hat{\mathbf{P}}$ and $\hat{\mathbf{R}}$, respectively, the center-of-mass momentum and coordinate operators; N the total number of electrons, and $M = Nm$, with m the electron effective mass. The free-electron Hamiltonian in the relative coordinates is given by

$$\hat{H}_e' = \sum_{\mathbf{k}} E_{\mathbf{k}} \hat{c}_{\mathbf{k}}^{\dagger} \hat{c}_{\mathbf{k}} \tag{2.4}$$

where $E_{\mathbf{k}} = \hbar^2\mathbf{k}^2/2m$ with $\hbar\mathbf{k}$ the momentum of electron in $\mathbf{k}$ state. (we use the parabolic band approximation); $\hat{c}_{\mathbf{k}}^{\dagger}$ and $\hat{c}_{\mathbf{k}}$ are the electron creation and annihilation operators in the relative coordinates; spin indices have been omitted (but not ignored); $\hat{V}_{ee}'$ is the Coulomb interaction in the relative coordinates. The latter has the standard form and can be found in the literature.[16] The free Hamiltonian for phonons is given by

$$\hat{H}_p = \sum_{\mathbf{q},\lambda} \hbar\Omega_{q\lambda}\, \hat{b}^{\dagger}_{\mathbf{q}\lambda}\hat{b}_{\mathbf{q}\lambda}\,, \tag{2.5}$$

where $\hat{b}^{\dagger}_{\mathbf{q}\lambda}$ and $\hat{b}_{\mathbf{q}\lambda}$ are the phonon creation and annihilation operators, with wave vector $\mathbf{q}$ and in branch λ. The electron–impurity interaction is described by

$$\hat{V}_{ei} = \sum_{\mathbf{q},a} u(\mathbf{q}) e^{i\mathbf{q}\cdot(\mathbf{R}-\mathbf{R}_a)} \hat{\rho}_{\mathbf{q}}\,, \tag{2.6}$$

with $\mathbf{R}_a$ the coordinates of impurity, $u(\mathbf{q})$ the impurity potential and $\hat{\rho}_{\mathbf{q}} = \sum_{\mathbf{k}} \hat{c}^{\dagger}_{\mathbf{k}+\mathbf{q}}\hat{c}_{\mathbf{k}}$ the electron density operator. The electron-phonon interaction is described by

$$\hat{V}_{ep} = \sum_{\mathbf{q},\lambda} M(\mathbf{q},\lambda)(\hat{b}_{\mathbf{q}\lambda} + \hat{b}^{\dagger}_{-\mathbf{q}\lambda}) e^{i\mathbf{q}\cdot\hat{\mathbf{R}}} \hat{\rho}_{\mathbf{q}}\,, \tag{2.7}$$

with $M(\mathbf{q}, \lambda)$ the electron-phonon matrix element.

2.2. Equations for drift velocity of electrons

One can derive the kinetic equations for a set of macroscopic observables, $\{\gamma_i(t)\}$ from the quantum-mechanical Liouville equation. Here we choose the following observables: $\{\gamma_i(t)\} = \{\mathbf{v}(t),\ f_{\mathbf{k}}(t),\ n_{\mathbf{q}\lambda}(t)\}$. The corresponding operators are $\{\hat{\gamma}_i\} = \{\hat{\mathbf{P}}/M,\ \hat{c}^{\dagger}_{\mathbf{k}}\hat{c}_{\mathbf{k}},\ \hat{b}^{\dagger}_{\mathbf{q}\lambda}\hat{b}_{\mathbf{q}\lambda}\}$, which are connected with observables by

$$\gamma_i(t) = \mathrm{Tr}\, \hat{\gamma}_i \hat{\rho}(t)\,, \tag{2.8}$$

with $\hat{\rho}(t)$ the statistical density matrix. The physical meaning of above observables is obvious: $\mathbf{v}(t)$ is the drift velocity of the electron system; $f_{\mathbf{k}}$ is the electron distribution function in relative coordinates; $n_{\mathbf{q}\lambda}$ is the phonon occupation number. If the Liouville equation is solved to the lowest order, the rate of change of any function of $\gamma(t)$, denoted by $F(\gamma(t))$, is given by

$$\frac{\partial F(\gamma(t))}{\partial t} = \frac{i}{\hbar} < [\, \hat{H}_0(t),\, F(\hat{\gamma}(t))\,] > + \frac{i}{\hbar} < [\hat{V}(t),\, F(\hat{\gamma}(t))\,] >$$

$$- (\frac{i}{\hbar}) \lim_{\varepsilon\to 0^+} \int_{-\infty}^{t} dt'\, e^{\varepsilon(t'-t)} < [\, \dot{F}(\hat{\gamma}(t)),\, \hat{V}(t')\,] >\,, \tag{2.9}$$

where $\dot{F}(\hat{\gamma}(t)) = -i/\hbar\, [F(\hat{\gamma}(t)),\, \hat{V}(t)]$, with $[\hat{A},\, \hat{B}]$ is the commutator of operators $\hat{A}$ and $\hat{B}$. In Eq. (2.9), $\hat{V} = \hat{V}'_{ee} + \hat{V}_{ep} + \hat{V}_{ei} + \hat{V}_{pp}$, and

$$\hat{V}(t') = \exp\left[\, i/\hbar \int_t^{t'} \hat{H}_0(s)ds \,\right] \hat{V}(t) \exp\left[\, -i/\hbar \int_t^{t'} \hat{H}_0(s)ds \,\right], \tag{2.10}$$

with $\hat{H}_0 = \hat{H}_c + \hat{H}'_e + \hat{H}_p$. $<.\quad> \equiv \mathrm{Tr}\{\hat{\rho}_0(\ .\ )\}$ with $\hat{\rho}_0$ the unperturbated density matrix.

We define a frictional force operator, $\hat{\mathbf{F}}$, as

$$\hat{\mathbf{F}} = \frac{-i}{\hbar}[\hat{\mathbf{P}}, \hat{V}] = -i\sum_{\mathbf{q},a} \mathbf{q}u(q)e^{i\mathbf{q}\cdot(\hat{\mathbf{R}}-\mathbf{R}_a)}\hat{\rho}_{\mathbf{q}} - i\sum_{\mathbf{q},\lambda} \mathbf{q}M(q,\lambda)e^{i\mathbf{q}\cdot\hat{\mathbf{R}}}(\hat{b}_{\mathbf{q}\lambda} + \hat{b}^{\dagger}_{-\mathbf{q}\lambda})\hat{\rho}_{\mathbf{q}} \,. \tag{2.11}$$

The operators in Eq. (2.9) are those in the interaction representation, since a lowest order expression is taken. In the interaction representation the motion of the mass-of-center $\hat{\mathbf{R}}$ is a "free" motion:

$$< \exp[i\mathbf{q}\cdot\hat{\mathbf{R}}(t)]\exp[-i\mathbf{q}\cdot\hat{\mathbf{R}}(t')] > = \exp[i\mathbf{q}\cdot\mathbf{v}(t)(t-t')] \,, \tag{2.12a}$$

One can see that in the expression of Eq. (2.12a) the memory effect has been lossed. The electron moves starting from a scatterer (at time t') along a straight line, with a constant drift velocity $\mathbf{v}(t)$, to the next scatterer (at time t). Using this expression for $\mathbf{R}$, we can directly derive to the usual Boltzmann transport equation.

We, however, use a different expression as follows:

$$< \exp[i\mathbf{q}\cdot\hat{\mathbf{R}}(t)]\exp[-i\mathbf{q}\cdot\hat{\mathbf{R}}(t')] > = \exp\left[i\mathbf{q}\cdot\int_{t'}^{t}\mathbf{v}(s)ds\right] \,, \tag{2.12b}$$

with $\mathbf{v}(s) = \mathbf{P}(s)/M$, the drift velocity of electron system. This approach was proposed by Ting and Nee.[17] Using this expression, the motion of mass-center of an electron is described by the "true" drift velocity $\mathbf{v}(s)$, which may rapidly change with time during the electron travels. Although Eq. (2.12b) appears natural and reasonable, it is non-trivial from a theoretical view point. Notice that Eq. (2.12b) is obtained by repeating use of Eq. (2.10) through infinitely small time intervals. This treatment implies a change of $\hat{\mathbf{R}}$ from the interaction representation to the Heisenberg representation and, hence, the approach is in some extent beyond the lowest perturbative theory. The validity of Eq. (2.12b) must be proven. A possible derivation to this form is obtained by the path integral approach (see Appendix). We first write an exact expression of the dynamic equation, Eq. (A9), in terms of Feymann-integral formula, where functional integral is taken over the variable of center-of-mass of electrons $\mathbf{R}$ and the variables of the relative motion of electrons, $c_{\mathbf{k}}$ and $c_{\mathbf{k}}^{\dagger}$ (see Eq. (A8)). This expression is equivalent to that all operators are in Heisenberg representation. We then keep only the classical path for center-of-mass variables of electrons $\mathbf{R}$, so the description of $\mathbf{R}$ is given by Eq. (2.12b). We, then, make a perturbative expansion only for the relative variables of electrons by replacing the exact density-density correlation function in Eq. (A9) by its zero-order approximation. Using this approach, we treat two kinds of variables in different ways: the variables of center-of-mass are in the Heisenberg representation but are "classical", while the variables of relative motion are "quantum" but are in the interaction representation. This leads to the present non-Boltzmann formulas.

Since the parameters described the distribution of electrons in the relative coordinates is treated in the lowest perturbative theory, these parameters are obtained by using the following assumption for $\hat{\rho}_0$:[18]

$$\hat{\rho}_0(t) = \frac{1}{\Xi} \exp\{ -\sum_i B_i[\gamma(t)] \ \hat{\gamma}_i \} , \tag{2.13}$$

where Ξ is a normalization constant and the functions B_i are determined by the requirement

$$\gamma_i(t) = \mathrm{Tr}\ \hat{\gamma}_i \hat{\rho}_0(t) . \tag{2.14}$$

By substituting Eq. (2.11), (2.12b)–(2.14) into Eq. (2.9), the following evolution equation for drift velocity of electrons, $\mathbf{v}(t)$, is obtained:

$$Nm\frac{\partial \mathbf{v}(t)}{\partial t} = -Ne\mathbf{E}(t) - i\sum_{\mathbf{q}} \mathbf{q} \lim_{\varepsilon \to 0^+} \int_{-\infty}^{t} dt' \ e^{\varepsilon(t'-t)} \exp\left(i\mathbf{q} \cdot \int_{t'}^{t} \mathbf{v}(s) ds \right)$$

$$\times \left(n_i \ | u(\mathbf{q}) |^2 \Pi(\mathbf{q}, \ t-t') + \sum_{\lambda} | M(\mathbf{q},\lambda) |^2 \Lambda^{(+)}(\mathbf{q}, \ \lambda, \ t-t') \right) ; \tag{2.15}$$

the last two terms in Eq. (2.15) are, respectively, related to scatttering with impurities and phonons:

$$\Pi(\mathbf{q}, \ t-t') \equiv \ < \frac{-i}{\hbar} \ [\hat{\rho}_{\mathbf{q}}(t), \ \hat{\rho}_{-\mathbf{q}}(t')] > = \frac{-i}{\hbar} \sum_{\mathbf{k}} [f_{\mathbf{k+q}}(t) - f_{\mathbf{k}}(t)] e^{\frac{i}{\hbar}(E_{\mathbf{k+q}} - E_{\mathbf{k}})(t-t')} , \tag{2.16}$$

$$\Lambda^{(+)}(\mathbf{q}, \ \lambda, \ t-t') \equiv \ < \frac{-i}{\hbar} \ [(\pm \hat{b}_{\mathbf{q}\lambda}(t) + \hat{b}^{\dagger}_{-\mathbf{q}\lambda}(t)) \hat{\rho}_{\mathbf{q}}(t), \ (\hat{b}_{-\mathbf{q}\lambda}(t') + \hat{b}^{\dagger}_{\mathbf{q}\lambda}(t')) \hat{\rho}_{-\mathbf{q}}(t')] >$$

$$= \mp \frac{i}{\hbar} \sum_{\mathbf{k}} [\ (1+n_{\mathbf{q}\lambda}(t)) f_{\mathbf{k+q}}(t)(1-f_{\mathbf{k}}(t)) - n_{\mathbf{q}\lambda}(t) f_{\mathbf{k}}(t)(1-f_{\mathbf{k+q}}(t)) \ ] e^{\frac{i}{\hbar}(E_{\mathbf{k+q}} - E_{\mathbf{k}} - \hbar\Omega_{\mathbf{q}\lambda})(t-t')}$$

$$+ \frac{i}{\hbar} \sum_{\mathbf{k}} [\ (1+n_{-\mathbf{q}\lambda}(t)) f_{\mathbf{k}}(t)(1-f_{\mathbf{k+q}}(t)) - (n_{-\mathbf{q}\lambda}(t) f_{\mathbf{k+q}}(t)(1-f_{\mathbf{k}}(t))] e^{\frac{i}{\hbar}(E_{\mathbf{k+q}} - E_{\mathbf{k}} + \hbar\Omega_{\mathbf{q}\lambda})(t-t')} \tag{2.17}$$

We will study the high frequency "steady state" conductivity. What the "steady state" means is that the drift velocity of electrons is described as follows, with the time independent amplitudes $\mathbf{v}_n$:

$$\mathbf{v}(t) = \mathbf{v}_0 + \sum_n \mathbf{v}_n e^{in\omega t} , \qquad (n = \pm 1, \ \pm 2, \ \ . \ \ ) \tag{2.18a}$$

here $\mathbf{v}_{-n} = \mathbf{v}_n^*$. Also, we assume the distribution function of electrons in the relative coordinates oscillates with the time independent amplitudes:

$$f_{\mathbf{k}}(t) = f_{\mathbf{k}}^{(0)} + \sum_j f_{\mathbf{k}}^{(j)} e^{ij\omega t} , \quad (j = \pm 1, \pm 2, \ldots .) \tag{2.18b}$$

with $f_{\mathbf{k}}^{(-j)} = f_{\mathbf{k}}^{(j)*}$. Let $\tau = t' - t$, we make a power expansion for following expression in Eq. (2.15):

$$\exp\left[i\mathbf{q}\cdot\int_{t'}^{t} \mathbf{v}(s)ds \right] = \exp[\,-i\mathbf{q}\cdot\mathbf{v}_0\tau + \sum_n \frac{\mathbf{q}\cdot\mathbf{v}_n}{n\omega} e^{in\omega t}(1 - e^{in\omega\tau})\,]$$

$$= e^{-i\mathbf{q}\cdot\mathbf{v}_0\tau}\left[1 + \sum_n \frac{\mathbf{q}\cdot\mathbf{v}_n}{n\omega} e^{in\omega t}(1 - e^{in\omega\tau}) + \right.$$

$$\left. + \frac{1}{m!}\sum_{n_1}\cdots\sum_{n_m} \frac{(\mathbf{q}\cdot\mathbf{v}_{n_1})\ldots(\mathbf{q}\cdot\mathbf{v}_{n_m})}{n_1\ldots n_m\omega^m} \prod_{k=1}^{m}(1 - e^{in_k\omega\tau})\exp\left[i\omega t\sum_{k=1}^{m} n_k \right] + \ldots . \right] \tag{2.19}$$

By substituting Eqs. (2.18)–(2.19) into Eq. (2.15), we can split the evolution equation (15) for drift velocity of electrons to a series of equations for different harmonic components of drift velocity. We have

$$0 = -Ne\mathbf{E}_0 + \mathbf{F}_0 ,$$

$$in\omega Nm\mathbf{v}_n = -Ne\mathbf{E}_n + \mathbf{F}_n , \tag{2.20}$$

here $\mathbf{F}_n$ is the n-th component of frictional force:

$$\mathbf{F}(t) = \mathbf{F}_0 + \sum_n \mathbf{F}_n e^{in\omega t} , \tag{2.21}$$

with $\mathbf{F}_{-n} = \mathbf{F}_n^*$. If only a few low harmonic components of electric field are applied, we assume the high harmonic current components are small enough and can be cutoff to some order. By expanding Eq. (2.19) up to 3-rd order and keeping $|n|$ and $|j|$ up to 2, we obtain that

$$\mathbf{F}_0 = \mathbf{F}_0^{(0)} + 2\mathrm{Re}\,\mathbf{F}_1^{(-1)} + 2\mathrm{Re}\,\mathbf{F}_2^{(-2)} ,$$

$$\mathbf{F}_1 = \mathbf{F}_1^{(0)} + \mathbf{F}_0^{(1)} + \mathbf{F}_2^{(-1)} + (\mathbf{F}_1^{(-2)})^* ,$$

$$\mathbf{F}_2 = \mathbf{F}_2^{(0)} + \mathbf{F}_1^{(1)} + \mathbf{F}_0^{(2)} . \tag{2.22}$$

In $\mathbf{F}_n^{(j)}$, lower index n is related the harmonic component of $\mathbf{v}(t)$, obtained by expansion of Eq. (2.19), and the upper index j is related to the harmonic component of $f_{\mathbf{k}}(t)$ in Eq. (2.18b). We have

$$\mathbf{F}_0^{(j)} = \sum_{\mathbf{q}} \left\{ \mathbf{\Gamma}_{\mathbf{q}}^{(j)}(0) - \frac{(\mathbf{q}\cdot\mathbf{v}_1)(\mathbf{q}\cdot\mathbf{v}_1^*)}{\omega^2}[2\mathbf{\Gamma}_{\mathbf{q}}^{(j)}(0) - \mathbf{\Gamma}_{\mathbf{q}}^{(j)}(\omega) - \mathbf{\Gamma}_{\mathbf{q}}^{(j)}(-\omega)] \right.$$

$$- \frac{(\mathbf{q}\cdot\mathbf{v}_2)(\mathbf{q}\cdot\mathbf{v}_2^*)}{4\omega^2}[(2\mathbf{\Gamma}_{\mathbf{q}}^{(j)}(0) - \mathbf{\Gamma}_{\mathbf{q}}^{(j)}(2\omega) - \mathbf{\Gamma}_{\mathbf{q}}^{(j)}(-2\omega)]$$

$$+ \frac{(\mathbf{q}\cdot\mathbf{v}_2)(\mathbf{q}\cdot\mathbf{v}_1^*)^2}{4\omega^3}[2\mathbf{\Gamma}_{\mathbf{q}}^{(j)}(\omega) - 2\mathbf{\Gamma}_{\mathbf{q}}^{(j)}(-\omega) - \mathbf{\Gamma}_{\mathbf{q}}^{(j)}(2\omega) + \mathbf{\Gamma}_{\mathbf{q}}^{(j)}(-2\omega)]$$

$$\left. - \frac{(\mathbf{q}\cdot\mathbf{v}_2^*)(\mathbf{q}\cdot\mathbf{v}_1)^2}{4\omega^3}[2\mathbf{\Gamma}_{\mathbf{q}}^{(j)}(-\omega) - 2\mathbf{\Gamma}_{\mathbf{q}}^{(j)}(\omega) + \mathbf{\Gamma}_{\mathbf{q}}^{(j)}(2\omega) - \mathbf{\Gamma}_{\mathbf{q}}^{(j)}(-2\omega)] \right\}, \quad (2.23)$$

$$\mathbf{F}_1^{(j)} = \sum_{\mathbf{q}} \left\{ \frac{\mathbf{q}\cdot\mathbf{v}_1}{\omega}[\mathbf{\Gamma}_{\mathbf{q}}^{(j)}(0) - \mathbf{\Gamma}_{\mathbf{q}}^{(j)}(\omega)] - \frac{(\mathbf{q}\cdot\mathbf{v}_2)(\mathbf{q}\cdot\mathbf{v}_1^*)}{2\omega^2}[\mathbf{\Gamma}_{\mathbf{q}}^{(j)}(0) + \mathbf{\Gamma}_{\mathbf{q}}^{(j)}(\omega) - \mathbf{\Gamma}_{\mathbf{q}}^{(j)}(2\omega) - \mathbf{\Gamma}_{\mathbf{q}}^{(j)}(-\omega)] \right.$$

$$- \frac{(\mathbf{q}\cdot\mathbf{v}_1)^2(\mathbf{q}\cdot\mathbf{v}_1^*)}{2\omega^3}[3\mathbf{\Gamma}_{\mathbf{q}}^{(j)}(0) - 3\mathbf{\Gamma}_{\mathbf{q}}^{(j)}(\omega) + \mathbf{\Gamma}_{\mathbf{q}}^{(j)}(2\omega) - \mathbf{\Gamma}_{\mathbf{q}}^{(j)}(-\omega)]$$

$$\left. - \frac{(\mathbf{q}\cdot\mathbf{v}_1)(\mathbf{q}\cdot\mathbf{v}_2)(\mathbf{q}\cdot\mathbf{v}_2^*)}{4\omega^3}[2\mathbf{\Gamma}_{\mathbf{q}}^{(j)}(0) - \mathbf{\Gamma}_{\mathbf{q}}^{(j)}(2\omega) - \mathbf{\Gamma}_{\mathbf{q}}^{(j)}(-2\omega) - 2\mathbf{\Gamma}_{\mathbf{q}}^{(j)}(\omega) + \mathbf{\Gamma}_{\mathbf{q}}^{(j)}(3\omega) + \mathbf{\Gamma}_{\mathbf{q}}^{(j)}(-\omega)] \right\},$$

(2.24)

$$\mathbf{F}_2^{(j)} = \sum_{\mathbf{q}} \left\{ \frac{\mathbf{q}\cdot\mathbf{v}_2}{2\omega}[\mathbf{\Gamma}_{\mathbf{q}}^{(j)}(0) - \mathbf{\Gamma}_{\mathbf{q}}^{(j)}(2\omega)] + \frac{(\mathbf{q}\cdot\mathbf{v}_1)^2}{2\omega^2}[\mathbf{\Gamma}_{\mathbf{q}}^{(j)}(0) - 2\mathbf{\Gamma}_{\mathbf{q}}^{(j)}(\omega) + \mathbf{\Gamma}_{\mathbf{q}}^{(j)}(2\omega)] \right.$$

$$- \frac{(\mathbf{q}\cdot\mathbf{v}_2)^2(\mathbf{q}\cdot\mathbf{v}_2^*)}{16\omega^3}[3\mathbf{\Gamma}_{\mathbf{q}}^{(j)}(0) - 3\mathbf{\Gamma}_{\mathbf{q}}^{(j)}(2\omega) + \mathbf{\Gamma}_{\mathbf{q}}^{(j)}(4\omega) - \mathbf{\Gamma}_{\mathbf{q}}^{(j)}(-2\omega)]$$

$$\left. - \frac{(\mathbf{q}\cdot\mathbf{v}_2)(\mathbf{q}\cdot\mathbf{v}_1)(\mathbf{q}\cdot\mathbf{v}_1^*)}{2\omega^3}[2\mathbf{\Gamma}_{\mathbf{q}}^{(j)}(0) - 2\mathbf{\Gamma}_{\mathbf{q}}^{(j)}(2\omega) + \mathbf{\Gamma}_{\mathbf{q}}^{(j)}(3\omega) - \mathbf{\Gamma}_{\mathbf{q}}^{(j)}(-\omega)] \right\}. \quad (2.25)$$

$\mathbf{\Gamma}_{\mathbf{q}}^{(j)}(n\omega)$ in Eqs. (2.23)–(2.25) is given by

$$\mathbf{\Gamma}_{\mathbf{q}}^{(j)}(n\omega) = \mathbf{q} \left[n_i \,|\, u(q)\,|^2 \sum_{\mathbf{k}}(f_{\mathbf{k+q}}^{(j)} - f_{\mathbf{k}}^{(j)})\zeta(\mathbf{k},\, \mathbf{q},\, n\omega - \mathbf{q}\cdot\mathbf{v}_0) + \right.$$

$$\sum_{\lambda} \,|\, M(q,\, \lambda)\,|^2 \left\{ \sum_{\mathbf{k}}[(1 + n_{\mathbf{q}\lambda})f_{\mathbf{k+q}}^{(j)}(1 - f_{\mathbf{k}}^{(0)}) - n_{\mathbf{q}\lambda}f_{\mathbf{k}}^{(j)}(1 - f_{\mathbf{k+q}}^{(0)})]\zeta(\mathbf{k},\, \mathbf{q},\, \Omega_{q\lambda} + n\omega - \mathbf{q}\cdot\mathbf{v}_0) \right.$$

$$- \sum_{\mathbf{k}}[(1+n_{-\mathbf{q}\lambda})f_{\mathbf{k}}^{(j)}(1-f_{\mathbf{k+q}}^{(0)}) - n_{-\mathbf{q}\lambda}f_{\mathbf{k+q}}^{(j)}(1-f_{\mathbf{k}}^{(0)})]\ \zeta(\mathbf{k},\mathbf{q},-\Omega_{q\lambda}+n\omega-\mathbf{q}\cdot\mathbf{v}_0)\Big]\Big], \quad (2.26)$$

where $\zeta(\mathbf{k}, \mathbf{q}, x)$, obtained by integrating exponential terms in Eq. (2.16), (2.17) and (2.19) over dt' (or $d\tau$), has the following form:

$$\zeta(\mathbf{k}, \mathbf{q}, x) = \frac{-2i}{(E_{\mathbf{k+q}} - E_{\mathbf{k}} - \hbar x) + i\varepsilon}, \quad (2.27)$$

where the 2 comes from spin degeneracy. In order to obtain Eq. (2.26) we have neglected the oscillating terms in the factor $(1 - f_{\mathbf{k}}(t))$, since we will use a non-degeneracy assumption later. $\mathbf{\Gamma}_{\mathbf{q}}^{(j)}(n\omega)$ has the following symmetry:

$$\mathbf{\Gamma}_{-\mathbf{q}}^{(-i)}(-n\omega) = \mathbf{\Gamma}_{\mathbf{q}}^{(j)}(n\omega)^* . \quad (2.28)$$

One can see from Eqs. (2.26) and (2.27) that the frequency of the electric field enters into the expression of energy conservation, Eq. (2.27), in a form of a multi-photon emission or absorption. This quantum effect is originated from Eq. (2.12b), which includes the memory effect in the theory. Comparing Eq. (2.12b) with Eq. (2.12a), that leads to the usual Boltzmann equation, we can see the latter only approximately valid when $t-t'$, which is related to the scattering time, less than 1/4 of the period of the applied HF field. This requirement greatly limits the frequency region, in which the Boltzmann equation valids. It shows the significance of introducing Eq. (2.12b), especially, in study the HF conductivity. Moreover, use of Eq. (2.12b) also benefits us, because we can easily separate each harmonic component, as shown before.

In the limit of low frequency, $\omega \to 0$, expressions in Eqs.(2.23)-(2.25) are not divergent, since the terms with ω^{-n} $(n > 0)$ in the Taylor expansion of ω (divided by denominator) completely cancel.

2.3. Equation for distribution of hot-electrons

To obtain a closed set of equations, we need equations for the electron distribution function in the relative coordinates, $f_{\mathbf{k}}^{(j)}$. To simplify our calculation, we use the drift-temparature model to deal with $f_{\mathbf{k}}^{(j)}$. We assume that owing to the frequent electron–electron collisions, the distribution function of electrons in the relative coordinates $f_{\mathbf{k}}(t)$ is a Maxwellian distribution at the temperature $T_e(t)$ (a typical relaxation time for electron-electron scattering is about 30-40 fsec, when the density of electrons is 10^{18} /cm^3):

$$f_{\mathbf{k}}(t) = \frac{N}{2}\left[\frac{2\pi\hbar^2\beta(t)}{m}\right]^{3/2} \exp[-E_{\mathbf{k}}\beta(t)] , \quad (2.29)$$

with $\beta(t) = 1/K_B T_e(t)$. We assume the electron temperature $T_e(t)$, or $\beta(t)$, oscillate in the following way:

$$T_e(t) = T_0 + \sum_j T_j e^{in\omega t} ,$$

$$\beta(t) = \beta_0 + \sum_j \beta_j e^{in\omega t} \tag{2.30}$$

By expanding to 3-th order and keeping $|n|$ up to 2 in Eq. (2.30), the relation between amplitudes T_n and β_n can be obtained:

$$K_B T_0 = \frac{1}{\beta_0} + \frac{1}{\beta_0^3}(2\,|\beta_1|^2 + 2\,|\beta_2|^2) - \frac{1}{\beta_0^4}(3\beta_2\beta_1^{*2} + 3\beta_2^*\beta_1^2) + \ldots ,$$

$$K_B T_1 = \frac{-\beta_1}{\beta_0^2} + \frac{2\beta_2\beta_1^*}{\beta_0^3} - \frac{1}{\beta_0^4}(3\,|\beta_1|^2\beta_1 + 6\,|\beta_2|^2\beta_1) + \ \ldots ,$$

$$K_B T_2 = \frac{-\beta_2}{\beta_0^2} + \frac{\beta_1^2}{\beta_0^3} - \frac{1}{\beta_0^4}(3\,|\beta_2|^2\beta_2 + 6\,|\beta_1|^2\beta_2) + \tag{2.31}$$

A similar expansion is made for $f_{\mathbf{k}}(t)$ in Eq. (2.29), we obtain the expressions of amplitudes $f_{\mathbf{k}}^{(j)}$ by β_n:

$$f_{\mathbf{k}}^{(0)} = A_{\mathbf{k}}[\,1 + C_{\mathbf{k}}(2\,|\beta_1|^2 + 2\,|\beta_2|^2) + D_{\mathbf{k}}(3\beta_2\beta_1^{*2} + 3\beta_2^*\beta_1^2)] + \ \ldots ,$$

$$f_{\mathbf{k}}^{(1)} = A_{\mathbf{k}}[\,B_{\mathbf{k}}\beta_1 + C_{\mathbf{k}}(2\beta_2\beta_1^*) + D_{\mathbf{k}}(3\,|\beta_1|^2\beta_1 + 6\,|\beta_2|^2\beta_1)] + \ldots ,$$

$$f_{\mathbf{k}}^{(2)} = A_{\mathbf{k}}[\,B_{\mathbf{k}}\beta_2 + C_{\mathbf{k}}(\beta_1^2) + D_{\mathbf{k}}(3\,|\beta_2|^2\beta_2 + 6\beta_2\,|\beta_1|^2)] + \quad , \tag{2.32}$$

with

$$A_{\mathbf{k}} = \frac{N}{2}\left(\frac{2\pi\hbar^2\beta_0}{m}\right)^{3/2} \exp(-E_{\mathbf{k}}\beta_0) ,$$

$$B_{\mathbf{k}} = \frac{3}{2\beta_0} - E_{\mathbf{k}} ,$$

$$C_{\mathbf{k}} = \frac{3}{8\beta_0^2} - \frac{3E_{\mathbf{k}}}{2\beta_0} + \frac{E_{\mathbf{k}}^2}{2} ,$$

$$D_{\mathbf{k}} = -\frac{1}{16\beta_0^3} - \frac{3E_{\mathbf{k}}}{8\beta_0^2} + \frac{3E_{\mathbf{k}}^2}{4\beta_0} - \frac{E_{\mathbf{k}}^3}{6} \tag{2.33}$$

The electron temperature $T_e(t)$ is related to the energy of electrons in the relative coordinates by $E_e' \equiv \langle \hat{H}_e' \rangle = C_v T_e$, with $C_v = 3NK_B/2$ the specific heat of electrons. The evolution equation for energy of electrons, $\partial E_e'(t)/\partial t$, can be derived in a similar way as that for drift velocity of electrons. We have

$$\frac{\partial E_e'(t)}{\partial t} \equiv \langle \dot{H}_e' \rangle = -Ne\mathbf{E}(t)\cdot\mathbf{v}(t) - Nm\mathbf{v}(t)\cdot\frac{d\mathbf{v}(t)}{dt} - \langle \dot{H}_{ph}(t)\rangle_{\text{e-p}}$$

$$= -\mathbf{F}(t)\cdot\mathbf{v}(t) - \langle \dot{H}_{ph}(t)\rangle_{\text{e-p}} \,. \tag{2.34}$$

The operator of change of energy of phonons via electron–phonon interaction is given by

$$\dot{H}_{ph} = \frac{-i}{\hbar}[\hat{H}_{ph}, \hat{V}_{ep}] = -i\sum_{\mathbf{q},\lambda} M(q,\lambda)\Omega_{q\lambda} e^{i\mathbf{q}\cdot\hat{\mathbf{R}}}(-\hat{b}_{\mathbf{q}\lambda} + \hat{b}^{\dagger}_{-\mathbf{q}\lambda})\hat{\rho}_{\mathbf{q}} \,. \tag{2.35}$$

The average energy change via electron–phonon interaction can then be written as

$$\langle \dot{H}_{ph}(t)\rangle_{\text{e-p}} = -i\sum_{\mathbf{q},\lambda}\Omega_{q\lambda}\lim_{\varepsilon\to 0^+}\int_{-\infty}^{t} dt' e^{\varepsilon(t'-t)}\exp\left[i\mathbf{q}\cdot\int_{t'}^{t}\mathbf{v}(s)ds\right] |M(q,\lambda)|^2\Lambda^{(-)}(\mathbf{q},\lambda,t-t')\,, \tag{2.36}$$

with $\Lambda^{(-)}(\mathbf{q}, \lambda, t-t')$ is defined in Eq. (2.17). By use of the expansion in Eq. (2.19), and substituting Eq. (2.36) into Eq. (2.34), similar to Eq. (2.20), we obtain a set of equations of harmonic components for the energy of electrons $E_e'(t)$ or the corresponding components for the electron temperature $T_e(t)$:

$$-(\mathbf{F}_0\cdot\mathbf{v}_0 + \mathbf{F}_1\cdot\mathbf{v}_1^* + \mathbf{F}_1^*\cdot\mathbf{v}_1 + \mathbf{F}_2\cdot\mathbf{v}_2^* + \mathbf{F}_2^*\cdot\mathbf{v}_2) - \langle \dot{H}_{ph}\rangle_0 = 0\,,$$

$$-(\mathbf{F}_2\cdot\mathbf{v}_1^* + \mathbf{F}_1\cdot\mathbf{v}_0 + \mathbf{F}_0\cdot\mathbf{v}_1 + \mathbf{F}_1^*\cdot\mathbf{v}_2) - \langle \dot{H}_{ph}\rangle_1 = i\omega\left[\frac{3N}{2}K_BT_1\right],$$

$$-(\mathbf{F}_2\cdot\mathbf{v}_0 + \mathbf{F}_1\cdot\mathbf{v}_1 + \mathbf{F}_0\cdot\mathbf{v}_2) - \langle \dot{H}_{ph}\rangle_2 = i2\omega\left[\frac{3N}{2}K_BT_2\right], \tag{2.37}$$

here $\mathbf{F}_0$, $\mathbf{F}_1$, and $\mathbf{F}_2$ is given by Eqs. (2.22)–(2.26), and $\langle \dot{H}_{ph}\rangle_n$ is obtained from corresponding $\mathbf{F}_n$ by changing $\Gamma_{\mathbf{q}}^{(j)}(n\omega)$ in Eqs. (2.23)–(2.25) to $\Phi_{\mathbf{q}}^{(j)}(n\omega)$, with

$$\Phi_{\mathbf{q}}^{(j)}(n\omega) = \sum_{\lambda}\Omega_{q\lambda}|M(q,\lambda)|^2$$

$$\left[-\sum_{\mathbf{k}}[(1+n_{\mathbf{q}\lambda})f_{\mathbf{k+q}}^{(j)}(1-f_{\mathbf{k}}^{(0)}) - n_{\mathbf{q}\lambda}f_{\mathbf{k}}^{(j)}(1-f_{\mathbf{k+q}}^{(0)})]\zeta(\mathbf{k},\mathbf{q},\Omega_{q\lambda}+n\omega-\mathbf{q}\cdot\mathbf{v}_0)\right.$$

$$\left.-\sum_{\mathbf{k}}[(1+n_{-\mathbf{q}\lambda})f_{\mathbf{k}}^{(j)}(1-f_{\mathbf{k+q}}^{(0)}) - n_{-\mathbf{q}\lambda}f_{\mathbf{k+q}}^{(j)}(1-f_{\mathbf{k}}^{(0)})]\,\zeta(\mathbf{k},\mathbf{q},-\Omega_{q\lambda}+n\omega-\mathbf{q}\cdot\mathbf{v}_0)\right]. \tag{2.38}$$

$\Phi_{\mathbf{q}}^{(j)}(n\omega)$ also has the following symmetry:

$$\Phi_{-\mathbf{q}}^{(-j)}(-n\omega) = \Phi_{\mathbf{q}}^{(j)}(n\omega)^* \tag{2.39}$$

We neglect hot-phonon effects and assume the phonons are in equilibrium with a heat bath, the occupation numbers of phonons are then determined by

$$n_{\mathbf{q}\lambda} = \frac{1}{\exp(\hbar\Omega_{q\lambda}/K_B T_L) - 1}, \tag{2.40}$$

with T_L the temperature of the heat bath.

Eq. (2.20) and Eq. (2.37), together with Eqs. (2.31)–(2.32) are consisted of a set of equations for determining the magnitudes and phases of the harmonic components of the drift velocity, $\mathbf{v}_n$, and the electron temperature T_j (or β_j).

If degeneracy is neglected, namely, $(1 - f_{\mathbf{k}}^{(0)})$ is replaced by 1, the summation over $\mathbf{k}$ in Eqs. (2.26) and (2.38), then, can be worked out. Typical terms in $\sum_{\mathbf{k}} f_{\mathbf{k}}^{(j)}\zeta(\mathbf{k},\mathbf{q},x)$ in Eq. (2.26) are obtained as follows:

$$\mathrm{Re}\left[\sum_{\mathbf{k}} E_{\mathbf{k}}^n \exp(-\beta_0 E_{\mathbf{k}})\zeta(\mathbf{k},\mathbf{q},x)\right] = \frac{-m^2}{2\pi\hbar^4 q}(-1)^n \frac{\partial^n}{\partial\beta_0^n}\left[\frac{1}{\beta_0}\exp(-\eta^2)\right] \tag{2.41}$$

with

$$\eta = \hbar\left(\frac{\beta_0}{2m}\right)^{1/2}\left(\frac{q}{2} - \frac{mx}{\hbar q}\right) \tag{2.42}$$

and

$$\mathrm{Im}\left[\sum_{\mathbf{k}} E_{\mathbf{k}}^n \exp(-\beta_0 E_{\mathbf{k}})\zeta(\mathbf{k},\mathbf{q},x)\right] = \frac{-4\pi^{1/2} m^2}{(2\pi)^2\hbar^4 q}(-1)^n \frac{\partial^n}{\partial\beta_0^n}\left[\frac{1}{\beta_0} e^{-\eta^2}\int_0^{\eta} e^{x^2} dx\right] \tag{2.43}$$

The integral in the above equation is related to Dawson's integral.[19]

3. LINEAR HIGH FREQUENCY CONDUCTIVITY

We first briefly discuss that a weak HF electric field is applied, without or with a strong dc electric field.

3.1. Cooling electron case

If we apply only a weak HF field (along x direction) on the sample:

$$\mathbf{E}(t) = \mathbf{E}_1(e^{i\omega t} + e^{-i\omega t}), \tag{3.1}$$

the electron distribution in the relative coordinates is assumed in equilibrium with the heat bath, and has only the component $f_{\mathbf{k}}^{(0)} = A_{\mathbf{k}}$ in Eq. (2.33), with $\beta_0 = 1/K_B T_L$. By taking only the linear term in Eq. (2.24), our formulas reduce to a simple equation given by Lei, Horing, and Zhang.[5]

$$i\omega N m \mathbf{v}_1 = -Ne\mathbf{E}_1 + \sum_{\mathbf{q}} \frac{\mathbf{q}\cdot\mathbf{v}_1}{\omega}[\boldsymbol{\Gamma}_{\mathbf{q}}^{(0)}(0) - \boldsymbol{\Gamma}_{\mathbf{q}}^{(0)}(\omega)] \,, \tag{3.2}$$

where $\boldsymbol{\Gamma}_{\mathbf{q}}^{(0)}$ is given by Eq. (2.26). Eq. (3.2) directly leads to the well known expression of linear conductivity, $\sigma(\omega)$, in terms of a memory function: (Götze and Wöfle [3])

$$\sigma(\omega) = i\frac{Ne^2}{m}\frac{1}{\omega + M(\omega)} \,, \tag{3.3}$$

with the memory function $M(\omega)$ given by

$$M(\omega) = \sum_{\mathbf{q}} \frac{-iq_x}{Nm\omega}[\Gamma_{q_x}^{(0)}(0) - \Gamma_{q_x}^{(0)}(\omega)] \tag{3.4}$$

3.2. Hot electron case

If a weak Hf field is applied together with a strong dc field (both are along the x direction), the electrons are heated by the dc field. The hot electron effect on the high frequency conductivity has previously been studied previously based on the Boltzmann transport equation.[6-9] We keep only the zero and first harmonic components and cut off the expansion beyond linear order because of the weak HF field. Two sets of equations then are obtained. One set of equations for the zero harmonic components is given by

$$-Ne\mathbf{E}_0 + \mathbf{F}_0^{(0)} = 0 \,,$$

$$-\mathbf{F}_0^{(0)}\cdot\mathbf{v}_0 - \langle \dot{H}_{ph}\rangle_0^{(0)} = 0 \,, \tag{3.5}$$

with $\mathbf{F}_0^{(0)} = \sum_{\mathbf{q}} \boldsymbol{\Gamma}_{\mathbf{q}}^{(0)}(0)$ and $\langle \dot{H}_{ph}\rangle_0^{(0)} = \sum_{\mathbf{q}} \Phi_{\mathbf{q}}^{(0)}(0)$. This set of equations is independent of the first harmonic components, and can be solved as if only a static field exists. The solutions of $\mathbf{v}_0$ and β_0 are parameters for further calculation of the linear differential conductivity for a weak HF field. Another set of equations for $\mathbf{v}_1$ and T_1 (or β_1) is given by

$$in\omega N m\mathbf{v}_1 = -Ne\mathbf{E}_1 + \mathbf{F}_1^{(0)} + \mathbf{F}_0^{(1)} \,,$$

$$-i\omega\frac{3\beta_1 N}{2\beta_0^2} = -[\, (\mathbf{F}_1^{(0)} + \mathbf{F}_0^{(1)})\cdot\mathbf{v}_0 + \mathbf{F}_0^{(0)}\cdot\mathbf{v}_1 \,] - \langle \dot{H}_{ph}\rangle_1^{(0)} - \langle \dot{H}_{ph}\rangle_0^{(1)} \,, \tag{3.6}$$

with

$$\mathbf{F}_1^{(0)} = \frac{\mathbf{v}_1}{\omega}\sum_{\mathbf{q}}\mathbf{q}[\boldsymbol{\Gamma}_{\mathbf{q}}^{(0)}(0) - \boldsymbol{\Gamma}_{\mathbf{q}}^{(0)}(\omega)] \,,$$

$$\langle \dot{H}_{ph}\rangle_1^{(0)} = \frac{\mathbf{v}_1\cdot}{\omega}\sum_{\mathbf{q}}\mathbf{q}[\Phi_{\mathbf{q}}^{(0)}(0) - \Phi_{\mathbf{q}}^{(0)}(\omega)] \,, \tag{3.7}$$

$$\mathbf{F}_0^{(1)} \approx \sum_{\mathbf{q}} \Gamma_{\mathbf{q}}^{(1)}(0)\,,$$

$$<\dot{H}_{ph}>_0^{(1)} \approx \sum_{\mathbf{q}} \Phi_{\mathbf{q}}^{(1)}(0)\,. \tag{3.8}$$

The superscript (1) in Eq. (3.8) is related to the first harmonic component of distribution function, $f_{\mathbf{k}}^{(1)} \approx A_{\mathbf{k}} B_{\mathbf{k}} \beta_1$ in Eq. (2.32), with $A_{\mathbf{k}}$ and $B_{\mathbf{k}}$ given by Eq. (2.33). The equation (3.6), therefore, is a set of linear equations for $\mathbf{v}_1$ and β_1 provided by the bias parameters $\mathbf{v}_0$ and β_0. The structure of these sets of equations is similar to that obtained by Das and Ferry[7]. Since they derived their formulas from a Boltzmann transport equation, the coefficients in the Eq. (3.6) of this paper are different from those of Das and Ferry. In Ref. 7, the coefficients in the linear equations corresponding to our equation (3.6) are obtained based on analytical expressions of the derivative of the relaxation times. Our formulas do not require a special analytical expression, and therefore, are easy to generalize to more complicate cases, such as two dimensional electrons in a quantum well.

4. NONLINEAR HF CONDUCTIVITY

We have used the approach shown in Sec. 2 to calculate the nonlinear conductivity in an n-GaAS bulk sample when a strong HF electric field is applied. The strength of HF field is applied up to 1 kV/cm, and the effect of higher valleys, which are important when the strength of the field reaches 3-4 kV/cm, are neglected in the present calculation. A weak dc electric field $E_0 = 10$ V/cm is also applied. The base frequency of HF electric field is chosen in the microwave region, $\nu = 35$ GHz. The lattice temperature $T_L = 300$K. The electron-longitudinal optical (LO) phonon scattering through polar interaction is the dominate scattering mechanism at this temperature, and we neglect electron scattering by impurities and other kinds of phonons. The electron-LO phonon interaction is described by the Fröhlich model: $M(\mathbf{q}) = \alpha/q^2$, with the Fröhlich coupling constant, $\alpha = [2\pi e^2 \hbar\Omega_{LO}(1/\varepsilon_\infty - 1/\varepsilon_0)]^{1/2}$, here the energy of the LO phonon $\hbar\Omega_{LO} = 36.2\,\mathrm{meV}$, the static dielectric constant $\varepsilon_0 = 12.91$ and the high frequency dielectric constant $\varepsilon_\infty = 10.91$. The effective mass of an electron is $m = 0.067\,m_0$, with m_0 the free electron mass. The density of electrons is chosen as $N = 10^{17}/\mathrm{cm}^3$, which is irrelevant in the calculation as long as a Maxwell distribution is assumed. The screening effect is neglected in present calculation, since the dynamic screening is weak for electron-LO phonon scattering.

We have calculated the dc velocity υ_0 and the harmonic components of velocity $\upsilon_n(t) = |\upsilon_n| \cos(n\omega + \phi_{\upsilon_n})$, $(n = 1, 2)$ and corresponding components of the electron temperature $T(t)$, as function of strength of the applied HF electric field, $E_1\cos(\omega t)$ and $E_2\cos(2\omega t)$. This calculation requires us to solve a set of 10 coupled nonlinear equations. We use an optimization program of the AT&T Bell laboratories Mathematical Subroutine library Port[20] based on the algorithm of a improved Newton's method. The accuracy generally is $10^{-8}-10^{-9}$. The results are shown in Fig. 5 – Fig. 10.

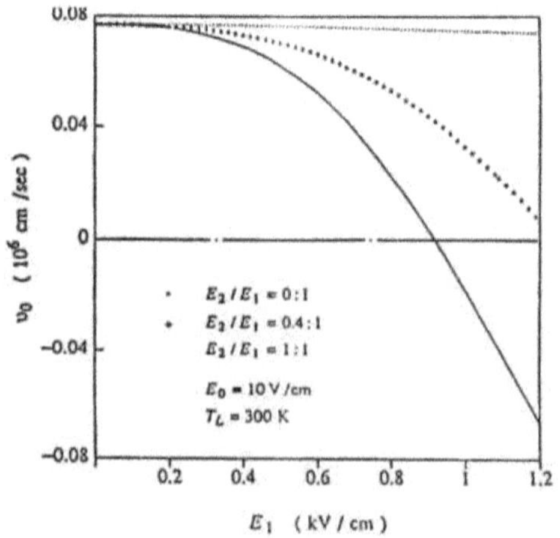

Fig. 5. The direct current component of the electron drift velocity, υ_0, as function of the strength of high frequency electric field, E_1. $E_1\cos(\omega t)$ and $E_2\cos(2\omega t)$ are applied together with (1): $E_2:E_1=0$ (dotted curve), (2): $E_2:E_1=0.4:1$ (cross curve), and (3): $E_2:E_1=1:1$ (solid curve). The frequency $\nu=\omega/2\pi=35$ GHz.

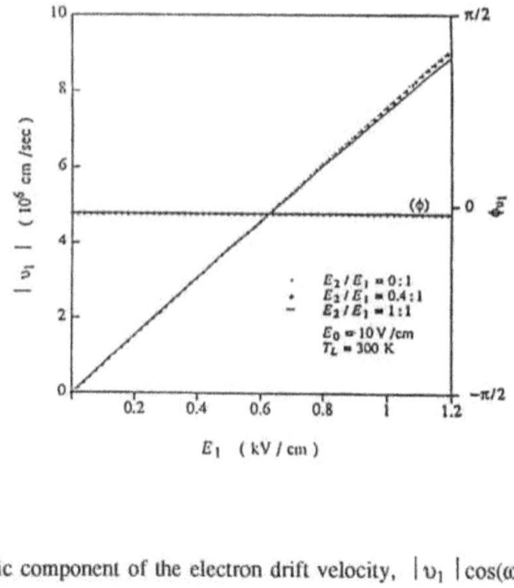

Fig. 6. The first harmonic component of the electron drift velocity, $|\upsilon_1|\cos(\omega t+\phi_{\upsilon_1})$, as function of E_1. The lines labeled by (ϕ) are that for phase ϕ_{υ_1}.

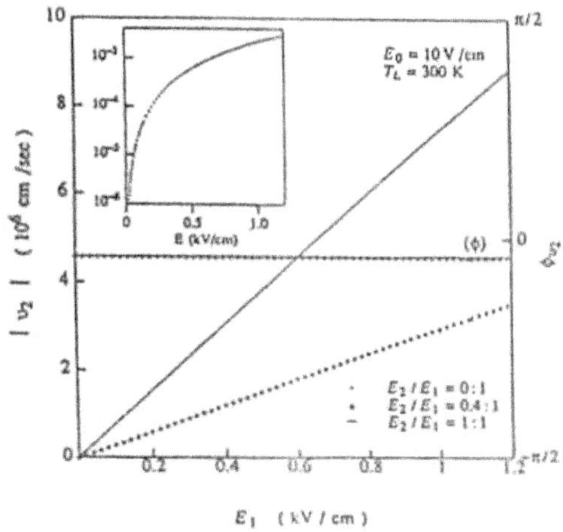

Fig. 7. The second harmonic component of the electron drift velocity, $|\upsilon_2|\cos(2\omega t+\phi_{\upsilon_2})$, as function of E_1. The lines labeled by (ϕ) are that for phase ϕ_{υ_2} Insert: $|\upsilon_2|$ as function of E_1 for $E_2:E_1=0:1$.

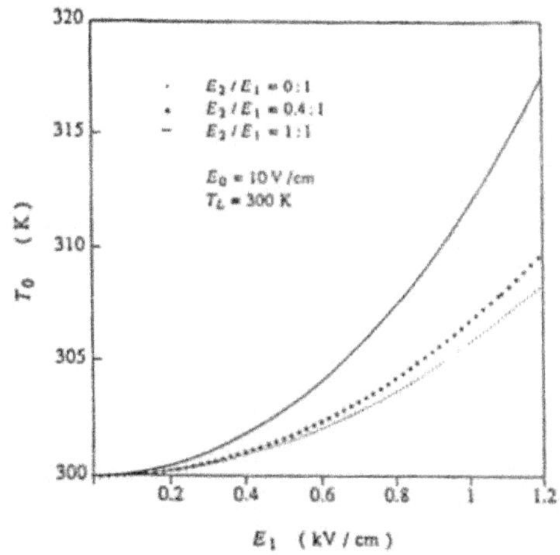

Fig. 8. The average electron temperature, T_0 as function of E_1.

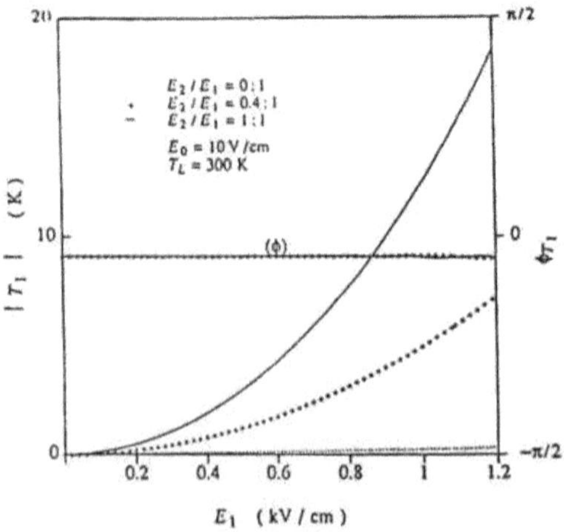

Fig. 9. The first harmonic component of the electron temperature, $|T_1|\cos(\omega t+\phi_{T_1})$, as function of E_1. The lines labeled by (ϕ) are that for phase ϕ_{T_1}.

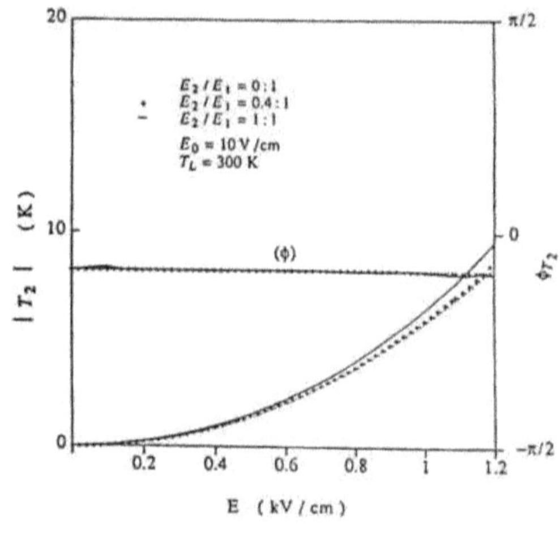

Fig. 10. The second harmonic component of the electron temperature, $|T_2|\cos(2\omega t+\phi_{T_2})$, as function of E_1. The lines labeled by (ϕ) are that for phase ϕ_{T_2}.

Three cases are considered: (1) $E_2:E_1=0:1$ (dotted curves), (2) $E_2:E_1=0.4:1$ (cross curves), and (3) $E_2:E_1=1:1$ (solid curve). Fig. 5 shows that the dc conductivity only reduces about 5% from its static value, up to $E_1=1$ kV/cm if only E_1 is applied. Surprisingly, if a second harmonic electric field is applied together with the first harmonic one (with the same phase), the dc conductivity decreases dramatically with increasing the HF field. As shown by the solid curve in Fig. 5, if $E_2:E_1=1:1$, the dc drift velocity definitely becomes negative when E_1 and E_2 are near 1 kV/cm, namely, a weak positive dc potential drop can produce a negative dc current. Fig. 6 and Fig. 7 show that the HF components of the electron drift velocity increase approximately linearly with increase of the HF electric field in the region of HF field $E_1<1$ kV/cm. The second harmonic components $|\upsilon_2|$ induced by the pure E_1 field is less than 10^{-3} of value of $|\upsilon_1|$, shown in Fig. 7 (dotted curve). The retarded phases, respectively, $\phi_{\upsilon_1}\approx-3.7^\circ$ and $\phi_{\upsilon_2}\approx-7.3^\circ$, remain constant with increasing HF field. Fig. 8 shows the heating effect of the HF electric field. The average temperature T_0 increases with increasing HF field, roughly proportional to E_1^2 as anticipated. Fig. 9 and Fig. 10 show the HF components of the electron temperature as functions of E_1 and E_2. We see that the second harmonic component T_2 induced by E_1 is large, but the first harmonic component T_1 is small (see dotted curves). A large T_2 appears because the input power to the electron system in the relative coordinates is $-Ne\mathbf{E}(t)\cdot\mathbf{v}(t)-M\mathbf{v}(t)\cdot d\mathbf{v}(t)/dt$, and a large E_1 and υ_1 should produce a large second harmonic component in the power input to the energy of electrons in the relative coordinates. On the other hand, the terms such as $E_1\upsilon_0$, $E_1\upsilon_2$ . . ., which contribute to the first harmonic component T_1, are relatively small. Fig. 9 and Fig. 10 also show that the amplitudes of oscillating components, $|T_1|$ and $|T_2|$ usually have the same order of magnitude as that of the average increase of electron temperature T_0-T_L in the microwave region, and should not be neglected. The retarded phases are $\phi_{T_1}\approx-7.8^\circ$ and $\phi_{T_2}\approx-17^\circ$.

The appearance of the negative dc conductivity was explained by Pozhela[10], owing to the formation of the high-field-domain in the sample. This leads that the electric field exceeds the threshold of the negative differential mobility, which is about 3 4 kV/cm for GaAs. They also showed, in a qualitative argument, that the HF rectification effect in this region of negative differential mobility may lead to a negative dc conductivity. Our calculation shows another possible origin of the negative DC conductivity, that is production and amplification of (applied or induced) higher harmonic components of the field. Since this possible higher harmonic field does not require formation of a high-field-domain, it may occur more easily. Further experiments and theoretical study are needed to understand the origin of this negative dc conductivity induced by an HF field. Our calculation also warns that in the measurement of the drift velocity of electrons by use of the integral microwave technique, it is important to avoid the appearance of the high harmonic components, since the existence of higher harmonic component greatly affect the average dc current.

5. DISCUSSION

In summary, we have developed a new analytical method, which is tractable, although is not easy, to study the carrier transport in a strong HF electric field. In view of the derivation of our formulas, our theory could be valid in the higher frequency region where the Boltzmann theory is not available. We have calculated the electron transport in a n-GaAs sample in a microwave field. Our results show an important mixing frequency effect which produces a negative DC conductivity. Finally, we conclude this paper with a few remarks.

(1) It is obvious that our approach is a balance equation (BE) method. Recently, there have been some arguments about the validity of the balance equation method.[21,22] It has been shown that in the weak field limit, the conductivity based on the balance equation method disagrees with that obtained from the Boltzmann equation or the Kubo formula, when the lattice temperature is nonzero. This is a so called " $<1/\tau> \neq 1/<\tau>$ " paradox.[21,23,24] We notice that this paradox appears because both the BE method and the Boltzmann theory are compared in the realm of a non-interacting electron theory. In order to simplify the infinite freedom problem of the dynamics of carriers, people often introduce a finite number of parameters to describe the carrier system, and use a finite number of balance equations to evaluate these parameters. One of the BE methods is based on the assumption that the strong electron-electron interaction equilibrates electrons, so that the term $(\partial f_{\mathbf{k}}/\partial t)_{e-e}$ in the dynamical equation becomes zero. This BE method is an approximate approach for treating interacting electrons. In the weak field limit, if only electron-impurity scattering are considered, we know that from the Boltzmann equation for non-interacting electrons, the distribution function of electrons can be obtained as $f_{\mathbf{k}} = f_{\mathbf{k}}^0 + f_{\mathbf{k}}^1$, with $f_{\mathbf{k}}^0$ the equilibrium distribution of electrons without an electric field and

$$f_{\mathbf{k}}^1 = f_{\mathbf{k}}^0 \beta \frac{e\hbar}{m} \mathbf{k}\cdot\mathbf{E}\tau_k \tag{5.1}$$

In Eq. (5.1) τ_k is the well known energy dependent scattering time, the expression of which is given, for example, in Eq. (5) of Ref. 24; $\beta = 1/(K_B T_L)$. On the other hand, if we assume the distribution function of electrons is a drift Maxwellian distribution due to strong electron-electron interaction (the change of electron temperature is proportional to $\mathbf{E}^2$, and is zero in the weak field limit), by expanding the distribution function to linear term of velocity, we have

$$f_{\mathbf{k}}^1 = f_{\mathbf{k}}^0 \beta \frac{e\hbar}{m} \mathbf{k}\cdot\mathbf{E}\bar{\tau} , \tag{5.2}$$

with $\bar{\tau} = <1/\tau_k>^{-1}$ We see that due to electron-electron interaction, the distribution function of electrons changes by taking a constant $\bar{\tau}$ instead of $\tau_{\mathbf{k}}$ and is more isotropic. This difference between Eqs. (5.1) and (5.2) immediately produces that the conductivity of electrons obtained by this BE method is different from that deduced by the Boltzmann equation and the Kubo formula for non-interacting electrons.[21,23,24]. This is reasonable and acceptable. In fact, comparing theories which derive this kind of balance equations, we find that there are some assumptions. For example, the distribution of electrons in the relative coordinates at $t = -\infty$ is assumed as Fermi-Dirac at temperature T_e[25]; and the thermal fluctuation-dissipation theorem with temperature T_e is used.[26] These assumptions reduce the complexity of solving a problem for the whole distribution function by solving two (momentum and energy) balance equations, and, in our opinion, means the BE approach equivalent to the drifted temperature model. It is interesting that the temperature model used in the realm of the Boltzmann theory[25], can also be inserted into a non-Boltzmann approach, so as to include the memory effect for drift motion[14] or the high orders of electron-impurity and electron-phonon interaction.[26] There are other methods, such as the classical moment expansion method of Chapman and Cowling to solve the Boltzmann equation,[27] or recent work by Su *et al*,[28] in which a Gaussian approximation is used. These methods, of course, is for non-interacting carriers, and should obtain the same result as that from the Kubo formula in the weak field limit. This view point was further proven by Chen, Ting and Horing using the method of closed-time-path Green's functions.[29]

(2) In our calculation, we keep terms up to second harmonic components for drift velocity and the electron temperature. We also make the expansion up to the 3-rd order. Although the series expansions in Eqs. (2.19), (2.31) and (2.32) are convergent, the higher harmonic components and the higher terms of expansions may be important with increasing of the applied electric field. There is no difficulty, in principle, to include higher order terms, however, it will require more computing time. It is also important to include the inter-valley scattering in the high field region, since it induces the Gunn effect. A multi-carriers model[30] of BE method is available, however, this will double the number of equations. Our theory could be easily extended to the quasi-two dimensional electron system, such as in a heterostructure. It is also interesting to study the nonlinear transport of carriers in a higher frequency field. These shall constitute a future subject for study.

ACKNOWLEDGMENT

The work in the City college of City University of New York was partly supported by the U. S. Army Research Office and the U. S. Department of Energy.

APPENDIX

As we pointed before, the formulas in the text, especially Eq. (2.12b), imply a generalization beyond the lowest perturbation theory. This generalization has not been proven. In this appendix we use the path integral formulation for closed-time-path Green's function (CTPGF)[26,28,31] to derive the formulas in the text. A clue to our derivation is similarity to that presented by Su *et al*[28] but the system we deal with is a system of many electrons rather than a single electron system in Ref. 28.

To simplify the description, we consider an electron-LO-phonon system. The total Lagrangian is given by

$$L(t) = \frac{1}{2}M\dot{\mathbf{R}}(t)^2 - eN\mathbf{E}(t)\cdot\mathbf{R}(t) + \sum_{\mathbf{k}} c_{\mathbf{k}}^{\dagger}(t)S(\mathbf{k})^{-1}c_{\mathbf{k}} + \frac{1}{2}\sum_{\mathbf{q}} B_{-\mathbf{q}}(t)\Delta(\mathbf{q})^{-1}B_{\mathbf{q}}$$

$$+ \sum_{\mathbf{q}} M(\mathbf{q})B_{\mathbf{q}}(t)e^{i\mathbf{q}\cdot\mathbf{R}(t)}\rho_{\mathbf{q}}(t) , \tag{A1}$$

where $B_{\mathbf{q}} = b_{\mathbf{q}} + b_{-\mathbf{q}}^{\dagger}$ and $\rho_{\mathbf{q}} = \sum_{\mathbf{q}} c_{\mathbf{k}+\mathbf{q}}^{\dagger}c_{\mathbf{k}}$; $S(\mathbf{k})$ and $\Delta(\mathbf{q})$ are, respectively, the free propagators of electron and phonon. Their expressions can be found, for example, in Eq. (2.21) and (2.26) of Ref. 31. The meanings of other notations are the same as that in the text.

The expectation of an arbitrary observable O can be expressed as the following functional integral which is defined on a closed-time path,

$$<<O>> = \int [d\mathbf{R}][dB_{\mathbf{q}}][dc_{\mathbf{k}}^{\dagger}][dc_{\mathbf{k}}]O[\mathbf{R},\ B_{\mathbf{q}},\ c_{\mathbf{k}}^{\dagger},\ c_{\mathbf{k}}]\exp(iI_p[\mathbf{R}, B_{\mathbf{q}}, c_{\mathbf{k}}^{\dagger}, c_{\mathbf{k}}]) . \tag{A2}$$

The action defined on the closed-time path is given by

$$I_p = \int_p dt\, L(t) = \int_p dt \Bigg[\frac{1}{2}M\dot{\mathbf{R}}(t)^2 - eN\mathbf{E}(t)\cdot\mathbf{R}(t) + \sum_{\mathbf{q}} M(\mathbf{q})B_{\mathbf{q}}(t)e^{i\mathbf{q}\cdot\mathbf{R}(t)}\rho_{\mathbf{q}}(t)$$

$$+ \frac{1}{2}\int_p dt' \sum_{\mathbf{q}} B_{-\mathbf{q}}(t)\Delta^{-1}(\mathbf{q};\ t,\ t')B_{\mathbf{q}}(t') + \int_p dt' \sum_{\mathbf{k}} C_{\mathbf{k}}^{\dagger}(t)S^{-1}(\mathbf{k};\ t,\ t')C_{\mathbf{k}}(t') \Bigg] , \tag{A3}$$

with

$$\int_p dt = \int_{-\infty}^{+\infty} dt_+ + \int_{+\infty}^{-\infty} dt_- = \int_{-\infty}^{+\infty} dt_+ - \int_{-\infty}^{+\infty} dt_- \tag{A4}$$

Since the $B_{\mathbf{k}}$ dependence of I_p is quadratic, the functional integration over the phonon variable $\int [dB_{\mathbf{q}}]$ in Eq. (A.2) can be exactly worked out. We obtain an effective action

$I_{eff}[\mathbf{R}, C_{\mathbf{k}}^{\dagger}, c_{\mathbf{k}}]$, which is defined by

$$\exp(iI_{eff}[\mathbf{R}, c_{\mathbf{k}}^{\dagger}, c_{\mathbf{k}}]) = \int [dB_{\mathbf{q}}] \exp(iI_p[\mathbf{R}, B_{\mathbf{q}}, c_{\mathbf{k}}^{\dagger}, c_{\mathbf{k}}]) \tag{A5}$$

We have

$$I_{eff}[\mathbf{R}, c_{\mathbf{k}}^{\dagger}, c_{\mathbf{k}}] = \int_p dt(\frac{1}{2}M\dot{\mathbf{R}}(t)^2 - eN\mathbf{E}(t)\cdot\mathbf{R}(t)) + \int_p dt \int_p dt' \sum_{\mathbf{k}} c_{\mathbf{k}}^{\dagger}(t) S^{-1}(\mathbf{k}; t, t') c_{\mathbf{k}}(t')$$

$$+ \frac{1}{2}\sum_{\mathbf{q}} |M(\mathbf{q})|^2 \int_p dt \int_p dt' \Delta(\mathbf{q}; t, t') e^{i\mathbf{q}\cdot[\mathbf{R}(t)-\mathbf{R}(t')]} \rho_{\mathbf{q}}(t)\rho_{-\mathbf{q}}(t') . \tag{A6}$$

Also, the following formula is useful for eliminating the phonon variables:

$$<<\cdots B_{\mathbf{q}}(t)>> = -\int_p ds M(\mathbf{q})^* \Delta(\mathbf{q}; t-s) <\cdots e^{-i\mathbf{q}\cdot\mathbf{R}(s)} \rho_{-\mathbf{q}}(s)> , \tag{A7}$$

where

$$<<\cdots>> = \int [d\mathbf{R}][dB_{\mathbf{q}}][dc_{\mathbf{k}}^{\dagger}][dc_{\mathbf{k}}] \cdots \exp(iI_p) ,$$

$$<\cdots> = \int [d\mathbf{R}][dc_{\mathbf{k}}^{\dagger}][dc_{\mathbf{k}}] \cdots \exp(iI_{eff}) \tag{A8}$$

From the Lagrangian in Eq. (A1), the operator equation for the center-of-mass motion of electrons is obtained as

$$\dot{\mathbf{P}}(t) = -eN\mathbf{E}(t) + i\sum_{\mathbf{q}} M(\mathbf{q})\mathbf{q} B_{\mathbf{q}}(t) \exp[i\mathbf{q}\cdot\mathbf{R}(t)] \rho_{\mathbf{q}}(t)$$

Taking the expectation value in the nonequilibrium ensemble, we have

$$<\dot{\mathbf{P}}(t)> = M\frac{d\mathbf{v}(t)}{dt} = -eN\mathbf{E}(t) + i\sum_{\mathbf{q}} <<M(\mathbf{q})\mathbf{q} \exp[i\mathbf{q}\cdot\mathbf{R}(t)] \rho_{\mathbf{q}}(t) B_{\mathbf{q}}(t)>>$$

$$= -eN\mathbf{E}(t) - i\sum_{\mathbf{q}} |M(\mathbf{q})|^2 \mathbf{q} \int_p ds <e^{i\mathbf{q}\cdot\mathbf{R}(t)} \rho_{\mathbf{q}}(t) e^{-i\mathbf{q}\cdot\mathbf{R}(s)} \rho_{-\mathbf{q}}(s)> \Delta(\mathbf{q}; t-s) \tag{A9}$$

Here $\mathbf{v}(t)$ is the drift velocity of electrons. Until now we did not make any approximation. In the following we assume that the center-of-mass motion of electrons is a classical motion due to its huge mass, therefore, we neglect the functional integration $\int [d\mathbf{R}]$ and take only the classical path:

$$\exp[i\mathbf{q}\cdot(\mathbf{R}(t) - \mathbf{R}(s))] \approx \exp[i\mathbf{q}\cdot<\mathbf{R}(t) - \mathbf{R}(s)>] = \exp\left[i\mathbf{q}\cdot\int_s^t \mathbf{v}(\tau)d\tau\right] , \tag{A10}$$

The Eq. (A9) is now written as

$$M\frac{d\mathbf{v}(t)}{dt} = -eN\mathrm{E}(t) - i\sum_{\mathbf{q}} |M(\mathbf{q})|^2 \mathbf{q}\int_p ds\, e^{i\mathbf{q}\cdot\int_s^t \mathbf{v}(\tau)d\tau}\Delta(\mathbf{q};\, t-s)\Pi(\mathbf{q};\, t,\, s)\,, \tag{A11}$$

with

$$\Pi(\mathbf{q};\, t,\, s) = \langle T_p\, \rho_{\mathbf{q}}(t)\, \rho_{-\mathbf{q}}(s)\rangle\,, \tag{A12}$$

where T_p is the time-ordering operator along the closed time path. $\langle\cdots\rangle$ is defined in Eq. (A.8) without integration $\int[d\mathbf{R}]$. $\Pi(\mathbf{q};\, t,\, s)$ is related to the density-density correlation function in the relative coordinates. It is difficult to calculate it completely (note: $c_{\mathbf{k}}{}^{\dagger}$ and $c_{\mathbf{k}}$ are Grassman numbers[31]). However, its lowest order (free) term is easily obtained by making a perturbative expansion of $\exp(iI_{eff})$ and taking only the term of zero order of $[M(\mathbf{q})]$. Then, following the standard closed-time-path integral method (see Sec. V and Appendix in Ref. 26), we directly obtain Eq. (2.15) and Eq. (2.17) in the text.

It is easy to include the electron-impurity interaction in our formulas. The term of electron-electron interaction can be introduced without difficulty in our derivation, since this term is only related to the electron variable in the relative coordinates. In the limit of very strong electron-electron scattering, electrons are in equilibrium with each other and the term reflecting electron-electron scattering in the dynamical equation disappear. The procedure of derivation of the dynamical equation in the strong electron-electron interaction limit is the same as that without the electron-electron interaction, however, the distribution function for these two cases are different from each other. The former in the relative coordinates should be Fermi-Dirac (or Maxwellian) distribution, described by temperatue; the latter is nonequilibrium and is described by infinite parameters.

The difference between this derivation and the usual perturbative expansion is that we first take the classical approximation for the center-of-mass motion of electrons and make the perturbation expansion only for the electron variable in the relative coordinates. The usual perturbation theory, on the other hand, makes perturbative expansion for both mass-of-center and relative variables of electrons to the lowest order, therefore, $\langle\mathbf{R}(t)-\mathbf{R}(s)\rangle = \mathbf{v}(t)(t-s)$ in that case, which induces to the Boltzmann theory.

[1]Y. K. Pozhela, "*Hot-electron transport in semiconductors*", Edited by L. Reggiani (Springer-Verlag 1985) Chapter 4. 113; J. Pozhela, "*Plasma and Current Instabilities in Semiconductors*", Intern. Ser. Sci. Solid State, Vol. 18 (Pergamon, Oxford 1981).

[2]A. Ron and N. Tzoar, Phys. Rev. 131, 1943 (1963).

[3]W. Götze and P. Wöfle, Phys. Rev. B 6, 1226 (1976).

[4]N. Tzoar and C. Zhang, Phys. Rev. B 32, 1149 (1985); 33, 2642 (1986); 35, 7596 (1987); 36, 7906 (1987).

[5]X. L. Lei, N. J. Horing, and J. Q. Zhang, Phys. Rev. 35, 2834 (1987).

[6]H. D. Rees, Solid State Commu., 7, 267 (1969).

[7]P. Das and D. K. Ferry, Solid State Electronics, 19, 851 (1976).

[8]P. A. Lebwohl, J. Appl. Phys., 44 1744 (1973).

[9]Y. Pozhela and A. Reklaitis, Sov. Phys. Semi, 13, 660 (1979).

[10]T. J. Banys, I. V. Parsheliunas, and Y. K. Pozhela, Liet. Fis. Rinkinys 11, 1013, (1971); Fis. Tekh. Polupr 5, 1990 (1971).

[11]K. Ashida, M. Inoue, and J. Shirafuji, J. Phys. Soc. Jpn. 37, 408 (1974).

[12]J. R. Barker, J. Phys. C 6, 1663 (1973).

[13]F. S. Khan, J. H. Davies, and J. W. Wilkins, Phys. Rev. B 36, 2578 (1987).

[14]W. Cai, P. Hu, T. F. Zheng, B. Yudanin, and M. Lax, Phys. Rev. B 40, 7671 (1989).

[15]C. S. Ting, S. C. Ying, and J. J. Quinn, Phys. Rev. B 16, 5394 (1977).

[16]For example, "*Physics of Nonlinear Transport in Semiconductors*", edited by D. K. Ferry *et al.* (Plenum, New York, 1980).

[17]C. S. Ting and T. W. Nee, Phys. Rev. 33, 7056 (1986); D. Y. Xing and C. S. Ting, Phys. Rev. 35, 3971 (1987).

[18]D. N. Zubarev, "*Nonequilibrium statistical Thermodynamics*", (Consultants Bureau, New York, 1974).

[19]"*Handbook of Mathematical Functions*", No. 55 of Applied Mathematics, National Bureau of Standards, Edited by M. Abramowitz and I. A. Stegun (U. S. GPO, Washington, D. C., 1964).

[20]P. A. Fox, A. D. Hall, and N. L. Schryer, "*The Port Mathematical Subroutine Library*", Computing Science Technical Report #47, (1976).

[21]R. S. Fishman and G. D. Mahan, Phys. Rev. B 39, 2990 (1989); 1994 (1989).

[22]P. A. Argyres, Phys. Rev. B 39, 2982 (1989).

[23]N. J. M. Horing, X. L. Lei, and H. L. Cui, Phys. Rev. B 33, 6929 (1986).

[24]W. Y. Lai and C. S. Ting, Phys. Rev. B 24, 7206 (1981).

[25]X. L. Lei and C. S. Ting, Phys. Rev. B 32, 1112 (1985).

[26]Z. B. Su, L.Y. Chen, and J. L. Birman, Phys. Rev. B 35, 9744 (1987).

[27]S. Chapman and T. G. Cowling "*Mathematical theory of Non-uniform Gases*", Cambridge University Press, New York, 1951.

[28]Z. B. Su, L. Y. Chen, and C. S. Ting, Phys. Rev. Lett. 60, 2323 (1988); Phys. Rev. B 38, 3886 (1988).

[29]L. Y. Chen, C. S. Ting, and N. J. M. Horing, Phys. Rev. B 40, 3756 (1989); 42, 1140 (1990).

[30]D. Y. Xing, M. Liu, and C. S. Ting, Phys. Rev. B 37, 10283 (1988).

[31]K. C. Chou, Z. B. Su, B. L. Hao, and L. Yu, Phys. Rep. 118, 1, (1985).

International Journal of Modern Physics B, Vol. 6, No. 7 (1992) 1037–1057

NONEQUILIBRIUM STATISTICAL OPERATOR IN HOT-ELECTRON TRANSPORT THEORY

D.Y. Xing
Department of Physics, Nanjing University, Nanjing 210008, China,
and International Center for Theoretical Physics, Trieste, Italy

M. Liu
Department of Physics, Southeast University, Nanjing 210018, China,
and International Center for Theoretical Physics, Trieste, Italy

The Nonequilibrium Statistical Operator method developed by Zubarev is generalized and applied to the study of hot-electron transport in semiconductors. The steady-state balance equations for momentum and energy are derived to the lowest order in the electron-lattice coupling. We show that the derived balance equations are exactly the same as those obtained by Lei and Ting. This equivalence stems from the fact that to the linear order in the electron-lattice coupling, two statistical density matrices have identical effect when they are used to calculate the average value of a dynamical operator. The application to the steady-state and transient hot-electron transport in multivalley semiconductors is also discussed.

1. Introduction

Since the early pioneering measurement[1] of nonlinear transport behavior in Ge and the later discovery of the Gunn effect[2] in GaAs, hot-electron transport in semiconductors has received a great deal of attention[3] Rapid development of submicrometer semiconductor devices, which may be employed in high-speed computers and telecommunication systems, further enhances the importance of hot-electron transport, since miniaturization of devices has led to high-field strengths well outside the linear-response region, where Ohm's law holds, for any reasonable voltage signal. For example, applying a potential difference of 1 V across a field effect transistor of length 0.25 μm gives a field of 40 KV/cm which far exceeds the typical fields at room temperature required to produce hot electrons.

Physical understanding of most of the microscopic processes which underlie the performances of semiconductor devices at high electric fields is provided by research into hot-electron phenomena[4] Although there exists a number of theoretical approaches to this problem, most of the realistic calculations concern themselves with the solution of the Boltzmann transport equation[3–5] and numerical methods based on the iterative procedure[6] or the Monte Carlo simulation[7,8] In particular, with the aid of modern large and fast computers, it would become possible to obtain exact numerical solutions of the Boltzmann equation for microscopic physical models.

At the same time, analytical methods for studying the hot-electron transport under the action of strong electric fields are also desirable to obtain physical insights and for the interpretation of new phenomena. Recently an analytical balance-equation approach to hot-electron transport has been developed by Lei and Ting[9], based on a separation of the center-of-mass motion from the relative motion of electrons[10]. Concise balance equations, written in terms of a electron density correlation function, provide quite a good description for the steady-state and transient transport in many-body systems and have been applied to a large class of experimental situations[11] It is further argued that this analytical approach has the capacity to go beyond the semiclassical Boltzmann approach if the density correlation functions in the balance equations are calculated beyond the random-phase approximation[12].

It is well known that hot-electron transport is a typical irreversible process of dissipative dynamic systems. It is very important to construct a statistical density matrix adequate to describe irreversible processes of the nonequilibrium systems. The balance-equation approach[9] is based on the use of a equilibrium distribution for a decoulped electron-lattice system as an initial condition for the Liouville's equation. This way of constructing a statistical density matrix seems to be different from the Nonequilibrium Statistical Operator (NSO) method[13], but similar to the early Mori's approach based on the solution of the Liouville's equation with the initial condition of local equilibrium[14] Then it is necessary to make a careful comparison between the balance-equation approach and the NSO method. An agreement between them will provide a powerful backing for the theoretical basis of the balance-equation approach.

The NSO method is an advanced generalization of statistical methods based on Boltzmann and Gibbs fundamental ideas, and it is considered as a powerful formalism that offers an elegant and concise way for an analytical treatment in the theory of irreversible processes. Among various types of NSO the elegant treatment of Zubarev[15,16] deserves particular attention. So we resort in this work to the use of the Zubarev's method to study steady-state and transient hot-electron transport properties. The NSO method was applied to the study of transport properties in solid-state systems first by Kalashnikov[17] Since he made a weak-field assumption that the electron drift velocity v_e is much smaller than the mean velocity of an eletron, the balance equations obtained by Kalashnikov is valid only to the linear order in v_e. They may be suitable to so-called warm-electron transport, but can not correctly describe nonlinear transport behavior in the hot-electron region. Recently, new applications of the NSO method to the hot-electron transport has been made by authors and their co-workers[18–21], the derived balance equations being identical to those obtained by Lei and Ting[9] Although their equivalence was shown[18] in the case of neglecting the electron-electron interaction, here we shall show that the inclusion of the Coulombic many-body effect does not change this conclusion. It then follows that these two approaches must have something in common. For the steady-state case, the statistical density matrices, used in these two approaches, will be pointed out to have identical effect on the calculation of averages to the lowest order in the electron-lattice coupling.

When the NSO method and balance-equation approach are applied to the transient hot-electron transport problem, superficially, there is a small difference between them, the former including seemingly more memory or intra-collisional field effect. In reality, their difference can be shown to be of the high order of magnitude in the electron-lattice coupling. Consequently, in second-order approximation in the electron-lattice coupling, the same evolution equations for the transient hot-electron transport can be obtained by the NSO method and by the extension of the balance-equation approach. It has been shown that for electric fields with moderate strengths, the memory effect included in the evolution equations has little influence upon the solutions of evolution equations[22], and the calculated results for the evolution equations without memory effect coincide well with those obtained from the Monte Carlo simulation[20,21].

In Section 2 we give the Hamiltonian of an electron-lattice system including Coulombic many-body effect, in terms of a separation of the center-of-mass motion from the relative motion of electrons. In Sections 3 we proceed to construct a NSO in the relative electron-lattice system along the line proposed by Zubarev. Section 4 is devoted to the derivation of the steady-state balance equations for momentum and energy based on the NSO method. There we show the equivalence of the derived balance equations to the Lei-Ting equations[9] and trace this equivalence to its source. Sections 5 and 6 are devoted to the application of the NSO method to the steady-state and transient hot-electron transport in multivalley semiconductors. Finally concluding remarks are included in Section 7.

2. Hamiltonian

Let us consider an electron-lattice system, which consists of N conduction electrons interacting with each other, in an applied uniform electric field $\mathbf{E}$. The electrons are accelerated by the applied field and scattered by phonons and n_i randomly distributed impurities, forming an electric current with drift velocity v_e. The total Hamiltonian of this system can be written in terms of the center-of-mass variables and electron variables in the relative coordinates[9]:

$$H = H_c + H_e + H_{ph} + H_{el}, \tag{1}$$

where H_c and H_e stand for the center-of-mass and relative parts of the electron Hamiltonian, respectively. H_{ph} is the free-phonon Hamiltonian, and H_{el} is the interaction of carriers with the lattice.

For hot-electron transport problems, it is necessary to separate the motion of the center of mass from the relative motion of the electrons. The former can be regarded as a mechanical movement of a single body with mass Nm, which obeys the classical equation of motion. Its drift velocity v_e depends on the external drifting force and internal resistance due to the electron-lattice coupling. The Hamiltonian H_c includes the kinetic energy of the center of mass and its potential energy in the external field:

$$H_c = \frac{\mathbf{P}^2}{2Nm} - Ne\mathbf{E}\cdot\mathbf{R}, \tag{2}$$

with $\mathbf{P}$ and $\mathbf{R}$ as the momentum and coordinate for the center of mass, respectively.

The relative electron system, which consists a large number of interacting particles, should satisfy the quantum statistics. It was in the statistical ensemble of the relative electron systems that the effective temperature T_e of hot electrons can be defined. In the representation of second quantization,

$$H_e = \sum_{\mathbf{p}} \varepsilon_{\mathbf{p}} c_{\mathbf{p}}^{\dagger} c_{\mathbf{p}} + (1/2) \sum_{\mathbf{q}} v_c(\mathbf{q}) \rho_{\mathbf{q}} \rho_{-\mathbf{q}}. \tag{3}$$

Here $c_{\mathbf{p}}^{\dagger}$ and $c_{\mathbf{p}}$ are the Fermi operators in the state with energy $\varepsilon_{\mathbf{p}}$ and momentum $\mathbf{p}$, $\varepsilon_{\mathbf{p}} = \mathbf{p}^2/2m$ for a parabolic energy band. $\rho_{\mathbf{q}} = \sum_{\mathbf{q}} c_{\mathbf{p}+\mathbf{q}}^{\dagger} c_{\mathbf{q}}$ is the electron density operator, and $v_c(\mathbf{q})$ being the Coulomb potential between electrons.

$$H_{ph} = \sum_{\mathbf{q},\lambda} \Omega_{\mathbf{q}\lambda} b_{\mathbf{q}\lambda}^{\dagger} b_{\mathbf{q}\lambda}. \tag{4}$$

$b_{\mathbf{q}\lambda}^{+}$ and $b_{\mathbf{q}\lambda}$ in the above equation are the phonon field operators in the state with momentum $\mathbf{q}$, energy $\Omega_{\mathbf{q}\lambda}$, and polarization index λ. Two types of electron-lattice coupling will be taken into account, one is the electron-phonon interaction, the other is the interaction with randomly distributed impurities:

$$H_{el} = \sum_{\mathbf{q},a} u(\mathbf{q}) e^{i\mathbf{q}(\mathbf{R}-\mathbf{R}a)} \rho_{\mathbf{q}} + \sum_{\mathbf{q},\lambda} M(\mathbf{q},\lambda) e^{i\mathbf{q}\mathbf{R}} (b_{\mathbf{q}\lambda} + b_{-\mathbf{q}\lambda}^{\dagger}) \rho_{\mathbf{q}}, \tag{5}$$

where $u(\mathbf{q})$ is the impurity potential in momentum space, and $M(\mathbf{q},\lambda)$ is the electron-phonon matrix element, which satisfies $M(\mathbf{q},\lambda) = M^*(-\mathbf{q},\lambda)$.

We wish to point out that the electron-lattice coupling plays an important role in the hot-electron transport theory. On the one hand, it is considered a small perturbation so that the frequency of electron-lattice collisions τ_{el}^{-1} is much smaller than that of mutual collisions of electrons τ_{ee}^{-1}. Then it is possible to assume the nonequilibrium distribution of electrons to be dependent on a set of macroparameters such as effective temperature and chemical potential of hot electrons. Further, the effective temperature of hot electrons may be much higher than the lattice temperaure. On the other hand, the electron-lattice interaction is also indispensable to irreversible transport processes since all dissipative mechanisms are included in it. Therefore, the transport equations are usually set up to the lowest order in H_{el}.

3. Nonequilibrium Statistical Operator

We shall consider the nonequilibrium ensemble of the systems with the Hamiltonian in (1). Let us assume that the nonequilibrium state of the ensemble may be described by a basic set of average values $< P_m >^t = Tr\{P_m \rho(t)\}$ of some operators P_m. Here $\rho(t)$ is the NSO which will be defined later. As the first step of constructing the NSO, which satisfies Liouville's equation and can describe irreversible transport processes, a quasi-equilibrium statistical operator is introduced as

$$\rho_l = exp\{-\phi - \sum_m P_m F_m(t)\} \equiv exp[-S(t,0)], \tag{6}$$

with

$$\phi = lnTr\{exp[-\sum_m P_m F_m(t)]\},$$

where S(t,0) is the entropy operator of the system. $F_m(t)$ are the thermodynamic parameters conjugate to the average values of P_m, and they are determined from the conditions[16]:

$$< P_m >_l^t \equiv Tr\{P_m \rho_l\} = Tr\{P_m \rho(t)\} \equiv < P_m >^t . \tag{7}$$

In the application of the NSO method to a specific problem of the statistical mechanics of irreversible processes, one of key steps is to choose operators P_m and their thermodynamically conjugated parameters $F_m(t)$. For the hot-electron transport with strong electric field and weak electron-lattice coupling, the energy which relative electrons obtain from the external field is redistributed among them in such a way that the relative electron system becomes almost in equilibrium within itself, but not in equilibrium with respect to the lattice. So ρ_l should be defined in both the relative electron and free phonon systems. A most forthright choice of P_m and $F_m(t)$ is[19,22]

$$P_m = \{H_e;\ H_{ph}\},$$

and

$$F_m(t) = \{\beta_e(t);\ \beta\}, \tag{8}$$

so that

$$\rho_l = exp\{-\phi - \beta_e(t)H_e - \beta H_{ph}\}, \tag{9}$$

where $\beta_e(t)$ is the inverse effective temperature of hot electrons, and β the inverse equilibrium temperature of the lattice. This statistical operator is close in form to the equilibrium Gibbsian distribution but with different parameters β and β_e which depends on time. Its physical meaning is easily understood. Imagine that the electron-lattice interaction, together with the electric field, is mathematically cut off at the time t, so that the center of mass, the relative electrons, and the phonon system are decoupled from each other. Then the center of mass moves with a constant velocity $v_e(t)$, and the relative electrons and free phonons, as two isolated systems, attain their thermal equilibrium states very quickly, respectively, with hot-electron temperature $T_e(t)$ and lattice temperature T. The above picture requires that there exist strong collisions among electrons ($\tau_{ee} << t$, t being the characteristic time of variation of the external field) so that they can be thermalized immediately after the field is applied. The $\rho_l(t)$ given by (9) is a generalized Gibbsian distribution of the decoupled systems in their frozen equilibrium states, and it can be used to determine instantaneous values of the macrovariables from the conditions (7).

In the application of the NSO method to hot-electron transport theory, Kalashnikov[17] chose a more complicate form for P_m and $F_m(t)$. They are

$$P_m = \{H_e';\ \mathbf{P};\ N;\ \mathbf{P}_l;\ H_{ph} + H_{el}\},$$

and

$$F_m(t) = \{\beta_e(t);\ -v_e(t)\beta_e(t);\ -\beta_e(t)[\mu(t) - mv_e^2(t)/2];\ 0;\ \beta\}. \tag{10}$$

Here H_e' is the electron Hamiltonian in the laboratory system, N is the number-of-electron operator, and $\mathbf{P}_l$ is the momentum operator of the lattice. $\mu(t)$ is the nonequilibrium chemical potential of hot electrons.

$\rho_l(t)$, which is obtained by substituting (10) into (6), differs superficially from that defined by (9). In reality, it can be shown that they are almost identical. First, the complex structure of (10) stems mainly from the fact that the $\rho_l(t)$ is defined in the laboratory system. By means of a canonical transformation from the relative electron system into the laboratory one[16], the expression (9) for ρ_l becomes

$$\rho_l(t) = exp\{-\phi - \beta_e(t)[H'_e - v_e(t)P - Nmv_e^2(t)/2] - \beta H_{ph}\}, \tag{11}$$

from which one can understand the origin of the operators H'_e, $\mathbf{P}$ and N which appear in (10). Second, the introduction of operators N and $\mathbf{P}_l$ into the set of P_m in (10) doesn't seem to be necessary. This is because a statistical ensemble of the relative electron systems with constant electron number can be described by a cononical distribution, and the average momentum of the lattice is usually assumed to be zero. Finally, we wish to point out that if introducing the electron-lattice interaction H_{el} into the set of P_m , as used in (10), one has some trouble in choosing its thermodynamically conjugated parameter F_m. The trouble is that it is difficult to determine whether the F_m corresponding to H_{el} is chosen to be β of the lattice, or $\beta_e(t)$ of the hot electrons, or an unknown function of both. Fortunately it has been shown[18] that in the approximation to the lowest order of H_{el}, which is usually used in hot-electron transport theory, the choice of this F_m is irrelevent to the final expression of the balance equations in steady-state hot-electron transport. So it is quite convenient to exclude H_{el} from the set of P_m from the outset. Therefore, the two choices of P_m and $F_m(t)$ are approximately equivalent to each other. Since P_m and $F_m(t)$ given by (8) have a much simpler form, in this work, we shall apply them to construct the NSO, and then to derive the balance equations for hot-eletron transpot. If Kalashnikov's choice in the form (10) is used, the same result for the balance equations can also be obtained as long as all the correlation functions which appear in the balance equations are correctly calculated[18].

The density matrix of the real system, of course, deviates from the quasi-equilibrium distribution because the ρ_l can neither satisfy Liouville's equation nor describe irreversible processes. However, it can be used to formulate boundary conditions to the Liouville's equation[16], in the same way as the free-particle wave function is used to formulate boundary conditions to the Schrodinger equation in the quantum theory of scattering.

It is well known that the Liouville's equation is symmetric under the time-reversal transformation (in the quantum case, this means the replacement $(t, iL) \Rightarrow (-t, -iL^+)$, where L is the Liouville's operator of the system). However, the macroscopic evolution of the system with dissipative structures must be irreversible. This poses the question of how to obtain irreversible behavior in the macroscopic description of the state of the system generated by a statistical operator satisfying Liouville's equation. Zubarev's approach to this question is to introduce into Liouville's equation an infinitesimally small source that breaks time-reversal symmetry[16]. Then the NSO for the relative electron-lattice system has the form

$$\rho(t) = exp\{-\varepsilon \int_{-\infty}^{0} dt' e^{\varepsilon t'} S(t + t', t')\}, \tag{12}$$

with

$$S(t, t') = e^{iHt'} S(t, 0) e^{-iHt'},$$

$$S(t,0) = \phi(t) + \beta_e(t)H_e + \beta H_{ph}. \tag{13}$$

where ε is a positive number and will tend to zero after the thermodynamic limiting process in the calculation of averages. As shown in (13), the first argument of $S(t,t')$ denotes the time dependence though the parameters $F_m(t)$, and the second denotes the Heisenberg picture time dependence of operators $P_m(t)$. Integrating by part, the operator $\rho(t)$ can be written as

$$\rho(t) = exp\{-S(t,0) + \int_{-\infty}^{0} dt' e^{\varepsilon t'} \dot{S}(t+t',t')\}, \tag{14}$$

where

$$\dot{S}(t,t') = e^{iHt'}\dot{S}(t,0)e^{-iHt'},$$

with

$$\dot{S}(t,0) = \beta_e(t)\dot{H}_e + \beta\dot{H}_{ph} + \dot{\beta}_e(t)(H_e - < H_e >^t), \tag{15}$$

as the entropy production operator. In the limit of E=0 and $\beta_e(t) = \beta$, the second term in the exponent of (14) can be shown to be $-\beta H_{el}$ when the entropy production operator is evaluated to the lowest order in the H_{el}. Then the density matrix reduces to the well-known form $\rho(t) = exp(-\phi - \beta H)$ for the equilibrium state. It has been shown[22] that the first two terms on the right hand side of (15) are of the first order in H_{el} whereas the last term is at least second order of magnitude in H_{el}. In the hot-electron transport theory, as already mentioned, the H_{el} is usually regarded as a small perturbation since the energy exchange between the electron and lattice systems is generally assumed to be small. If we confine $\rho(t)$ to terms of the first order in H_{el}, the last term in (15) may be neglected. Further, using the following operator equation in the approximation linear in B:

$$exp(A+B) \simeq \{1 + \int_0^1 d\tau e^{A\tau}Be^{-A\tau}\}e^A, \tag{16}$$

where A and B stand for arbitrary operators, and replacing A anb B by -S(t,0) and $\int_{-\infty}^{0} dt' e^{\varepsilon t'}\dot{S}(t+t',t')$, respectively, we obtain the expansion of $\rho(t)$ to the first-order approximation in H_{el}:

$$\rho(t) = \{1 + \int_{-\infty}^{0} dt' e^{\varepsilon t'} \int_0^1 d\tau[\beta_e(t+t')\dot{H}_e(t',i\tau) + \beta\dot{H}_{ph}(t',i\tau)]\}\rho_l. \tag{17}$$

where

$$\dot{H}_e(t',i\tau) = e^{-\tau S(t,0)}e^{iHt'}\dot{H}_e e^{-iHt'}e^{\tau S(t,0)}, \tag{18}$$

and

$$\dot{H}_e = -i[H_e,H] = -\sum_{\mathbf{q},a} u(\mathbf{q})e^{i\mathbf{q}(\mathbf{R}-\mathbf{R}a)}\dot{\rho}_{\mathbf{q}} - \sum_{\mathbf{q},\lambda} M(\mathbf{q},\lambda)e^{i\mathbf{q}\mathbf{R}}(b_{\mathbf{q}\lambda} + b^\dagger_{-\mathbf{q}\lambda})\dot{\rho}_{\mathbf{q}}, \tag{19}$$

with

$$\dot{\rho}_{\mathbf{q}} = -i[\rho_{\mathbf{q}},H] = -i[\rho_{\mathbf{q}},H_e]. \tag{20}$$

The $\dot{H}_{ph}(t', i\tau)$ has the same dependences of t' and τ as $\dot{H}_e(t', \tau)$ given by (18), but

$$\dot{H}_{ph} = i \sum_{\mathbf{q},\lambda} M(\mathbf{q},\lambda)\Omega_{\mathbf{q}\lambda} e^{i\mathbf{q}\mathbf{R}}(b_{\mathbf{q}\lambda} - b^{\dagger}_{-\mathbf{q}\lambda})\rho_{\mathbf{q}}. \tag{21}$$

By means of the approximate expression (17) for $\rho(t)$, one can obtain the balance equations in second-order approximation in the electron-lattice interaction.

4. Balance Equations For Momentum and Energy

We now focus our attention on steady-state hot-electron transport. In steady state, the center of mass moves with a constant v_e along the electric field direction. At the same time, the total energy of the relative electron system must be constant, the energy supplied per unit time by the electric field to the electron system being equal to the energy transfer rate from the electron system to the lattice. Therefore, the steady-state hot-electron transport is determined by the following balance equations for momentum and energy[17]:

$$Nm\dot{v}_e = <\dot{\mathbf{P}}>^t = 0, \tag{22}$$

$$c_e\dot{T}_e = <\dot{H}_e>^t = 0, \tag{23}$$

where c_e is the heat capacity of the relative electron system, and the averages with the NSO will be approximately calculated by means of (17).

If the external electric field $\mathbf{E}$ is assumed to be along the x-direction, the ceter-of-mass momentum operator $\mathbf{P}$ has only x-component P_x and its derivative is equal to

$$\dot{P}_x = NeE + \dot{P}_{el}, \tag{24}$$

with

$$\dot{P}_{el} = -i\sum_{\mathbf{q},a} q_x u(\mathbf{q}) e^{i\mathbf{q}(\mathbf{R}-\mathbf{R}_a)}\rho_{\mathbf{q}} - i\sum_{\mathbf{q},\lambda} q_x M(\mathbf{q},\lambda) e^{i\mathbf{q}\mathbf{R}}(b_{\mathbf{q}\lambda} + b^{\dagger}_{-\mathbf{q}\lambda})\rho_{\mathbf{q}}, \tag{25}$$

Substituting (17) into (22) and (23), we obtain

$$NeE + \int_{-\infty}^{0} dt' e^{\epsilon t'} \int_0^1 d\tau \{\beta_e < \dot{P}_{el}\dot{H}_e(t', i\tau) >_l + \beta < \dot{P}_{el}\dot{H}_{ph}(t', i\tau) >_l\} = 0, \tag{26}$$

$$\int_{-\infty}^{0} dt' e^{\epsilon t'} \int_0^1 d\tau \{\beta_e < \dot{H}_e\dot{H}_e(t', i\tau) >_l + \beta < \dot{H}_e\dot{H}_{ph}(t', i\tau) >_l\} = 0. \tag{27}$$

Here all the correlation functions should be calculated in the relative electron-lattice system by means of the quasi-equilibrium density matrix ρ_l given by (9). Since in the steady state the ρ_l is independent of time t, for simplicity, we have omitted the superscripts of all the correlation functions. The solutions of the balance equations (26) and (27) determine the steady-state values v_e and β_e as functions of the external field.

The next step is to calculate the correlation functions which appear in the balance equations (26) and (27). In Ref. 17, a weak-field approximation was made under the consideration of warm-electron transport. On the basis of this approximation, only the diagonal-correlation functions like $< \dot{P}_{el}\dot{P}_{el}(t', i\tau) >_l$ and $< \dot{H}_e\dot{H}_e(t', i\tau) >_l$ were retained, while all the off- diagonal-correlation functions in the balance equations were neglected[17] In the region of hot-electron transport, however, this kind of approximation will not work. It has been shown[18] that both the diagonal- and off-diagonal-correlation functions play important roles in hot-electron transport theory, and the latter can vanish only when $\beta_e = \beta$.

Up to this point, we have considered the general case of including the Coulomb interaction between carries. In the following we shall calculate the correlation functions in (26) and (27) in two separate cases. First we assume that the H_e has a simple form of a free-electron Hamiltonian. This corresponds to the case in which the influence of mutual scattering of electrons may be taken into account by a normalization of an electron-lattice coupling. Then the effect due to the electron-electron interaction, which is explicitly included in (3), will also be considered.

For free-electron and phonon operators, there are the following relationships[17,18]:

$$c_k(t, i\tau) = c_k exp(\tau\beta\varepsilon_k - it\varepsilon_k), \tag{28}$$

$$b_{q\lambda}(t, i\tau) = b_{q\lambda} exp(\tau\beta\Omega_{q\lambda} - it\Omega_{q\lambda}), \tag{29}$$

from which the t and τ dependences of the correlation functions can be taken from the calculation of averages and the integrations over τ and t' in (26) and (27) are easily done[18] Taking the energy balance equation (27) as an example, we have

$$\begin{aligned}
&\int_0^1 d\tau\{\beta_e < \dot{H}_e\dot{H}_e(t', i\tau) >_l + \beta < \dot{H}_e\dot{H}_{ph}(t', i\tau) >_l\} \\
&= n_i \sum_{\mathbf{q},\mathbf{k}} | u(\mathbf{q}) |^2 \omega_{\mathbf{kq}}[\exp(\beta_e\omega_{\mathbf{kq}}) - 1] \exp[-\mathbf{i}\mathbf{t}'(\omega_{\mathbf{kq}} + \mathbf{q_x v_e})] \\
&\times < c^\dagger_{\mathbf{k+q}}c_{\mathbf{k}}c^\dagger_{\mathbf{k}}c_{\mathbf{k+q}} >_l + \sum_{\mathbf{K},\mathbf{q},\lambda} | M(\mathbf{q},\lambda) |^2 \omega_{\mathbf{kq}} < c^\dagger_{\mathbf{k+q}}c_{\mathbf{k}}c^\dagger_{\mathbf{k}}c_{\mathbf{k+q}} >_l \\
&\times\{< b_{\mathbf{q}\lambda}b^\dagger_{\mathbf{q}\lambda} >_l [\exp(\beta_e\omega_{\mathbf{kq}} - \beta\Omega_{\mathbf{q}\lambda}) - 1] \exp[-it'(\omega_{\mathbf{kq}} + q_x v_e - \Omega_{\mathbf{q}\lambda})] \\
&+ < b^\dagger_{-\mathbf{q}\lambda}b_{-\mathbf{q}\lambda} >_l [\exp(\beta_e\omega_{\mathbf{kq}} + \beta\Omega_{\mathbf{q}\lambda}) - 1] \exp[-it'(\omega_{\mathbf{kq}} + q_x v_e + \Omega_{\mathbf{q}\lambda})]\},
\end{aligned} \tag{30}$$

where $\omega_{\mathbf{kq}} = \varepsilon_{\mathbf{k+q}} - \varepsilon_{\mathbf{k}}$. In deriving of (30) we have used the relation $\mathbf{q}\cdot[\mathbf{R}(t') - \mathbf{R}(0)] = q_x v_e t'$ for the steady-state transport under a constant field along the x-direction. Further, taking into account the following relations

$$< c^\dagger_{\mathbf{k+q}}c_{\mathbf{k}}c^\dagger_{\mathbf{k}}c_{\mathbf{k+q}} >_l = f(\beta_e\varepsilon_{\mathbf{k+q}})[1 - f(\beta_e\varepsilon_{\mathbf{k}})], \tag{31}$$

$$< b_{\mathbf{q}\lambda}b^\dagger_{\mathbf{q}\lambda} >_l = 1 + < b^\dagger_{\mathbf{q}\lambda}b_{\mathbf{q}\lambda} >_l = 1 + n(\beta\Omega_{\mathbf{q}\lambda}), \tag{32}$$

with f(x) and n(x) being the usual Fermi and Bose functions, respectively, and

$$\int_{-\infty}^0 dt' \exp[-it'(\omega_{\mathbf{kq}} + q_x v_e \pm \Omega_{\mathbf{q}\lambda}) + \varepsilon t'] = i[\omega_{\mathbf{kq}} + q_x v_e \pm \Omega_{\mathbf{q}\lambda} + i\varepsilon]^{-1}, \tag{33}$$

with real part being $\pi\delta(\omega_{\mathbf{kq}} + q_x v_e \pm \Omega_{\mathbf{q},\lambda})$, we finally obtion for the energy balance equation

$$2\sum_{\mathbf{q},\lambda} | M(\mathbf{q},\lambda) |^2 (q_x v_e + \Omega_{\mathbf{q}\lambda})\Pi_2^0(\mathbf{q}, q_x v_e + \Omega_{\mathbf{q}\lambda})[n(\beta\Omega_{\mathbf{q}\lambda}) - n(\beta_e(q_x v_e + \Omega_{\mathbf{q}\lambda}))]$$

$$+ n_i \sum_{\mathbf{q}} | u(\mathbf{q}) |^2 q_x v_e \Pi_2^0(\mathbf{q}, q_x v_e) = 0, \qquad (34)$$

where

$$\Pi_2^0(\mathbf{q},\omega) \equiv Im\Pi^0(\mathbf{q},\omega) = 2\pi \sum_{\mathbf{k}} [f(\beta_e \varepsilon_{\mathbf{k}}) - f(\beta_e \varepsilon_{\mathbf{k+q}})]\delta(\omega - \omega_{\mathbf{k+q}}), \qquad (35)$$

is the imaginary part of the density-density correlation function in the absence of the Coulomb interaction between electrons.

Using a same procedure to that outlined above, we can obtain for the momentum balance equation

$$NeE + 2\sum_{\mathbf{q},\lambda} | M(\mathbf{q},\lambda) |^2 q_x \Pi_2^0(\mathbf{q}, q_x v_e + \Omega_{\mathbf{q}\lambda})[n(\beta\Omega_{\mathbf{q}\lambda}) - n(\beta_e(q_x v_e + \Omega_{\mathbf{q}\lambda}))]$$

$$+ n_i \sum_{\mathbf{q}} | u(\mathbf{q}) |^2 q_x \Pi_2^0(\mathbf{q}, q_x v_e) = 0. \qquad (36)$$

Equations (35) and (36) are the main results derived from the NSO method. Comparing them with the results in Ref. 17, we find that two sets of balance equations are quite different in the region of hot-electron transport. The momentum balance equations agree with each other only when $\beta_e = \beta$, while the energy balance equations are the same in the appoximation linear in v_e[18]

To see clear the difference between two sets of balance equations, we make a comparison between them by means of studying numerically their solutions for v_e and T_e for p-type Ge. We assume that the transport is due to heavy holes with a parabolic band. The scattering mechanisms include both acoustic and nonpolar optical phonons[9] The parameters[23,24] we employ are given as follows. The heavy-hole mass is chosen as $m = 0.31m_e$, with m_e as the free-electron mass. The reason for such a choice is that the contribution to the current by the light holes is compensated by adjusting the mass of heavy-hole band. The density of the lattice is $d = 5.32g/cm^3$. The effective acoustic deformation potential E=4.34eV and the longtitudinal acoustic phonon velocity is $v_s = 5.4 \times 10^5 cm/sec$. The optical phonon deformation field $D = 8.86 \times 10^8 eV/cm$ and the optical phonon frequency is $\Omega_0 = 430K$. In Fig.1, v_e and T_e are depicted as functions of electric field strengths E at a given lattice temperature T=77K. The solid curves are calculated according to (34) and (36), while the dashed curves according to the balance equations obtained by Kalashnikov[17] The results have been also compared with the experimental data of Reggiani et.al.[23] In the linear region ($E < 100V/cm$) the drift velocities v_e for both theories agree with each other and also with experimental data. But for $E > 200V/cm$ the discrepancy in v_e between the solid and dashed curves begins to show up. It clearly indicates that the present result (solid curve) is consistent with experimental data in high-field region. We also would like to emphasize here that the present result for v_e as a function of electric field agrees with that of Monte Carlo simulation[23] for a single isotropic hole band.

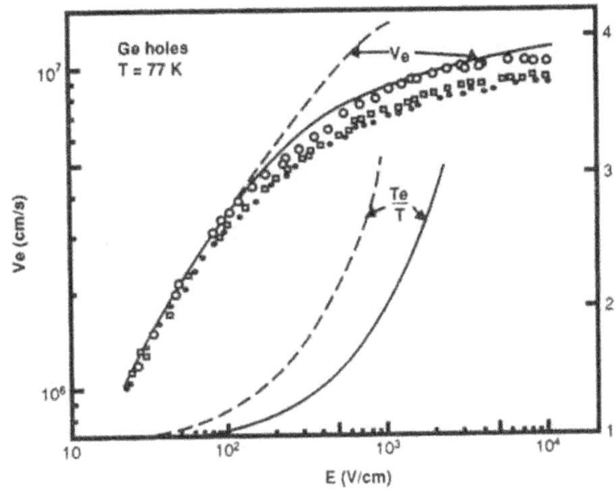

Fig.1. Hole drift velocity v_e and carrier temperature T_e as functions of electric field E at lattice temperature 77K. The solid curves present the results of Eqs. (34) and (36) and the dashed curves present the results of Kalashnikov[17] The circles, squares, and dots refer to the experimental data from Ref. 23 for the direction of E along different orientations of the crystal; the circles indicate $\mathrm{E} \| < 100 >$, the squares are for$\mathrm{E} \| < 110 >$, and the dots for $\mathrm{E} \| < 111 >$.

It is easy to see that the balance equations (34) and (36) obtained above are identical to the Lei-Ting equations in the absence of the Coulomb interaction between carriers. We now study the effect of the electron-electron interaction on the balance equations. Let us consider the relative electron Hamiltonian in the form (3). Since the Coulomb interaction has been explicitly included in H_e, the simple relations (28) and (31), which are used in deriving balance equations in the preceding section, no longer hold. So the derivation for the balance equations can not be straightforward as done above.

In what follows we shall generally show that in the steady state, statistical average of a dynamical operator, calculated with the approximate expression (17) for the NSO, coincides with its average over the statistical density matrix used by Lei and Ting[9]. So the balance equations for momentum and energy, derived from (22) and (23), must be identical with theirs.

For the steady-state hot-electron transport under consideration, both β_e and β are independent of time so that the entropy operator S(t,0) does not depend explicitly on time. In this section, for the sake of simplicity, we omit to write the arguments of the entropy operator S and rewrite the entropy production operator as

$$\begin{aligned}\dot{S}(t,i\tau) &= \beta_e\dot{H}_e(t,i\tau) + \beta\dot{H}_{ph}(t,i\tau)\\ &= (-i)e^{-\tau S}e^{iHt}[S,H_{el}]e^{-iHt}e^{\tau S},\end{aligned} \tag{37}$$

where we have utilized the property that both H_e and H_{ph} commutes with the operator S. Then the NSO in the form (17) can be rewitten as

$$\rho(t) = \{1 + \int_{-\infty}^{0} dt' e^{\varepsilon t'} \int_0^1 d\tau \dot{S}(t', i\tau)\}e^{-S}. \tag{38}$$

The average of a dynamical operator A over the statistical density matrix (38) is

$$< A >^t = < A >_l + \int_{-\infty}^{0} dt' e^{\varepsilon t'} \int_0^1 d\tau < A\dot{S}(t', i\tau) >_l, \tag{39}$$

where the second term, including the time correlation function, will be calculated in terms of the spectral representation for the correlation function.

Let $| m >$ and S_m be the eigenfunctions and eigenvalues of the operator S. Substituting (37) into (39), we write explicitly the statistical averaging for the time correlation function:

$$\begin{aligned}&< A\dot{S}(t', i\tau) >_l\\ &= (-i)\sum_m < m \mid Ae^{-\tau S}e^{iHt'}(SH_{el} - H_{el}S)e^{-iHt'}e^{(\tau-1)S} \mid m > .\end{aligned} \tag{40}$$

Between the operators S and H_{el}, we insert another complete set of states, which is unity

$$1 = \sum_m | m >< m |,$$

which gives

$$\begin{aligned}< A\dot{S}(t', i\tau) >_l = (-i)\sum_{m,n} &< m \mid A \mid n >< n \mid H_{el}(t') \mid m >\\ &(S_n - S_m)\exp\{\tau(S_m - S_n) - S_m\}.\end{aligned} \tag{41}$$

Therefore,

$$\begin{aligned}&\int_0^1 d\tau < A\dot{S}(t', i\tau) >_l\\ &= (-i)\sum_{m,n} < m \mid A \mid n >< n \mid H_{el}(t') \mid m > \{\exp(-S_m) - \exp(-S_n)\},\\ &= (-i)\sum_m < m \mid \{AH_{el}(t') - H_{el}(t')A\}e^{-S} \mid m >\\ &= (-i) < [A, H_{el}(t')] >_l .\end{aligned} \tag{42}$$

Substituting (42) into (39), we obtain

$$< A >^t = < A >_l + \int_{-\infty}^{\infty} dt' << A \mid H_{el}(t') >>, \tag{43}$$

where

$$<< A(t) \mid H_{el}(t') >> \equiv (-i)\theta(t-t') < [A(t), H_{el}(t')] >_l \tag{44}$$

It is easily seen that the present result for the average value of any operator A is just the same as that in Ref. 9. This means that in the calculation of steady-state averages, the approximate NSO (17) is equivalent to the statistical density matrix used in Ref. 9, even though they differ greatly from each other in the form. It then follows from this equivalence that starting from (17), we can obtain exactly the same balance equations for momentum and energy as those given by Lei and Ting[9], the Coulombic many-body effect being included. It has been shown[9] that the main effect of the Coulomb interaction between carries on the balance equations is that all the retarded density-density correlation fuctions in (34) and (36) need to be renormalized by their form in the random-phase approximation. Another important effect of electron-electron interaction is to make the charge carriers to reach fast thermalization[25] after an electric field is applied. Therefore, we have shown that in the presence of the electron-electron interaction, the steady-state balance equations, derived from the NSO method, is still identical with the Lei-Ting equations.

5. Application to Multivalley Semiconductors

Up to now the NSO approach to the steady-state hot-electron transport is based on a single valley model with spherical equienergetic surfaces. However, realistic semiconductors usually have more complicated band structures, especially many-valley structures. As a typical example, n-type GaAs has a conduction band with a Γ-L-X valley ordering, i.e., there are one Γ valley, four equivalent L valleys and three equivalent X valleys in order of energy minimum. For n-type Si, the electrons which contribute to transport are those in the six equivalent valleys which are around the six minima of the conduction band along the < 100 > directions. Since the equienergetic surfaces of each valley are near ellipsoids whose major axis is along the crystallographic directions, the effective mass of hot electrons for each valley is anisotropic. Therefore, for an external electric field along any direction, except the < 111 > crystallographic directions, the six valleys may be not equivalent to each other due to different effective masses along the field direction. In multivalley semiconductors the electron transfer between nonequivalent valleys plays a very important role in determining their hot-electron transport properties. Both the wellknown Gunn effect[2] in n-type GaAs and the anisotropic effect[3] for the drift velocity in n-type Si originate from the redistribution of hot electrons between nonequivalent valleys via intervalley transfer.

In the single valley case, as already shown, a pair of balance equations, which determine v_e and T_e as functions of the external field strengths, are quite enough for describing the steady-state hot-electron transport. In the case of multivalley semiconductors, the balance equations for momentum and energy must be written for each valley together with an additional balance equation for population since the hot-electron population for each valley is no longer a constant in the presence of intervalley coupling.

We now proceed to extend the NSO method to the study in the multivalley case and put stress on discussing how to construct a NSO adequate to a multivalley system.

First, take each valley as an electron subsystem and separate its center-of-mass motion from the relative motion of electrons for this subsystem. The whole electron system can be regarded as consisting of various subsystems which are coupled with each other due to intervalley electron-lattice and electron-electron interactions. Then the total Hamiltonian of an electron-lattice system with n valleys can be written as

$$H = \sum_{\alpha=1}^{n} [H_{c\alpha} + H_{e\alpha} + H_{el\alpha}] + H_{ph} + H_{inter}, \tag{45}$$

where $H_{c\alpha}$ and $H_{e\alpha}$ are the center-of-mass and relative parts of the electron Hamiltonian for the αth valley, respectively, and $H_{el\alpha}$ is the intravalley electron-lattice interaction. They have the same expressions as those given by (2), (3) and (5), but it is necessary to mark all the operators and physical quantities for each valley by a valley index, since they may be different for nonequivalent valleys. For example, the single-electron energy for the αth valley is $\varepsilon_{\alpha\mathbf{p}} = \mathbf{p}_\alpha^2/(2m_\alpha) + \varepsilon_{\alpha 0}$ where the energy minimum $\varepsilon_{\alpha 0}$ and the effective mass m_α may generally have different values for different α. The phonon Hamiltonian H_{ph} retains its original expression (4). All the intervalley couplings are included in H_{inter}:

$$H_{inter} = \sum_{\alpha\neq\gamma} \{ \sum_{\mathbf{k},\mathbf{q}} u_{\alpha\gamma}(\mathbf{q}) exp[(\mathbf{k}+\mathbf{q})\mathbf{R}_\alpha - \mathbf{k}\mathbf{R}_\gamma] c^\dagger_{\alpha,\mathbf{k}+\mathbf{q}} c_{\gamma,\mathbf{k}}$$

$$+ \sum_{\mathbf{k},\mathbf{q},\lambda} M_{\alpha\gamma}(\mathbf{q},\lambda) exp[(\mathbf{k}+\mathbf{q})\mathbf{R}_\alpha - \mathbf{k}\mathbf{R}_\gamma](b_{\mathbf{q}\lambda} + b^\dagger_{-\mathbf{q}\lambda}) c^\dagger_{\alpha,\mathbf{k}+\mathbf{q}} c_{\gamma,\mathbf{k}}$$

$$+ \sum_{\mathbf{q}} v_c(\mathbf{q}) \rho_{\alpha,\mathbf{q}} \rho_{\gamma,-\mathbf{q}} \}, \tag{46}$$

with

$$\rho_{\alpha,\mathbf{q}} = \sum_{\mathbf{k}} c^\dagger_{\alpha,\mathbf{k}+\mathbf{q}} c_{\gamma,\mathbf{k}}.$$

Here the first two terms in the summation over α and γ are the intervalley electron-impurity and electron-phonon interactions, and the third term is the Coulomb interaction between electrons for different valleys.

In the Hamiltonian (45) we consider $H' = \sum_\alpha H_{el\alpha} + H_{inter}$ as a small perturbation. If H' is neglected in the zeroth approximation, as in the construction of the quasi-equiliblium statistical operator, all the electron subsystems, together with the lattice system, are decoupled from each other. So the effective temperatures of hot electrons for various valleys may be different from each other, and also from the lattice temperature. In view of the above-mentioned consideration, we choose the following basic sets of P_m and $F_m(t)$:

$$\{P_m\} = \{H_{e1}; \;......;\; H_{en}; \; N_1; \;......;\; N_n; \; H_{ph}\}, \tag{47}$$

and

$$\{F_m(t)\} = \{\beta_1(t); \;......;\; \beta_n(t); \; -\beta_1(t)\mu_1(t); \;......;\; -\beta_n(t)\mu_n(t); \; \beta\}, \tag{48}$$

where $\beta_\alpha(t)$ and $\mu_\alpha(t)$ are the inverse effective temperature and chemical potential of hot electrons for the αth valley, respectively. Since the population of electrons for each valley is not constant due to intervalley transfer, the population operators N_α should be included in the set of P_m.

Starting from (47) and (48), it is straightforward to construct the NSO and obtain its approximate expression to the first order in H'. Using a similar procedure to that in the single valley case, as already outlined in Section 4, we can obtain the analytic results for statistical averages of the time derivatives of the center-of-mass momentum $\mathbf{p}_\alpha$, relative electron energy $H_{e\alpha}$ and population N_α for various valleys. Their explicit expressions have been given by Ref. 20 and will not presented here. Letting them be equal to zero for the steady state, we have 3n balance equations:

$$< \dot{\mathbf{p}}_\alpha >^t = 0, \tag{49}$$

$$< \dot{H}_{e\alpha} >^t = 0, \tag{50}$$

and

$$< \dot{N}_\alpha >^t = 0, \tag{51}$$

with $\alpha = 1, 2, ..., n$. They form a complete set of equations to determine the 3n steady-state values of v_α, T_α and N_α as functions of the electric field strengths at a given lattice temperature. The chemical potential μ_α ($\alpha = 1, 2, ...n$) of n valleys, which appear in the balance equations (49)-(51), are self-consistently determined by the following relations:

$$< N_\alpha >^t = \sum_{\mathbf{k}} 1/\{exp[\beta_\alpha(\varepsilon_{\mathbf{k}\alpha} - \mu_\alpha)] + 1\}. \tag{52}$$

Finally, the drift velocity of the whole electron system is given by

$$v_e = \sum_{\alpha=1}^{n} (< N_\alpha >^t /N) v_\alpha, \tag{53}$$

with N the total carrier number.

For the same reason as that discussed in the single valley case, the set of balance equations derived from the NSO method should be identical with that obtained by the Lei-Ting approach. This equivalence in the multivalley case has been presented by authors and their co-workers[20,26], At the same time, we have compared the numerical solutions for the set of balance equations (49)-(51) with those of Monte Carlo calculations and experimental data of n-type Si and GaAs. For n-type Si, without any adjustable parameters, our calculated results[20] are in good agreement with those of Monte Carlo simulation[27] in wide ranges of temperature and field strength. Further, the results calculated from the balance equations also agree with the experimental data for hot-electron transport in n-type GaAs[26]

6. Extension to Transient Hot-Electron Transport

It is straightforward to extend the NSO method to the study of transient hot-electron transport. The problem is to study the motion of an ensemble of electrons in a uniform bulk sample subject to various time configurations of a uniform electric field. Thus an ensemble average will enable us to obtain the time dependence of all the dynamic parameters such as the mean carrier energy and the mean carrier velocity. Here for simplicity, we will confine our discussion to the single valley case. The application to the case of multivalley semiconductors has been made by Ref.21.

For the transient transport, a pair of evolution equations

$$Nm\dot{v}_e = <\dot{\mathbf{p}}>^t, \tag{54}$$

$$c_e\dot{T}_e = <\dot{H}_e>^t, \tag{55}$$

are needed to determine the transient drift velocity $v_e(t)$ and the effective temperature $T_e(t)$ as functions of time resulting from a time-dependent configuration of the electric field. Further, it should to check whether or not the statistical averages in (54) and (55) are same as those in (22) and (23) since the NSO used for the transient transport seems to be a bit different from that in the steady-state case. To make this point clear let us extend the derivation in Section 6 to the case for transient transport. We again start from the approximate NSO in the form (17), over which the average of a dynamical operator A is

$$<A>^t = <A>_l^t + \int_{-\infty}^0 dt' e^{\epsilon t'} \int_0^1 d\tau <Ae^{-\tau S(t,0)}\dot{S}(t+t',t')e^{\tau S(t,0)}>_l^t . \tag{56}$$

Here

$$\dot{S}(t+t',t') = (-i)e^{iHt'}\{\beta_e(t+t')[H_e,H_{el}] + \beta[H_{ph},H_{el}]\}e^{-iHt'}, \tag{57}$$

and

$$<A>_l^t = Tr(Ae^{-S(t,0)}), \tag{58}$$

is calculated with the quasi-equilibrium statistical operator ρ_l at time t. Since the entropy operator $S(t,0)$ depends explicitly on time, the average over ρ_l must be time dependent. To calculate the time correlation function in (56) let $|\, m>$ be the exact eigenstates of H_e and H_{ph}, E_m and Ω_m be the corresponding eigenvalues of H_e and H_{ph}. We can repeat the same calculation of the spectral representation for the time correlation function as done in Section 6, yielding

$$\int_0^1 d\tau <Ae^{-\tau S(t,0)}\dot{S}(t+t',t)e^{\tau S(t,0)}>_l^t = (-i)\sum_{m,n} <m\,|\,A\,|\,n><n\,|\,H_{el}(t')\,|\,m>$$

$$\times \{exp[-S_m(t)] - exp[-S_n(t)]\}\gamma_{mn}(t+t',t), \tag{59}$$

where

$$S_m(t) = \beta_e(t)E_m + \beta\Omega_m,$$

and

$$\gamma_{mn}(t+t',t) = [S_m(t+t') - S_n(t+t')]/[S_m(t) - S_n(t)],$$

is a factor including some memory effect. Substituting(59) into (56) and integrating by parts, we obtain

$$<A>^t = <A>_l^t - i\int_{-\infty}^0 dt' e^{\epsilon t'} \sum_{m,n} <m\,|\,A\,|\,n><n\,|\,H_{el}(t')\,|\,m>$$

$$\times\{exp[-S_m(t)] - exp[-S_n(t)]\} + i\int_{-\infty}^0 dt' \dot{\gamma}_{mn}(t+t',t)\int_{-\infty}^{t'} ds e^{\epsilon s}$$

$$\times \sum_{m,n} < m \mid A \mid n >< n \mid H_{el}(s) \mid m > \{exp[-S_m(t)] - exp[-S_n(t)]\}, \tag{60}$$

where

$$\dot{\gamma}_{mn}(t+t',t) = \dot{\beta}_e(t+t')(E_m - E_n)/[S_m(t) - S_n(t)].$$

It is obvious that the second term on the right-hand side of (60) is of the first order in H_{el} while the third term is of the third order of magnitude in H_{el} because $\dot{\beta}_e$ is directly proportional to the square of H_{el} , as will be shown later. Therefore, when we construct the evolution equations in second order approximation in H_{el}, the third term in (60) should be neglected. Further, making the variable transformation $t' \to t' - t$ for the integral over t' in (60) and using the relation, valid in the first approximation in H_{el}

$$< m \mid A \mid n >< n \mid H_{el}(t'-t) \mid m > = < m \mid A(t) \mid n >< n \mid H_{el}(t') \mid m >,$$

we have

$$< A >^t = < A >_l^t - i \int_{-\infty}^{t} dt' e^{\varepsilon(t'-t)} \sum_{m,n} < m \mid A(t) \mid n >< n \mid H_{el}(t') \mid m >$$

$$\times \{exp[-S_m(t)] - exp[-S_n(t)]\}. \tag{61}$$

On the analogy of (42), we can write (61) as

$$< A >^t = < A >_l^t + \int_{-\infty}^{\infty} dt' << A(t) \mid H_{el}(t') >>^t, \tag{62}$$

where

$$<< A(t) \mid H_{el}(t') >>^t = -i\theta(t-t') < [A(t), H_{el}(t')] >_l^t, \tag{63}$$

is a retarded double-time Green's function. Equation (62) seems to have the same form as (43) for the steady-state case, but they have two major differences. One is the time-dependent statistical density matrix, as already mentioned, the other is that the center-of-mass coordinate, $R(t) = R(t') + \int_{t'}^{t} v_e(s)ds$ with $v_e(t)$ as the drift velocity, can not simply be expressed by $R(t') + v_e(t-t')$, as in the steady state, since v_e is time dependent in the transient transport case.

To obtain the evolution equations for $v_e(t)$ and $T_e(t)$ we substitute (19) and (24) into (62) and calculate the retarded Green's function. Omitting the tedious calculations we write down the final result only:

$$Nm\dot{v}_e = NeE(t) - i\sum_{\mathbf{q}} q_x \int_{-\infty}^{\infty} dt' exp[q_x \int_{t'}^{t} v(s)ds]\{n_i \mid u(q) \mid^2 << \rho_q(t) \mid \rho_{-q}(t') >>^t$$

$$+ \sum_{\lambda} \mid M(q,\lambda) \mid^2 [<< \rho_q(t) b_{q\lambda}(t) \mid \rho_{-q}(t') b^{\dagger}_{q\lambda}(t') >>^t$$

$$+ << \rho_q(t) b^{\dagger}_{-q\lambda}(t) \mid \rho_{-q}(t') b_{-q\lambda}(t') >>^t]\}, \tag{64}$$

$$c_e\dot{T}_e = -\sum_{\mathbf{q}} \int_{-\infty}^{\infty} dt' exp[q_x \int_{-t'}^{t} v(s)ds]\{n_i \mid u(q) \mid^2 << \dot{\rho}_q(t) \mid \rho_{-q}(t') >>^t$$

$$+\sum_{\lambda} | M(q,\lambda) |^2 [<< \dot{\rho}_q(t) b_{q\lambda}(t) \mid \rho_{-q}(t') b^{\dagger}_{q\lambda}(t') >>^t$$
$$+ << \dot{\rho}_q(t) b^{\dagger}_{-q\lambda}(t) \mid \rho_{-q}(t') b_{-q\lambda}(t') >>^t]\}. \tag{65}$$

Here we wish to point out that the evolution equations (63) and (64), derived from the NSO method, are exactly the same as those obtained by extending the Lei-Ting approach to the transient hot-electron transport [22]. This shows that the conclusion that two statistical density matrices in the linear approximation in H_{el} has identical effect on the calculation of averages is applicable not only to the steady-state transport, but also to the transient hot-electron transport. It is worth to mention that the evolution equations (63) and (64) contain the memory effect of the drift velocity, which is included in their exponent factors. When the evolution equation (63) is linearized, we can obtain a Langevin-like equation for the drift velocity. It has been shown that the memory effect or the intra-collisional field effect has little influence upon the solutions of evolution equations, at least for applied electric fields with moderate strengths[22]. Further, the evolution equations without memory effect have been applied to the calculations of transient hot-electron transport in p-type Ge and n-type GaAs under electric fields with several time-dependent configurations[21,22]. The calculated results coincide well with those obtained by interative procedure and Monte Carlo simulation.

7. Concluding Remarks

We have applied the NSO method to study the steady-state and transient hot-electron transport, including Coulombic many-body effect. After making a careful comparison between the NSO method and the balance-equation approach [9], we conclude that they yield the identical balance or evolution equations in the second-order approximation in H_{el}, which is usually used in hot-electron transport theory. This identity stems from the fact that even though they belong to two different statistical methods, their statistical density matrices to the lowest order in H_{el} have the same effect on evaluating the averages. It is conjucturable that such identity may not extend beyond this approximation; if so the balance or evolution equations obtained by the NSO method and the balance-equation approach may differ from each other only by high-order terms in H_{el}.

The essential feature of the balance-equation approach is the separation of the center-of-mass motion from the relative motion of electrons in the Hamiltonian and in the density matrix. After the separation the external electric field acts only on the center of mass, and the electrons which move relative to the center of mass do not directly experience the external field. They couple with the center of mass and with the lattice only through the electron-lattice interaction H_{el}. Taking into account the inequality $\tau_{el} >> \tau_{ee}$ for the electron-lattice system under consideration, we can regard H_{el} as a small perturbation. In the zeroth order approximation the relative electrons decouple themselves from the center of mass and from the lattice. So for time intervals larger than τ_{ee} but smaller than τ_{el} the hot-electron system in a nonequilibrium state may be described by an effective temperature T_e which indicates the mean energy of the relative motion of hot electrons, and the center-of-mass motion described by the average value of the momentum operator of the center of mass. On the other hand, the perturbation operator H_{el} provides dissipative mechanism for irreversible transport processes, so both steady-state balance equations and transient evolution equations are set up in terms of the perturbation expansion of H_{el}, usually in the second-order approximation in H_{el}. Therefore, the "separation" idea provides a very clear physical picture, as well as a simple mathematical treatment, for the hot-electron transport. This is the reason why we had absorbed this advanced idea at beginning of the present work.

A difficult problem in the nonequilibrium dissipative theory is how to obtain irreversible behavior in the macroscopic description of the state of the system generated by a statistical operator $\rho(t)$, if it satisfies Liouville's equation with time-reversal invariance. This can be obtained in terms of Prigogine's dynamic condition for dissipativity [27]:

$$\frac{\partial \rho(t)}{\partial t} + i\Lambda(t)\rho(t) = 0. \tag{66}$$

Here Λ is a modified Liouville's operator composed of an odd and an even part under time-reversal, and $\rho(t)$ is a transformed NSO now having irreversibility built in it. All available approaches to the nonequilibrium statistical problems introduce at some point in their derivations condition analogous to Prigogine's principle. As already mentioned, the Zubarev's approach is to introduce into Liouville's equation an infinitesimally small source that break time-reversal symmetry. The modified Liouville's equation is[16]

$$\frac{\partial \ln \rho(t)}{\partial t} - i[\ln \rho(t), H] = -\varepsilon(\ln \rho(t) - \ln \rho_l), \tag{67}$$

where $\varepsilon \rightarrow +0$ after the thermodynamic limiting process. In construting the above equation which is linear in $\ln \rho(t)$, it has been considered that the logarithm of a statistical operator satisfying Liouville's equation also satisfies Liouville's equation. It is easily shown that the NSO (12) is just the solution of (67). Although this point is not explicitly put forward in the balance-equation approach by Lei and Ting[9], when they select the retard solution of Liouville's equation and use the retarded Green's functions, an infinitesimally small source which violates the time-reversal symmetry has been implicitly introduced into Liouvelle's equation.

In this work we restrict ourselves to consider the spatially homogeneous distribution of hot electrons interacting with thermal-equilibrium scatterers in semiconductors. This approach may be easily extended to take into account the spatial inhomogeneities of the system and the deviation of the scatterers from thermal equilibrium such as phonon drag effect. Besides hot-electron transport problems, the NSO method has been also applied to different problems in the theory of irreversible processes[13] The successes on various applications show that the NSO method may constitute a very powerful theoretical tool in the study of dissipative dynamic systems.

8. Acknowledgment

The authors wish to thank professor C.S. Ting for enlightening discussions. This work is in part supported by the Natural Science Foundation of China.

References

1. W. Shockley, *Bell Syst. Tech. J.* **30**(1951)990.
2. J.B. Gunn, *Solid State Commun.* **1**(1963)88.
3. E.M. Conwell, *High Field Transport in Semiconductors* (Academic, New York, 1967).
4. For a general review, see L. Reggiani, *Hot-Electron Transport in Semiconductors* (Springer-Verlag, Berlin, 1985).
5. C.J. Stanton, H.U. Baranger, and J.W. Wilkins, *Appl. Phys. Lett.* **49**(1986)176.
6. H.F. Budd, *J. Phys. Soc. Jpn. Suppl.* **21**(1966)420; *Phys. Rev.* **158**(1967)798.
7. T. Kurosawa, *J. Phys. Soc. Jpn. Suppl.* **21**(1966)424.
8. C. Jacoboni and L. Reggiani, *Rev. Mod. Phys.* **55**(1983)645.
9. X.L. Lei, and C.S. Ting, *Phys. Rev.* **B32**(1985)1112.
10. C.S.Ting, S.C. Ying, and J.J. Quinn, *Phys. Rev.* **B14**(1976)4439.
11. X.L. Lei, and N.J.M. Horing, in this review volume.
12. X.L. Lei, and C.S. Ting, *Phys. Rev.* **B36**(1987)8162.
13. For a recent review, see R. Luzzi, and A.P. Vasconcellos, *Fortschr. Phys.* **38**(1990)887.
14. H. Mori, *J. Phys. Soc. Jpn.* **11**(1956)1029; *Phys. Rev.* **112** (1958)1829; **115**(1959)298.
15. D.N. Zubarev, *Sov. Phys. -Doklady* **6**(1962)776.
16. D.N. Zubarev, *Neravnovesnaia Statisticheskaia Termodinamika* (Izd. Nauka, Moskwa, 1971) [English Transl.: *Nonequilibrium Statistical Thermodynamics* (Consultants Bureau, New York,1974)].
17. V.P. Kalashnikov, *Physica* **48**(1970)93.
18. D.Y. Xing, P. Hu, and C.S. Ting, *Phys. Rev.* **B35**(1987)6379.
19. C.S. Ting, and D.Y. Xing, *Chin. J. Phys.* **25**(1987)85.
20. M. Liu, D.Y. Xing, C.S. Ting, and W.T. Xu, *Phys. Rev.* **B37** (1988)2997.
21. M. Liu, D.Y. Xing, and C.S. Ting, *J. Phys.: Condens. Matter* **1**(1989)417.
22. D.Y. Xing and C.S. Ting, *Phys. Rev.* **B35**(1987)3971.
23. L. Reggiani, C. Canali, F. Nava, and G. Ottaviani, *Phys. Rev.* **B16**(1977)2781.
24. J.P. Nougier and M. Rolland, *Phys. Rev.* **B8**(1973)5728.

25. L.Y. Chen and C.S. Ting, *Phys. Rev.* **B42**(1990)1129; C.S.Ting and L.Y. Chen, in this review volume.

26. X.L. Lei, D.Y. Xing, M.Liu, C.S.Ting, and J.L. Birman, *Phys. Rev.* **B36**(1987)9134.

27. C. Canali, C. Jacoboni, F. Nava, G. Ottaviani, and A. Alberigi-Quaranta, *Phys. Rev.* **B12**(1975)2265.

28. I. Prigogine, *Nature* **246**(1973)67.

International Journal of Modern Physics B, Vol. 6, No. 7 (1992) 1059–1078

Path Integral Study of Polaron Transport under High Electric Field

Z. B. Su
Institute of Theoretical Physics, Academia Sinica
Beijing, 100080, China

With help of the nonequilibrium Green's function technique and the Feynman path integral theory, we present a self-consistent approach to the Thornber-Feynman model of a polaron under high electric field. The applied field and the electron-phonon interaction both are treated non-perturbatively so that the high-field effect as well as the quantum interference effect have been taken into account. A set of coupled equations are derived to consistently describe the drift motion and the fluctuation of the electron. By solving these equations numerically we obtain (1) the nonlinear relation between the drift velocity and the applied electric field; (2) the effective electron temperature, the nonequilibrium noise, and the distribution functions.

I Introduction

Modern fabrication technology has brought various microstructures which evidently exhibit the quantum interference and the nonequilibrium statistics. As the size of devices is becoming smaller and the electric field affecting the transporting electrons becomes very strong, more and more researchers are getting involved in the nonlinear transport problem of solids (see, *e.g.*, Refs.1-18). To exploit the basic physics for the nonlinear transport, Thornber and Feynman[2], in a pioneering work at early seventies, investigated an interesting model of a single electron drifting in a thermal crystal with an applied high electric field. It is the first *ab initio* derivation of the quantum force balance equation for a single electron system, *i.e.*, in a steady transport state, the electron's momentum gain from the applied field should be balanced by its momentum loss to the phonon reservoir. Very importantly, they treated the applied field and the electron-phonon interaction both nonperturbatively. In an approximate treatment for the quantum force balance equation, they derived transparently the electron drift velocity as a function of the applied field. Especially, they found a threshold field beyond which no steady state can exist. Unfortunately, such an elegant treatment, as realized by the authors themselves, cannot be consistent with the energy balance equation. A decade later, Peeters and Devreese[4] also studied this model system and worked in the framework of Heisenberg's equation of motion. They found that the distribution function corresponding to the Thornber-Feynman theory is a drifted Maxwellian with an electron temperature which is independent of the applied electric field. This result is unphysical and needs to be reexamined more carefully.

In 1988, We also investigated this interesting model from the point of view of non-equilibrium quantum statistical mechanics and developed a consistent treatment of the electron's drift motion and fluctuations.[13,14] The basic physics coming out of this self-consistent approach is: to sustain a non-equilibrium stationary state (*i.e.*, the steady transport state), we need not only mechanically the balance between electron's momentum gain from the applied field and its loss due to phonon friction but also statistically the balance between the drift friction (dissipating drifting energy) and the fluctuation. These two balance conditions are closely coupled to each other and should be treated on the same footing. The former balance directly implies E-v (electric field E, drift velocity v) relation while the latter balance represents the nonequilibrium fluctuation dissipation theorem (FDT). There the energy balance condition would be guaranteed as a natural consequence from the FDT. Technically, this approach is based upon a path integral formulation[19-21] for the Schwinger-Keldysh non-equilibrium Green's functions,[22-26] *i.e.*, the closed-time-path Green's function, so that not only the applied field and the electron phonon interaction can be treated nonperturbatively but also the quantum statistical information could be incorporated automatically into the generalized Feynman's disentangling technique of this approach.

Our calculated E-v curve agree well with that of Thornber and Feynman. This confirms the existence of the steady-state transport threshold which has first been shown by these authors. The quantum fluctuation for the electron's velocity deviates away from the equilibrium distribution (in thermal equilibrium with the phonon reservoir) gradually as the applied field increases. If we introduce the effective electron temperature $T_e = 2\varepsilon/3$ to characterize the electron's fluctuationg energy ε, and the classical limit temperature (its definition will be given in Eq.(3.24)) to characterize the low frequency behavior of the current fluctuation, these two kinds of electron "temperature" have quite different features. Especially, the latter exhibits a sort of "cooling effect" but the former does not. Besides, the power spectrum of noise current is also calculated.

In this review article, we present a detailed discussion of the self-consistent approach to the Thornber-Feynman model. In section II we present the description of the system considered and derive the kinetic equations for the drift motion and the fluctuation. In section III we apply the coupled drift-fluctuation equations to the steady state transport and deduce various physical expectations from the solution of the coupled equations. In section IV the energy-balance equation is specifically emphasized and proven to be consistent with the momentum balance equation. In section V numerical calculation will be conducted for an electron-LO-phonon system. In the five appendices, we provide more details of some calculational steps.

II Nonequilibrium Green's Function Description of the Drift and the Fluctuation

The system we consider is an electron drifting in a thermal crystal in the presence of an applied electric field. The electron is described by its position $\mathbf{x}(t)$ and momentum $\mathbf{p} = m\dot{\mathbf{x}}$, while the crystal is characterized by an equilibrium phonon system of temperature $1/\beta$. So the total Lagrangian of the system is

$$L(t) = \frac{1}{2}m\dot{\mathbf{x}}(t)^2 + \frac{1}{2V}\sum_{\mathbf{k}} B_{-\mathbf{k}}\Delta^{-1}B_{\mathbf{k}} + \frac{1}{V}\sum_{\mathbf{k}} g_{\mathbf{k}}e^{i\mathbf{k}\cdot\mathbf{x}(t)}B_{\mathbf{k}}(t) + e\mathbf{E}\cdot\mathbf{x}(t), \tag{2.1}$$

where

$$B_{\mathbf{k}} = b_{\mathbf{k}} + b^{\dagger}_{-\mathbf{k}} \tag{2.2}$$

in which $b_{\mathbf{k}}$ and $b^{\dagger}_{\mathbf{k}}$ are the destruction and creation operators of the phonon with momentum $\mathbf{k}$. In Eq.(2.1), $\mathbf{E}$ is the applied electric field and Δ is the free propagator of the phonon system, the third term in the right hand side describes the electron-phonon interaction with $g_{\mathbf{k}}$ the coupling strength and V the volume of the crystal.

In the path integral formulation of CTPGF, the expectation of an arbitrary operator can expressed as a functional integral defined on a closed time path as shown in Fig.1,

$$\langle\langle O\rangle\rangle = \int[d\mathbf{x}][dB_{\mathbf{k}}]O[\mathbf{x}, B_{\mathbf{k}}]\exp(iI_p[\mathbf{x}, B_{\mathbf{k}}]), \tag{2.3}$$

where $[d\mathbf{x}]$ and $[dB_{\mathbf{k}}]$ are, respectively,

$$d\mathbf{x}_-(t_0)d\mathbf{x}_-(t_1)\cdots d\mathbf{x}_-(t_N)d\mathbf{x}_+(t_N)\cdots d\mathbf{x}_+(t_1)d\mathbf{x}_+(t_0)$$

and

$$\prod_{\mathbf{k}}\left[dB_{\mathbf{k}-}(t_0)dB_{\mathbf{k}-}(t_1)\cdots dB_{\mathbf{k}-}(t_N)dB_{\mathbf{k}+}(t_N)\cdots dB_{\mathbf{k}+}(t_1)dB_{\mathbf{k}+}(t_0)\right],$$

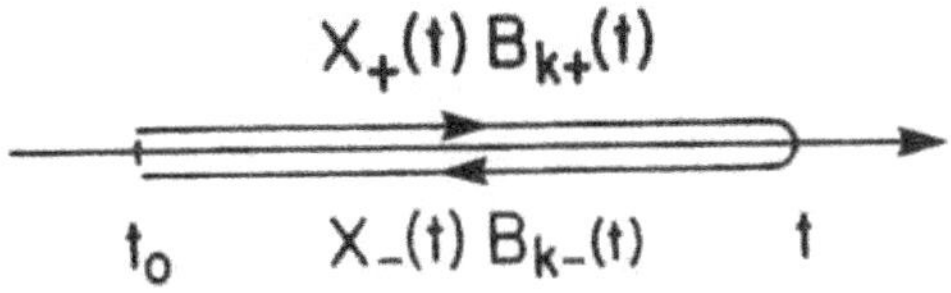

Fig.1. The closed time path.

with $-\infty = t_0 < t_1 < \cdots < t_N = \infty$. $\mathbf{x}_+(t)$ is defined on the positive branch (+) and $\mathbf{x}_-(t)$ on the negative branch (−). The action defined on the closed-time path is expressed as

$$I_p = \int_p dtL(t) = \int_p dt\left[\frac{1}{2}m\dot{\mathbf{x}}(t)^2 + e\mathbf{E}\cdot\mathbf{x}(t) + \frac{1}{V}\sum_{\mathbf{k}} g_{\mathbf{k}}e^{i\mathbf{k}\cdot\mathbf{x}(t)}B_{\mathbf{k}}(t)\right.$$
$$\left. + \frac{1}{2}\int_p dt'\frac{1}{V}\sum_{\mathbf{k}} B_{-\mathbf{k}}(t)\Delta^{-1}(\mathbf{k}, t-t')B_{\mathbf{k}}(t')\right], \tag{2.4}$$

with

$$\int_p dt = \int_{-\infty}^{+\infty} dt_+ + \int_{+\infty}^{-\infty} dt_- = \int_{-\infty}^{+\infty} dt_+ - \int_{-\infty}^{+\infty} dt_-. \tag{2.5}$$

It should be remarked that in Eq.(2.3) all the information of the density matrix for the thermal phonons has been incorporated into the closed-time path action I_p, so the Δ in Eq.(2.4) is the thermal propagator of the phonons.[20,21] Since the $B_{\mathbf{k}}$ dependence of I_p is quadratic, we can eliminate the phonon coordinates exactly by carrying out the functional integration $\int[dB_{\mathbf{k}}]$ in Eq.(2.3). Then we obtain an effective action $I_{eff}[\mathbf{x}(t)]$ for only the electron variable with all the phonon contributions being taken into account implicitly. $I_{eff}[\mathbf{x}(t)]$ is introduced through

$$\exp(iI_{eff}[\mathbf{x}]) = \int [dB_{\mathbf{k}}]\exp(iI_p[\mathbf{x}, B_{\mathbf{k}}]),$$

so that

$$I_{eff}[\mathbf{x}] = \int_p dt\left(\frac{1}{2}m\dot{\mathbf{x}}^2 + e\mathbf{E}\cdot\mathbf{x}\right)$$
$$+\frac{1}{2V}\sum_{\mathbf{k}}|g_{\mathbf{k}}|^2\int_p dt\int_p dt'\Delta(\mathbf{k}, t-t')e^{i\mathbf{k}\cdot[\mathbf{x}(t)-\mathbf{x}(t')]}.$$

Moreover, we can also derive without difficulty the following useful formula for eliminating the phonon variables:

$$\langle\langle\cdots B_{\mathbf{k}}(t)\rangle\rangle = -\int_p ds g^*_{\mathbf{k}}\Delta(\mathbf{k}, t-s)\langle\cdots e^{-i\mathbf{k}\cdot\mathbf{x}(s)}\rangle, \tag{2.6}$$

where

$$\begin{aligned}\langle\langle\cdots\rangle\rangle &\equiv \textstyle\int[d\mathbf{x}][dB_{\mathbf{k}}]\cdots\exp(iI_p),\\ \langle\cdots\rangle &= \textstyle\int[d\mathbf{x}]\cdots\exp(iI_{eff}).\end{aligned} \tag{2.7}$$

For convenience we also introduce the time-branch index with Greek subscript running over the positive (+) and negative (−) branches. For example, $\Delta_{\alpha\beta}(t,t')$ implies that t is on the αth branch and t' on the βth branch. Furthermore, we introduce two symbols,

$$\xi_\alpha = 1 \quad \text{for } \alpha = +,-,$$
$$\eta_\alpha = \begin{cases} 1 & \text{for } \alpha = +, \\ -1 & \text{for } \alpha = -, \end{cases}.$$

With these notations, the integral along the closed-time path (Eq.(2.5)) can be written as

$$\int_p dt = \eta_\alpha\int_{-\infty}^{\infty} dt_\alpha, \tag{2.8}$$

and the expression in Eq.(2.6) becomes

$$\langle\langle\cdots B_{\mathbf{k}\alpha}(t)\rangle\rangle = -\int ds g^*_{\mathbf{k}}\eta_\beta\Delta_{\alpha\beta}(\mathbf{k}, t-t')\langle\cdots\exp[-i\mathbf{k}\cdot\mathbf{x}_\beta(s)]\rangle. \tag{2.9}$$

Here and in the rest of this article the Einstein summation rule is assumed for all the repeated subscript and superscripts unless they are noted as exceptions.

Now we start to derive the quations for the drift motion and fluctuation. From the Lagrangian given in Eq.(2.1) we can immediately derive the operator equation for the electron motion,

$$\dot{\mathbf{p}}(t) = e\mathbf{E} + \mathbf{f}(t) = e\mathbf{E} + \frac{i}{V}\sum_{\mathbf{k}} g_{\mathbf{k}}\exp[i\mathbf{k}\cdot\mathbf{x}(t)]B_{\mathbf{k}}(t)\mathbf{k}. \tag{2.10}$$

Taking the expectation value of the nonequilibrium ensemble, we have

$$\langle\dot{\mathbf{p}}(t)\rangle = e\mathbf{E} + \frac{i}{V}\sum_{\mathbf{k}} g_{\mathbf{k}}\mathbf{k}\langle\langle\exp[i\mathbf{k}\cdot\delta\mathbf{x}(t)]B_{\mathbf{k}}(t)\rangle\rangle\exp[i\mathbf{k}\cdot\langle\mathbf{x}(t)\rangle], \tag{2.11}$$

where $\langle\mathbf{x}(t)\rangle$ and $\langle\mathbf{p}(t)\rangle$ describe the electron's drift motion and $\delta\mathbf{x}(t)$ represents the fluctuation:

$$\langle\mathbf{p}(t)\rangle = m\mathbf{v}(t) = m\langle\dot{\mathbf{x}}(t)\rangle, \tag{2.12}$$
$$\delta\mathbf{x}(t) = \mathbf{x}(t) - \langle\mathbf{x}(t)\rangle. \tag{2.13}$$

Here the drift velocity $\mathbf{v}$ is a C-number. By applying Eq.(2.9) to Eq.(2.1) we arrive at a momentum equation in which the drift motion is closely coupled to the electric field as well as the fluctuation,

$$\langle\dot{\mathbf{p}}(t)\rangle = e\mathbf{E} - \frac{i}{2V}\sum_{\mathbf{k}}|g_{\mathbf{k}}|^2\mathbf{k}\int ds\xi_\alpha\eta_\beta\Delta_{\alpha\beta}(\mathbf{k}, t-s)F_{\beta\alpha}(\mathbf{k}, t, s)\lambda(t, s), \tag{2.14}$$

where

$$\lambda = \exp[i\mathbf{k} \cdot \langle \mathbf{x}(t) - \mathbf{x}(s) \rangle] = \exp\left[i\mathbf{k} \cdot \int_s^t d\tau \mathbf{v}(\tau)\right]. \tag{2.15}$$

In Eq.(.14) the fluctuating effects are included in the nonequilibrium Green's functions

$$F_{\alpha\beta}(\mathbf{k}, t, s) = \langle T_p \exp\{i\mathbf{k} \cdot [\delta\mathbf{x}_\alpha(t) - \delta\mathbf{x}_\beta(s)]\}\rangle, \tag{2.16}$$

where T_p is the time-ordering parameter along the closed time path. It is identical to the usual time-ordering operator T on the positive branch, is identical to the anti-chronological operator $\tilde{T}$ on the negative branch, and orders any operator at the negative branch to stand left to that at the positive branch.

In order to determine the fluctuating effects we introduce the following two kinds of nonequilibrium Green's functions. For the elctron position fluctuation we define

$$\begin{aligned} iP^{ij}_{\alpha\beta}(t,t') &= \left\langle T_p \delta x^i_\alpha(t) \delta x^j_\beta(t')\right\rangle \\ &= \theta_{\alpha\beta}(t-t')\langle \delta x^i(t)\delta x^j(t')\rangle + \theta_{\beta\alpha}(t'-t)\langle \delta x^j(t')\delta x^i(t)\rangle, \end{aligned} \tag{2.17}$$

and for the velocity fluctuation,

$$\begin{aligned} iD^{ij}_{\alpha\beta}(t,t') &= \langle T_p \delta\dot{x}^i(t)\delta\dot{x}^j(t')\rangle \\ &= \theta_{\alpha\beta}(t-t')\langle \delta\dot{x}^i(t)\delta\dot{x}^j(t')\rangle + \theta_{\beta\alpha}(t'-t)\langle \delta\dot{x}^j(t')\delta\dot{x}^i(t)\rangle. \end{aligned} \tag{2.18}$$

Hereafter we use the Latin alphabet to denote the vector components. In Eqs.(2.17) and (2.18) the generalized step function $\theta_p(t-t')$ is defined on the closed time path as

$$\theta_p(t-t') = \begin{bmatrix} \theta_{++}(t-t') & \theta_{+-}(t-t') \\ \theta_{-+}(t-t') & \theta_{--}(t-t') \end{bmatrix} = \begin{bmatrix} \theta(t-t') & 0 \\ 1 & \theta(t'-t) \end{bmatrix} \tag{2.19}$$

Differentiating this expression with respect to time t, we immideiately get the generalized δ function

$$\delta_p(t-t') = \partial_t\theta_p(t-t') = \begin{bmatrix} \delta(t-t') & 0 \\ 0 & -\delta(t'-t) \end{bmatrix}, \tag{2.20}$$

which satisfies

$$\int_p dt \delta_p(t-t') = 1. \tag{2.21}$$

Now differentiating Eq.(2.17) twice and utilizing Eqs.(2.10), (2.11), and (2.20) and the simultaneous commutator

$$[x^i(t), \dot{x}^j(t)] = \frac{i}{m}\delta^{ij}, \tag{2.22}$$

we obtain

$$\begin{aligned} -m\partial_t^2 P^{ij}_{\alpha\beta}(t,t') = \delta^{ij}\delta_{\alpha\beta}(t-t') - \frac{1}{V}\sum_{\mathbf{k}} g_{\mathbf{k}} k^i \langle\langle \exp[i\mathbf{k}\cdot\delta\mathbf{x}_\alpha(t)] B_{\mathbf{k}\alpha}(t)\delta x^j_\beta(t')\rangle\rangle \\ \times \exp[i\mathbf{k}\cdot\langle \mathbf{x}(t)\rangle]. \end{aligned} \tag{2.23}$$

Using the expression in Eq.(2.9) once again for eliminating the phonon variables, we have

$$\begin{aligned} -m\partial_t^2 P^{ij}_{\alpha\beta}(t,t') = \delta^{ij}\delta_{\alpha\beta}(t-t') + \frac{1}{V}\sum_{\mathbf{k}} |g_{\mathbf{k}}|^2 k^i \int ds \eta_\delta \Delta_{\alpha\delta}(\mathbf{k}, t-s)\lambda \\ \times\langle \exp\{i\mathbf{k}\cdot[\delta\mathbf{x}_\alpha(t) - \delta\mathbf{x}_\delta(s)]\} x^j_\beta(t')\rangle. \end{aligned} \tag{2.24}$$

So far we have derived exactly the equations both for the electron drift motion (Eq.(2.14)) and for its fluctuation (Eq.(2.24)). Now we adopt the Gaussian approximation to make this set of equations closed. In the Gaussian approximation for the classical stochastic processes we keep only the first and second cumulants with all the higher order cumulants being assumed zero. In the same way for the quantum statistical system, the Gaussian approximation formulated in terms of the CTPGF is assumed as the following: Keep the expectation of variable $\mathbf{x}(t)$ and the two-point connected CTPGF $\langle T_p \delta x(t) \delta x(t') \rangle$ with the assumption that all higher-order connected CTPGF's vanish. Therefore the fluctuation functions F in Eq.(2.16) can be expressed in terms of the two-point GF's as (see Appendix A for the proof)

$$\begin{aligned} F_{\alpha\beta}(\mathbf{k},t,s) &= \exp\left(-\frac{1}{2}\left\langle T_p\{\mathbf{k}\cdot[\delta\mathbf{x}_\alpha(t)-\delta\mathbf{x}_\beta(s)]\}^2\right\rangle\right) \\ &= \exp\left[-\frac{i}{2}k^i k^j [P^{ij}_{\alpha\alpha}(t,t)+P^{ij}_{\beta\beta}(s,s)-2P^{ij}_{\alpha\beta}(t,s)\right]. \end{aligned} \tag{2.25}$$

Moreover, we can also prove (see Appendix B) a useful expression

$$\left\langle T_p \exp\{i\mathbf{k}\cdot[\delta\mathbf{x}_\alpha(t)-\delta\mathbf{x}_\delta(s)]\}\delta x^j_\beta(t')\right\rangle = -k^i[P^{ij}_{\alpha\beta}(t,t')-P^{ij}_{\delta\beta}(s,t')]F_{\alpha\delta}(\mathbf{k},t,s). \tag{2.26}$$

Then the GF equation (2.24) becomes

$$\begin{aligned} -m\partial_t^2 P^{ij}_{\alpha\beta}(t,t') &= \delta^{ij}\delta_{\alpha\beta}(t-t') + \frac{1}{V}\sum_{\mathbf{k}} |g_{\mathbf{k}}|^2 k^i k^l \int ds \eta_\delta \Delta_{\alpha\delta}(\mathbf{k},t-s)\lambda \\ &\times \left\{P^{lj}_{\delta\beta}(s,t')-P^{lj}_{\alpha\beta}(t,t')\right\} F_{\alpha\beta}(\mathbf{k},t,s), \end{aligned} \tag{2.27}$$

which, together with the momentum equation (2.14), provides a complete description of our nonequilibrium transport problem with regard to the Gaussian approximation Eq.(2.25).

For the convenience of the steady state discussions we may use the velocity GF: D's instead of the above position GF's. We can express the fluctuation function F in terms of D's and derive a kinetic equation for those velocity GF's from the above GF equations (2.27). In appendix C it is proven that

$$F_{\alpha\alpha}(\mathbf{k},t,s) = \exp\left[-\frac{i}{2}k^i k^j\left(\int_s^t d\tau \int_s^t d\tau' D^{ij}_{\alpha\alpha}(\tau,\tau') + \alpha\frac{\delta^{ij}}{m}|t-s|\right)\right], \alpha=+,-, \tag{2.28}$$

$$F_{-+}(\mathbf{k},t,s) = \theta(t-s)F_{++} + \theta(s-t)F_{--}, \tag{2.29}$$

$$F_{+-}(\mathbf{k},t,s) = \theta(s-t)F_{++} + \theta(t-s)F_{--}. \tag{2.30}$$

From the definitions Eqs.(2.17) and (2.18) it can be easily proven by noticing Eqs.(2.30) and (2.22) that the velocity GF's and position GF's are related as

$$D^{ij}_{\alpha\beta}(t,t') = -\frac{1}{m}\delta^{ij}\delta_{\alpha\beta}(t-t') + \partial_t\partial'_t P^{ij}_{\alpha\beta}(t,t'). \tag{2.31}$$

After an appropriate substitution of this relation into Eq.(2.27) we get the equation for the velocity GF's

$$\begin{aligned} m\partial^2_{t'} D^{ij}_{\alpha\beta}(t,t') &= \frac{1}{V}\sum_{\mathbf{k}} |g_{\mathbf{k}}|^2 k^i k^l \int ds \eta_\delta \Delta_{\alpha\delta}(\mathbf{k},t-s)\lambda F_{\alpha\delta}(\mathbf{k},t,s) \\ &\times \left[\frac{1}{m}\delta^{lj}[\delta_{\alpha\beta}(t-t')-\delta_{\delta\beta}(s-t')] + D^{lj}_{\alpha\beta}(t,t') - D^{lj}_{\delta\beta}(s,t')\right]. \end{aligned} \tag{2.32}$$

This GF equation together with the momentum equation (2.14) enable us to obtain a convenient self-consistent description of all the steady state transport properties in the following sections.

III Steady State Transport

When the applied electric field is constant the system may reach, after a transient process, a steady state if that state is stable. In this section we investigate the steady-state transport properties of our single-electron-phonon reservoir system without paying any attention to the transient behavior. We shall solve the coupled momentum equation and GF equations (2.14) and (2.32) derived in the last section and obtain a stationary solution for them to extract all the steady state transport properties. Intuitively, for establishing a nonequilibrium stationary state (NESS) in the presence of an applied field, we need not only the balance between the applied field and the phonon friction but also the "balance" between the drift friction and the fluctuation which is described as a generalized fluctuation-dissipation theorem (FDT) for the NESS. This is the basic physics of the coupled equations (2.14) and (2.32). Since we are able to solve the fluctuation consistently with the drift motion, the energy-balance condition as well as the distribution for the electron, as is going to be discussed below, will follow consistently. Therefore it is no longer necessary to introduce a parameterized *ad hoc* distribution for the electron.

According to the discussion of the last section our system can be viewed as a quantum stochastic process characterized by the electron velocity variable $\dot{\mathbf{x}}(t)$ with its first cumulant being the electron drift velocity, while the second cumulants are the nonequilibrium velocity GF's. So the stationarity of the process $\dot{\mathbf{x}}(t)$ means that all its cumulants satisfy the time translation invariant, *i.e.*, the electron drift velocity is constant and the velocity GF's depend only upon the time differences.

Since the drift velocity is constant, *i.e.*,

$$\langle \dot{\mathbf{p}}(t) \rangle = 0, \tag{3.1}$$

we have

$$\lambda(t,s) = \exp[-i\mathbf{k}\cdot\mathbf{v}(t-s)] = \lambda(t-s). \tag{3.2}$$

Due to the time translation invariance of the velocity GF's we have also the Fourier transformation

$$D^{ij}_{\alpha\beta}(t,t') = D^{ij}_{\alpha\beta}(t-t') = \int \frac{d\omega}{2\pi} e^{-i\omega(t-t')} D^{ij}_{\alpha\beta}(\omega). \tag{3.3}$$

Using these conditions it is not difficult to prove that the fluctuation function F in Eqs.(2.28)-(2.30) are also temporal-translation invariance, *i.e.*,

$$F_{\alpha\beta}(\mathbf{k},t,s) = F_{\alpha\beta}(\mathbf{k},t-s). \tag{3.4}$$

For simplicity it is natural to assume that the system is axially symmetric with its axis parallel to the applied field. Therefore the direction of the drift velocity is certainly along the electric field $\mathbf{E}$ and the GF: $D^{ij}_{\alpha\beta}$ for $i \neq j$ should vanish, *i.e.*,

$$D^{ij}_{\alpha\beta} = \delta^{ij} D^{(i)}_{\alpha\beta}. \tag{3.5}$$

If we choose the coordinate with its third axis along the $\mathbf{E}$ direction we will have $D^{(1)} = D^{(2)}$ as a result of the axial symmetry. So we will take the notation $\perp$ for $i = 1,2$ and $\parallel$ for $i = 3$.

With all these conditions discussed above, the momentum equation (2.14) now reads

$$eE = \frac{i}{V}\sum_{\mathbf{k}} |g_{\mathbf{k}}|^2 k_3 \int ds \lambda(s) \frac{1}{2}\xi_\alpha \eta_\beta \Delta_{\alpha\beta}(\mathbf{k}, s) F_{\alpha\beta}(\mathbf{k}, s), \tag{3.6}$$

which clearly shows the balance between the momentum gain from the electric field (left-hand side) and the loss to the phonon reservoir (right-hand side). Fourier transformation of the GF equation (2.32) gives

$$\begin{aligned} -m\omega^2 D^{(i)}_{\alpha\beta}(\omega) &= \frac{1}{V}\sum_{\mathbf{k}} |g_{\mathbf{k}}|^2 k^i k^j \int ds \lambda(s) \Delta_{\alpha\delta}(\mathbf{k}, s) \eta_\delta F_{\alpha\delta}(\mathbf{k}, s) \\ &\times \left[\frac{1}{m}(\delta_{\alpha\beta} - e^{i\omega s}\delta_{\delta\beta}) + D^{(i)}_{\alpha\beta}(\omega) - e^{i\omega s} D^{(i)}_{\delta\beta}(\omega)\right]. \end{aligned} \tag{3.7}$$

In Eqs.(3.5) and (3.7) and in what follows the repeated superscript i is not summed. In the above two equations (3.6) and (3.7) the drift velocity v, the GF's, and the electric field E are closely coupled together. We may take the v as a parameter and first solve Eq.(3.7) to obtain the nonequilibrium GF's and then calculate the E-v relation through Eq.(3.6). Solving Eq.(3.7) formally, we obtain the following expressions for the nonequilibrium Green's functions:

$$D^{(i)}_r(\omega) = -i\gamma_i(\omega)/m[\omega + i\gamma_i(\omega)], \tag{3.8}$$

$$D^{(i)}_a(\omega) = i\gamma_i(-\omega)/m[\omega - i\gamma_i(-\omega)], \tag{3.9}$$

$$D^{(i)}_c(\omega) = N_i(\omega)[D^{(i)}_r(\omega) - D^{(i)}_a(\omega)], \tag{3.10}$$

with $\gamma_i(\omega)$ and $N_i(\omega)$ satisfying the following self-consistent equations:

$$\gamma_i(\omega) = -\frac{1}{V}\sum_{\mathbf{k}} |g_{\mathbf{k}}|^2 k_i^2 \frac{1}{2m\omega}\xi_\alpha\eta_\beta \int dt (1 - e^{i\omega t})\Delta_{\alpha\beta}(\mathbf{k}, t) F_{\beta\alpha}(\mathbf{k}, t) e^{i\mathbf{k}\cdot\mathbf{v}t}, \tag{3.11}$$

$$N_i(\omega) = \frac{i}{V}\sum_{\mathbf{k}} |g_{\mathbf{k}}|^2 k_i^2 \frac{1}{2m\omega}\int dt \cos\omega t \Delta_{\alpha\beta}(\mathbf{k}t) F_{\beta\alpha}(\mathbf{k}, t)\frac{exp(i\mathbf{k}\cdot\mathbf{v}t)}{2Re\gamma_i(\omega)}. \tag{3.12}$$

D_r, D_a and D_c are, respectively, the nonequilibrium retarded, advanced, and correlation Green's functions which are related to the $D_{\alpha\beta}$ as[20]

$$D_r = \frac{1}{2}\xi_\alpha\eta_\beta D_{\alpha\beta}, \tag{3.13}$$

$$D_a = \frac{1}{2}\eta_\alpha\xi_\beta D_{\alpha\beta}, \tag{3.14}$$

$$D_c = \frac{1}{2}\xi_\alpha\xi_\beta D_{\alpha\beta}, \tag{3.15}$$

$$0 = \frac{1}{2}\eta_\alpha\eta_\beta D_{\alpha\beta}. \tag{3.16}$$

Now we can determine the solutions for the nonequilibrium Green's functions: D_r, D_a, and D_c by solving the coupled equations (3.11) and (3.12) with Eqs.(3.8)-(3.10) self-consistently. With these nonequilibrium GF's which have the drift velocity as a parameter we can easily derive all the transport properties of the system.

The nonlinear E-v relation can be obtained from Eq.(3.6) by a straightforward integration. The nonequilibrium noise, current fluctuation, which is defined as

$$\langle ne\delta\dot{x}^i(t) ne\delta\dot{x}^i(t')\rangle = (ne)^2 S_i(t - t'), \tag{3.17}$$

can be directly connected to the correlation Green's function as

$$S_i(t-t') = \frac{i}{2} D_c^{(i)}(t-t'). \tag{3.18}$$

So the power spectrum is

$$S^{(i)}(\omega) = \frac{i}{2} D_c^{(i)}(\omega). \tag{3.19}$$

We can also introduce the effective electron temperature as

$$T_e = \frac{2}{3}\varepsilon = \frac{i}{12} m \xi_\alpha \xi_\beta D^{ll}_{\alpha\beta}(t,t) \tag{3.20}$$

for characterizing the fluctuating energy ε. It is related to the correlation GF as

$$T_e = \frac{1}{6}\sum_{i=1}^{3} \int \frac{d\omega}{2\pi} m i D_c^{(i)}(\omega). \tag{3.21}$$

Noting that the right hand term $(D_r - D_a)$ of Eq.(3.10) describes the dissipation effect while the left hand side $D_c^{(i)}$ is simply the current fluctuation, we may, therefore, interpret this equation as a generalized FDT for the NESS in which the function $N_i(\omega)$ plays the role of a nonequilibrium distribution for the current fluctuation. When the applied electric field is zero the drift velocity vanishes. In this case we can verify without difficulty that our generalized FDT Eq.(3.10) goes back to the equilibrium FDT, *i.e.*, the solution of Eq.(3.12) now gives an equilibrium Bose distribution for the current fluctuation

$$N_i^{(0)}(\omega) = (e^{\beta\omega} + 1)/(e^{\beta\omega} - 1), \tag{3.22}$$

with the β exactly equal to the inverse temperature of the phonon reservoir. In the low-frequency limit $\omega \to 0$, the classical limit, the right-hand side of our generalized FDT becomes the product of temperature and conductivity while the left-hand side gives the diffusivity. Therefore we can arrive at the well-known Einstein relation[27,28] between conductivity σ and diffusivity d:

$$d \propto \sigma/\beta. \tag{3.23}$$

It is obvious that the low-frequency limit of an equilibrium Bose distribution defines the equilibrium electron temperature as

$$1/\beta = \lim_{\omega\to 0} \frac{1}{2}\omega N_i^{(0)}(\omega). \tag{3.24}$$

Similarly, we may take the limit of our nonequilibrium distribution $N_i(\omega)$ to define another kind of electron "temperature", the classical limit temperature (compared with the above-defined effective temperature)

$$T_e^{(i)} = \lim_{\omega\to 0} \frac{1}{2}\omega N_i(\omega), \tag{3.25}$$

to characterize the low-frequency behavior of the current fluctuation. For the NESS it would be expected that the classical limit temperature $T_e^{(i)}$ is different from the effective temperature T_e in Eq.(3.21) which carries the integrated information of the current fluctuation. The difference between $T_e^{||}$ and $T_e^{\perp}$ also describes, in a certain degree, the anisotropy of the system caused by the applied electric field. All these aboved-mentioned aspects give a full description for the distinguished features of the NESS in an applied electric field which is different from the usual equilibrium state.

IV Energy Balance Equation

The basic philosophy of above discussion is that a nonequilibrium stationary state is essentially a balance between the drift motion and the fluctuation. Accordingly, we derived the coupled equations for the drift motion and the velocity GF's which describe the fluctuation. The assumed temporal-translation invariance is the natural consequence for the stationary state. Therefore the energy-balance equation, which is inconsistent with the equation for the drift motion in the Thornber-Feynman approach, should also be consistent with the coupled equations for the drift motion and the fluctuation provided our approximation scheme, the Gaussian approximation in the CTPGF sense, is correct. In this section we would like to show the consistency between the energy-balance equation and our coupled equations by proving that the energy-balance condition is naturally guaranteed in our steady-state solution for the nonequilibrium GF's.

Taking derivative of the electron fluctuatiing energy

$$\varepsilon(t) = \frac{m}{2}[\langle \dot{x}^l(t)\dot{x}^l(t)\rangle - \langle \dot{x}^l(t)\rangle^2], \tag{4.1}$$

and using the dynamic equation (2.10), we get

$$\begin{aligned}\dot{\varepsilon}(t) &= \langle \dot{x}^l(t)\rangle(eE^l - \langle \dot{p}^l(t)\rangle) + \langle \dot{x}^l(t) f^l(t)\rangle \\ &= \langle \dot{x}^l(t) f^l(t)\rangle - \langle \dot{x}^l(t)\rangle\langle f^l(t)\rangle,\end{aligned} \tag{4.2}$$

where the operator $f^l(t)$, Eq.(2.10), represents the frictional force due to electron-phonon interation. In former expression of Eq.(4.2), the first part explicitly shows the energy gain from the applied field while the second part represents the loss due to phonon friction. In the latter expression, the energy equation is written as a correlation between the friction force and the velocity. For the convenience of the following calculation by utilizing Eqs.(2.10) and (2.11), we can rewrite this energy equation (4.2) in the closed-time path form with electron-phonon interaction shown explicitly,

$$\dot{\varepsilon}(t) = \xi_\alpha \frac{i}{V}\sum_{\mathbf{k}} g_{\mathbf{k}} k^l \langle\langle T_p \exp[i\mathbf{k}\cdot\mathbf{x}_\alpha(t)] B_{\mathbf{k}\alpha}(t)\delta\dot{x}^l_\alpha(t)\rangle\rangle. \tag{4.3}$$

Making use of Eq.(2.6) to eliminate the phonon variables, we then have

$$\begin{aligned}\dot{\varepsilon} &= \frac{1}{2}\xi_\alpha\eta_\beta\frac{-i}{V}\sum_{\mathbf{k}}|g_{\mathbf{k}}|^2 k^l \int ds \Delta_{\alpha\beta}(\mathbf{k}, t-s)\langle T_p \exp\{i\mathbf{k}\cdot[\mathbf{x}_\alpha(t) - \mathbf{x}_\beta(s)]\}\delta\dot{x}^l_\alpha(t)\rangle \\ &= \frac{1}{2}\xi_\alpha\eta_\beta\frac{-i}{V}\sum_{\mathbf{k}}|g_{\mathbf{k}}|^2 k^l \int ds \lambda(t-s)\Delta_{\alpha\beta}(\mathbf{k}, t-s) \\ &\qquad \times\left\langle T_p\delta\dot{x}^l_\alpha(t)\exp\left[i\mathbf{k}\cdot\int_{s_\beta}^{t_\alpha} d\tau\delta\dot{\mathbf{x}}(\tau)\right]\right\rangle.\end{aligned} \tag{4.4}$$

Now we introduce the Gaussian approximation to the fluctuation in Eq.(4.4). After a few steps similar to that for deriving Eq.(2.26), we get

$$\begin{aligned}\dot{\varepsilon} &= \frac{1}{2}\xi_\alpha\eta_\beta\frac{1}{V}\sum_{\mathbf{k}}|g_{\mathbf{k}}|^2 k^l k^i \int ds \lambda(t-s)\Delta_{\alpha\beta}(\mathbf{k}, t-s) F_{\alpha\beta}(\mathbf{k}, t-s) \\ &\qquad \times\left\langle T_p\delta\dot{x}^l_\alpha(t)\int_{s_\beta}^{t_\alpha} d\tau\delta\dot{x}^i(\tau)\right\rangle.\end{aligned} \tag{4.5}$$

Using Eqs.(D1)-(D4) proved in Appendix D and the relations Eqs.(3.13)-(3.16), we can express the above energy equation in term of the velocity GF's,

$$\dot{\varepsilon} = \frac{1}{V}\sum_{\mathbf{k}} |g_{\mathbf{k}}|^2 k^l k^l \int ds \lambda(t-s)\Delta_{\alpha\beta}(\mathbf{k}, t-s)F_{\alpha\beta}(\mathbf{k}, t-s)$$
$$\times \left\{ \frac{1}{4}\xi_\alpha\xi_\beta \left[\int_s^t d\tau D_r^{(l)}(t,\tau) + \frac{1}{m}\theta(t-s) \right] + \frac{1}{4}\xi_\alpha\xi_\beta \int_s^t d\tau D_c^{(l)}(t,\tau) \right\}. \quad (4.6)$$

By noticing the temporal-translation invariance property of the steady-transport state Eqs.(3.2)-(3.4), Eq.(4.6) can be written as

$$\dot{\varepsilon} = -\frac{i}{4V}\sum_{\mathbf{k}} |g_{\mathbf{k}}|^2 k^l k^l \int d\tau \lambda(\tau)\Delta_{\alpha\beta}(\mathbf{k},\tau)F_{\alpha\beta}(\mathbf{k},\tau)$$
$$\times \left\{ -\frac{1}{2}\xi_\alpha\xi_\beta \int \frac{d\omega}{2\pi} e^{i\omega\tau}[D_r^{(l)} - D_a^{(l)}]/\omega + \frac{1}{2}\xi_\alpha\eta_\beta \int \frac{d\omega}{2\pi} e^{i\omega\tau} D_c^{(l)}(\omega)/\omega \right.$$
$$\left. + \frac{1}{2}\theta(\tau)\xi_\alpha\xi_\beta \left[\int \frac{d\omega}{2\pi}[D_r^{(l)} - D_a^{(l)}]/\omega + \frac{i}{m} \right] \right\}, \quad (4.7)$$

where we have made use of the symmetry properties of the GF's: $D_r(-\omega) = D_a(\omega)$ and $D_c(-\omega) = D_c(\omega)$. Expressing Eq.(4.7) in terms of the damping factor γ_i and the nonequilibrium distribution $N_i(\omega)$ defined in Eqs.(3.11) and (3.12) and noticing the identity Eq.(E1) given in Appendix E, we obtain the final form of the energy equation

$$\dot{\varepsilon} = \int \frac{d\omega}{2\pi} 2iRe\gamma_i(\omega)\{D_c^{(i)}(\omega) - N_i(\omega)[D_r^{(i)}(\omega) - D_a^{(i)}(\omega)]\}. \quad (4.8)$$

Now it becomes obvious that the energy-balance condition

$$\dot{\varepsilon} = e\mathbf{E}\cdot\mathbf{v} + \langle \dot{x}^l f^l \rangle = 0, \quad (4.9)$$

i.e., the balance between the energy gain from the electric field and the loss due to phonon friction, is naturally guaranteed by our generalized FDT Eq.(3.10). And therefore we complete the proof for the self-consistency of our discussion.

V Numerical results

In this section we first rewrite the coupled integral equations (3.8)-(3.12) for the purpose of numerical calculation and then present our numerical results. In the derivation of the coupled equations (3.8)-(3.12), the only assumption we have made is the Gaussian approximation to truncate the GF equations, and we treat the phonon variables and the applied electric field both nonperturbatively. Moreover, we notice that the imaginary part of Eq.(3.11) contributes a renormalization effect to the electron mass. Without loss of essential physics, from now on, we would assume the renormalized electron (polaron) mass as a constant for simplicity. Therefore the damping factor (the real part of Eq.(3.11))

$$\gamma_i(\omega) = -\frac{1}{V}\sum_{\mathbf{k}} |g_{\mathbf{k}}|^2 \frac{k_i^2}{2m}\int dt \frac{\sin(\omega t)}{\omega}\cos(\mathbf{k}\cdot\mathbf{v}t)\xi_\alpha\eta_\beta\Delta_{\alpha\beta}(\mathbf{k},t)F_{\alpha\beta}(\mathbf{k},t) \quad (5.1)$$

becomes a real, even function of ω and the nonequilibrium distribution

$$N_i(\omega) = \frac{i}{V}\sum_{\mathbf{k}} |g_{\mathbf{k}}|^2 \frac{k_i^2}{2m}\int dt \frac{\cos(\omega t)}{2\omega\gamma_i(\omega)}\cos(\mathbf{k}\cdot\mathbf{v}t)\xi_\alpha\xi_\beta\Delta_{\alpha\beta}(\mathbf{k},t)F_{\alpha\beta}(\mathbf{k},t) \quad (5.2)$$

is a real, odd function of ω. With the help of Eqs.(3.8)-(3.10) the fluctuation function F in the above equations can be expressed in terms of $\gamma_i(\omega)$ and $N_i(\omega)$ by using Eqs.(2.28)-(2.30):

$$F_{++}(\mathbf{k},t) = \exp\left[-\frac{k^2}{m}(\alpha + i\phi)\right], \tag{5.3}$$

$$F_{--}(\mathbf{k},t) = \exp\left[-\frac{k^2}{m}(\alpha - i\phi)\right], \tag{5.4}$$

in which

$$\alpha \equiv \frac{1}{2}\int \frac{d\omega}{2\pi}\frac{\sin^2(\omega t/2)}{\omega^2/4}\sum_{i=1}^{3}\left[\frac{k^i}{k}\right]^2 \frac{\omega N_i(\omega)\gamma_i(\omega)}{\omega^2+\gamma_i(\omega)^2}, \tag{5.5}$$

$$\phi \equiv \frac{1}{2}|t| - \frac{1}{2}\int \frac{d\omega}{2\pi}\frac{\sin^2(\omega t/2)}{\omega^2/4}\sum_{i=1}^{3}\left[\frac{k^i}{k}\right]^2 \frac{\gamma_i(\omega)^2}{\omega^2+\gamma_i(\omega)^2}. \tag{5.6}$$

The thermal propagators of the phonon reservoir in Eqs.(5.1) and (5.2) have the form[20]

$$\Delta_{++}(\mathbf{k},t) = -\sin(\omega_{\mathbf{k}}|t|) - i[1+2n(\beta\omega_{\mathbf{k}})]\cos(\omega_{\mathbf{k}}t), \tag{5.7}$$

$$\Delta_{--}(\mathbf{k},t) = \sin(\omega_{\mathbf{k}}|t|) - i[1+2n(\beta\omega_{\mathbf{k}})]\cos(\omega_{\mathbf{k}}t), \tag{5.8}$$

$$\Delta_{-+}(\mathbf{k},t) = -\sin(\omega_{\mathbf{k}}t) - i[1+2n(\beta\omega_{\mathbf{k}})]\cos(\omega_{\mathbf{k}}t), \tag{5.9}$$

$$\Delta_{+-}(\mathbf{k},t) = \sin(\omega_{\mathbf{k}}t) - i[1+2n(\beta\omega_{\mathbf{k}})]\cos(\omega_{\mathbf{k}}t), \tag{5.10}$$

where $n(\beta\omega) = 1/[\exp(\beta\omega)-1]$ is the population of phonons with momentum $\mathbf{k}$ and energy $\omega_{\mathbf{k}}$ and $1/\beta$ is the phonon temperature. Substitute the expressions of the $\Delta's$ and $F's$ from Eqs.(5.3)-(5.10) into Eqs.(5.1) and (5.2) and we obtain explicit forms of self-consistent integral equations for the $\gamma_i(\omega)$ and $N_i(\omega)$,

$$\omega N_i(\omega) = \frac{2}{V}\sum_{\mathbf{k}}|g_{\mathbf{k}}|^2\frac{k_i^2}{m\gamma_i(\omega)}\int_0^\infty dt\cos(\omega t)\cos(\mathbf{k}\cdot\mathbf{v}t)\exp\left(-\frac{k^2}{m}\alpha\right)$$
$$\times\left[\sin(\omega_{\mathbf{k}}t)\sin\left(\frac{k^2}{m}\phi\right)+\coth\left(\frac{\beta\omega_{\mathbf{k}}}{2}\right)\cos(\omega_{\mathbf{k}}t)\cos\left(\frac{k^2}{m}\phi\right)\right], \tag{5.11}$$

$$\gamma_i(\omega) = \frac{2}{V}\sum_{\mathbf{k}}|g_{\mathbf{k}}|^2\frac{k_i^2}{m}\int_0^\infty dt\frac{1}{\omega}\sin(\omega t)\cos(\mathbf{k}\cdot\mathbf{v}t)\exp\left(-\frac{k^2}{m}\alpha\right)$$
$$\times\left[\sin(\omega_{\mathbf{k}}t)\cos\left(\frac{k^2}{m}\phi\right)-\coth\left(\frac{\beta\omega_{\mathbf{k}}}{2}\right)\cos(\omega_{\mathbf{k}}t)\sin\left(\frac{k^2}{m}\phi\right)\right], \tag{5.12}$$

with α and ϕ given in Eqs.(5.5) and (5.6). These equations can be solved numerically for a given phonon temperature $1/\beta$ and the electron-phonon interaction strength $g_{\mathbf{k}}$. Once the solution is obtained, all the transport properties can be easily evaluated. Substituting Eqs.(5.3)-(5.10) into Eq.(3.6), it becomes

$$eE = \frac{2}{V}\sum_{\mathbf{k}}|g_{\mathbf{k}}|^2 k_3\int_0^\infty dt\sin(\mathbf{k}\cdot\mathbf{v}t)\exp\left(-\frac{k^2}{m}\alpha\right)\left[\sin(\omega_{\mathbf{k}}t)\cos\left(\frac{k^2}{m}\phi\right)\right.$$
$$\left.-\coth\left(\frac{\beta\omega_{\mathbf{k}}}{2}\right)\cos(\omega_{\mathbf{k}}t)sin\left(\frac{k^2}{m}\phi\right)\right]. \tag{5.13}$$

Then we obtain the E-v curve by numerical integration of Eq.(5.13). Substituting the $\gamma_i(\omega)$ and $N_i(\omega)$ into Eq.(3.21), the effective electron temperature can be calculated by

$$T_e = \frac{1}{3}\sum_{i=1}^{3}\int_{-\infty}^{\infty}\frac{d\omega}{2\pi}\frac{\omega N_i(\omega)\gamma_i(\omega)}{\omega^2+\gamma_i(\omega)^2}. \tag{5.14}$$

Similarly the power spectrum of the nonequilibrium noise, the current fluctuation can also be calculated by

$$S^{(i)}(\omega) = 2\omega N_i(\omega)\gamma_i(\omega)/[\omega^2+\gamma_i(\omega)^2]. \tag{5.15}$$

The concrete numerical calculation proceeds as follows. We use the same model as that investigated by Thornber and Feynman[2] where

$$|g_{\mathbf{k}}|^2 = 4\pi\alpha\omega_0^{3/2}/\sqrt{2m}k^2, \tag{5.16}$$

in which ω_0 is the LO-phonon energy and m is the electron effective mass. We adopt the dimensionless coupling constant $\alpha = 3$ and the inverse phonon temperature $\beta = 5$. The Planck constant $\hbar = 1$. ω, $1/\beta$, k_BT_e and $S^{(i)}(\omega)$ are all scaled by ω_0. v, k, and eE are scaled by $\sqrt{\omega_0/m}$, $\sqrt{m\omega_0}$, and $\alpha\sqrt{m\omega_0^3}$, respectively. Moreover, we assume a continuous cutoff for the phonon momentum

$$\frac{1}{V}\sum_{\mathbf{k}} \Rightarrow \frac{2}{\sqrt{\pi}}\int\frac{d\mathbf{k}}{(2\pi)^3}\exp\left(-\frac{k^2}{k_M^2}\right), \tag{5.17}$$

with a maximum momentum $k_M = 1$ to simplify the numerical integrations involved.

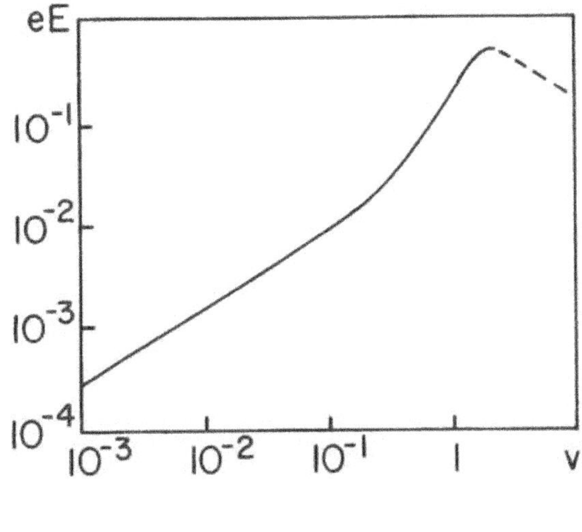

Fig.2. E-v curve.

It is interesting to see that our result for the E-v curve (Fig.2) qualitatively agrees with that of Thornber and Feynman[2], especially there is a threshold electric field strength E_{th} beyond which the steady state becomes unstable. The threshold field strength has its value with electron drifting energy approximately equal to the LO-phonon energy. When the electric field is small we have the usual linear E-v relations; the nonlinear behavior will be important only when the applied field approaches to the threshold. Although Thornber and Feynman's study is not self-consistent, their approximation is based upon a variational principle which minimizes the electron ground state energy, therefore their result would be expected to be quite valid for the low-temperature cases which is the right situation to which the above chosen parameters corresponded. When the temperature goes up, however, we would reasonably expect that the result of the present self-consistent treatment would show many more differences from that of Thornber-Feynman.

When the applied electric field is small, the response of the system is linear, the nonequilibrium distribution functions $N_{\|}(\omega)$ and $N_{\perp}(\omega)$ are both very close to the equilibrium Bose function, and the effective electron temperature is only at a little higher than the phonon reservoir. As the field increases, the response becomes more complicated. The effective electron temperature then goes up rapidly while the distributions are also driven farther away from the Bose function. The larger and larger difference between $N_{\|}(\omega)$ and $N_{\perp}(\omega)$ describes the increasing anisotropy of the system. The nonequilibrium distribution functions for a few typical values of the drift velocity are plotted in Fig.3. It seems interesting that when the electric field goes beyond the threshold and the steady state becomes unstable, the nonequilibrium distribution function $N_{\|}(\omega)$ begins to get an eminence structure near the phonon frequency.

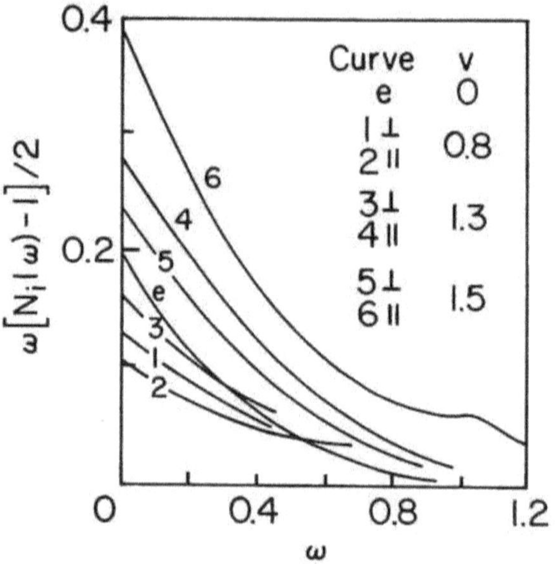

Fig.3. The nonequilibrium distribution functions.

The effective electron temperature T_e (shown in Fig.4), which describes the average behavior of the electron fluctuation, is always higher than the temperature of the phonon reservoir as long as the applied field is nonzero. Its field dependence is almost quadratic when E is not too large which coincides with the usual hot-electron studies. The classical limit temperatures $T_e^{\|}$ and $T_e^{\perp}$, which characterize the low-frequency behavior of the nonequilibrium distribution for the electron fluctuation (see also Fig.4) are lower than the phonon temperature in a certain range. This implies that the low-frequency current fluctuation is suppressed in presence of finite electric field. It should be remarked that T_e

and $T_e^{(i)}$ represent difference physics, which was ignored in previous literatures. The so-called "cooling effect", in the present model, appears only in the behavior of the classical limit temperature rather than in that of effective temperature. These three kinds of temperatures go back to the same limit: the phonon temperature when the applied field goes to zero.

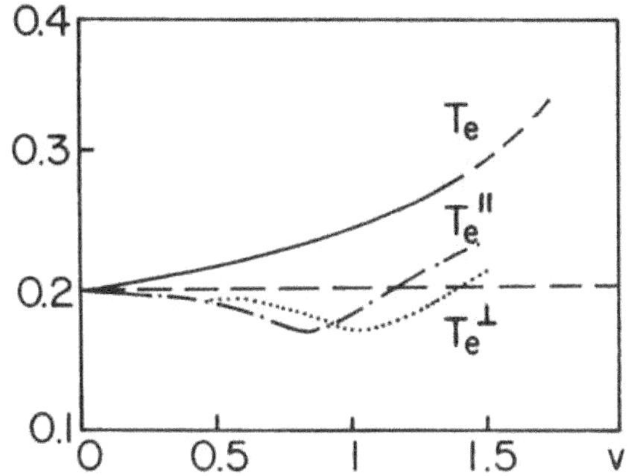

Fig.4. Electron temperatures.

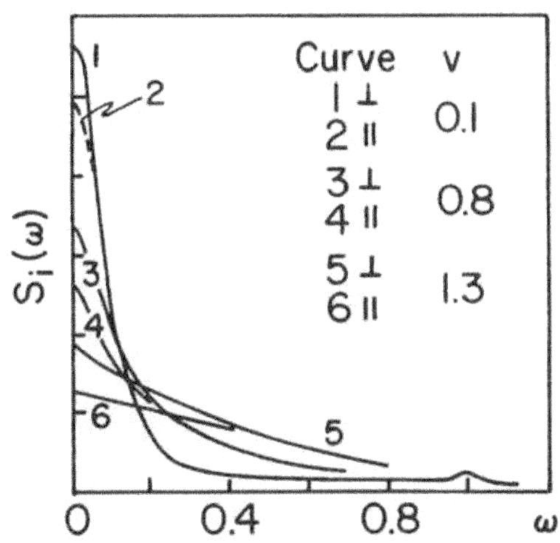

Fig.5. The noise current power spectrum.

Besides the E-v curve, electron temperature, and distribution, we also obtain the power spectra of the nonequilibrium current fluctuation which are plotted in Fig.5. Evidently, as the applied electric field increases, the low-frequency part of fluctuation is gradually suppressed. At the same time, the high-frequecy potion rises up. The overall fluctuation as represented by the area underneith a spectrum curve goes up with the field. Interestingly, there is also a little bump in a power spectrum near the phonon frequency.

Acknowledgements

The author wishes to thank Dr. L. Y. Chen and Professor C. S. Ting for fruitful collaboration and helpful discussions. This work was partially supported by the National Science Foundation of China and partially supported by a Grant No. MDA 972-88-G-0002 from U.S. Defense Advanced Research Projects Agency and the State of Texas.

Appendix A: Proof of Eq.(2.25)

Since $\delta\mathbf{x}(t)$ is a Gaussian process the composite operator

$$A = i\mathbf{k} \cdot [\delta\mathbf{x}_\alpha(t) - \delta\mathbf{x}_\beta(s)] \tag{A1}$$

is also Gaussian. Obviously,

$$\langle T_p A\rangle = 0. \tag{A2}$$

Moreover, all the connected GF's of order higher than 2 vanish, *i.e.*,

$$\langle T_p(AA\cdots A)_n\rangle_{connect} = 0 \quad \text{for} \quad n > 2. \tag{A3}$$

Using (A2) and the Gaussian condition (A3) we may conclude that all the odd-order GF's vanish and the even $2n$-th order GF can be expressed in terms of the only nonvanishing connected GF, $\langle T_p AA\rangle$ (the second order), as

$$\langle T_p(AA\cdots AA)_{2n}\rangle = \frac{(2n)!}{2^n n!}(\langle T_p AA\rangle\cdots\langle T_p AA\rangle)_n. \tag{A4}$$

Now expanding the exponential $\exp(A)$ in the Taylor series and using (A4), the proof of Eq.(2.25) is completed as

$$\begin{aligned}\langle\exp(A)\rangle &= \sum_{n=0}^{\infty}\frac{1}{n!}\langle T_p(AA\cdots A)_n\rangle \\ &= \sum_{m=0}^{\infty}\frac{1}{2^m m!}(\langle T_p AA\rangle\cdots\langle T_p AA\rangle)_m \\ &= \exp(\langle T_p AA\rangle/2).\end{aligned} \tag{A5}$$

Appendix B: Proof of Eq.(2.26)

Obviously, the left-hand side of Eq.(2.26) can be expressed as

$$\partial_J\langle T_p \exp\{i\mathbf{k}\cdot[\delta\mathbf{x}_\alpha(t) - \delta\mathbf{x}_\delta(s)] + J\delta x^j_\beta(t')\}\rangle|_{J=0}. \tag{B1}$$

The Gaussian approximation for the exponential in Eq.(B1) is similar to that in Appendix A and then Eq.(B1) becomes

$$\partial_J\exp(\langle T_p\{i\mathbf{k}\cdot[\delta\mathbf{x}_\alpha(t) - \delta\mathbf{x}_\delta(s)] + J\delta x^j_\beta(t')\}^2\rangle/2)|_{J=0}. \tag{B2}$$

Taking the derivative ∂_J and then setting $J = 0$, we get the desired equation (2.26) by noticing the definition of the GF in Eq.(2.17).

Appendix C: Proof of Eqs.(2.28)-(2.30)

We first prove Eqs.(2.29) and (2.30). If we introduce the following notation:

$$A^{ij}_{\alpha\beta}(t,s) = \langle T_p[\delta x^i_\alpha(t) - \delta x^i_\beta(s)][\delta x^j_\alpha(t) - \delta x^j_\beta(s)]\rangle, \tag{C1}$$

then the F in Eq.(2.25) becomes

$$F_{\alpha\beta} = \exp\left(-\frac{1}{2}k^i k^j A^{ij}_{\alpha\beta}(t,s)\right). \tag{C2}$$

It is not difficult to prove that

$$A^{ij}_{-+} = \theta(t-s)A^{ij}_{++} + \theta(s-t)A^{ij}_{--}, \tag{C3}$$

$$A^{ij}_{+-} = \theta(s-t)A^{ij}_{++} + \theta(t-s)A^{ij}_{--}. \tag{C4}$$

Substituting these two expressions into (C2) and noticing that

$$\exp[\theta(t-s)(\cdot) + \theta(s-t)(\cdot\cdot)] = \theta(t-s)\exp(\cdot) + \theta(s-t)\exp(\cdot\cdot), \tag{C5}$$

we have the proof of Eqs.(2.29) and (2.30).

Now we turn to the proof of Eq.(2.28). Express the operator $[\delta x_\alpha(t) - \delta x_\alpha(s)]$ of Eq.(2.25) in terms of the velocity operator $\delta\dot{\mathbf{x}}_\alpha(t)$, we obtain

$$F_{\alpha\alpha} = \exp\left[-\frac{1}{2}k^i k^j \left\langle T_p \int_s^t d\tau \int_s^t d\tau' \delta\dot{x}^i_\alpha(\tau)\delta\dot{x}^j_\alpha(\tau')\right\rangle\right]. \tag{C6}$$

Moving the time-ordering operator into the integral we can get an extra term due to the commutator between the operator $\delta x(t)$ and $\delta\dot{x}(t)$. After a careful, straightforward check we can express (C6) in the form of Eq.(2.28).

Appendix D: For the Proof of the Energy-balance Equation

In this appendix we derive the following four equations.

$$\left\langle T_p \delta\dot{x}^l_\alpha(t) \int_{\tau_\beta}^{t_\alpha} ds\,\delta\dot{x}^i(s)\right\rangle = i\delta^{il}\int_\tau^t ds\, D^{(i)}_{\alpha\beta}(t,s) + H_{\alpha\beta} \tag{D1}$$

where

$$H_{++} = \frac{i}{2m}[\theta(t-\tau) - \theta(\tau-t)], \tag{D2}$$

$$H_{--} = -\frac{i}{2m}[\theta(t-\tau) - \theta(\tau-t)], \tag{D3}$$

$$H_{-+} = \frac{i}{2m}, \tag{D4}$$

$$H_{+-} = -\frac{i}{2m}. \tag{D5}$$

In the following, we will prove Eq.(D2) as an example only, the other three equations (D3)-(D5), can be proven similarly without difficulty. First, we present the equal time GF's:

$$\langle T_p \delta x_+(t)\delta\dot{x}_+(t)\rangle = \frac{1}{2}\langle\delta x(t)\delta\dot{x}(t) + \delta\dot{x}(t)\delta x(t)\rangle, \tag{D6}$$

$$\langle T_p \delta x_-(t)\delta\dot{x}_-(t)\rangle = \frac{1}{2}\langle\delta x(t)\delta\dot{x}(t) + \delta\dot{x}(t)\delta x(t)\rangle, \tag{D7}$$

$$\langle T_p \delta x_-(t)\delta\dot{x}_+(t)\rangle = \langle\delta x(t)\delta\dot{x}(t)\rangle, \tag{D8}$$

$$\langle T_p \delta x_+(t)\delta\dot{x}_-(t)\rangle = \langle\delta\dot{x}(t)\delta x(t)\rangle, \tag{D9}$$

which are consistent with the CTPGF causality requirement. Then the left hand side of Eq.(D1) with $\alpha = \beta = +$ can be expressed as

$$\langle T\delta\dot{x}^l(t)[\delta x^i(t) - \delta x^i(\tau)]\rangle = \frac{1}{2}\langle \delta\dot{x}^l(t)\delta x^i(t) + \delta x^i(t)\delta\dot{x}^l(t)\rangle$$
$$-\theta(t-\tau)\langle\delta\dot{x}^l(t)\delta x^i(\tau)\rangle - \theta(\tau - t)\langle\delta x^i(\tau)\delta\dot{x}^l(t)\rangle, \tag{D10}$$

while the right-hand side of it can be written in the following form:

$$\delta^{il}\int_\tau^t ds\langle\theta(t-s)\delta\dot{x}^l(t)\delta\dot{x}^i(s) + \theta(s-t)\delta\dot{x}^i(s)\delta\dot{x}^l(t)\rangle$$
$$= \langle\theta(t-\tau)\delta\dot{x}^l(t)[\delta x^i(t) - \delta x^i(\tau)]$$
$$+\theta(\tau - t)[\delta x^i(t) - \delta x^i(\tau)]\delta\dot{x}^l(t)\rangle. \tag{D11}$$

Comparing Eqs.(D10) and (D11) and using the commutator Eq.(2.22), we complete the proof.

Appendix E: An Identity for Velocity GF

In this appendix, we derive the following identity for the velocity GF's:

$$\int\frac{d\omega}{2\pi}\frac{D_r(\omega) - D_a(\omega)}{\omega} = -\frac{i}{m}. \tag{E1}$$

For the purpose of proving the above identity we would first discuss the relation Eq.(2.31) between the velocity GF's $D_{\alpha\beta}(t_1, t_2)$ and the position GF's $P_{\alpha\beta}(t_1, t_2)$. Introducing $t = t_1 - t_2$ and $T = (t_1 + t_2)/2$ and expressing the t_1 and t_2 in terms of the t and T, we have

$$D_{\alpha\beta}(t) = -\frac{1}{m}\delta_{\alpha\beta}(t) + \left(\frac{1}{4}\partial_T^2 - \partial_t^2\right)P_{\alpha\beta}(t,T), \tag{E2}$$

where D is independent of T as required by the time translation invariance. Taking into account Eqs.(3.13)-(3.16), we can easily derive from Eq.(E2) that

$$D_r(t) - D_a(t) = \left(-\partial_t^2 + \frac{1}{4}\partial_T^2\right)[P_r(t,T) - P_a(t,T)]. \tag{E3}$$

Moreover, because $\partial_T[P_r(t,T) - P_a(t,T)]$ is independent of T, *i.e.*,

$$\begin{aligned} &i\partial_T[P_r(t,T) - P_a(t,T)] \\ &= \partial_T\langle[\delta x(T+t/2), \delta x(T-t/2)]\rangle \\ &= \langle[\delta\dot{x}(T+t/2), \delta x(T-t/2)]\rangle + \langle[\delta x(T+t/2), \delta\dot{x}(T-t/2)]\rangle \\ &= \int_{-t/2}^{t/2} d\tau\langle[\delta\dot{x}(T+\tau), \delta\dot{x}(T-t/2)] - [\delta\dot{x}(T+\tau), \delta\dot{x}(T-t/2)]\rangle \\ &= \int_{-t/2}^{t/2} d\tau i[D_r(\tau - t/2) - D_a(\tau - t/2) - D_r(\tau + t/2) + D_a(\tau + t/2)] \end{aligned} \tag{E4}$$

is dependent upon t only, we have

$$\partial_T^2[P_r(t,T) - P_a(t,T)] = 0. \tag{E5}$$

Therefore Eq.(E3) becomes

$$D_r(\omega) - D_a(\omega) = \omega^2[P_r(\omega,T) - P_a(\omega,T)], \tag{E6}$$

where we have made the Fourier transformation. With the help of Eq.(E6) and the commutator Eq.(2.22) we are now able to complete the proof of Eq.(E1) as

$$\begin{aligned}\int \frac{d\omega}{2\pi}\frac{D_r(\omega)-D_a(\omega)}{\omega} &= \int \frac{d\omega}{2\pi}\omega[P_r(\omega,T)-P_a(\omega,T)] \\ &= i\partial_t \int \frac{d\omega}{2\pi} e^{-i\omega t}[P_r(\omega,T)-P_a(\omega,T)]|_{t=0} \\ &= i\partial_t[P_r(t,T)-P_a(t,T)]|_{t=0} \\ &= \partial_t\langle[\delta x(T+t/2),\delta x(T-t/2)]\rangle|_{t=0} \\ &= \langle[\delta\dot{x}(T),\delta x(T)]\rangle \\ &= -i/m. \end{aligned} \tag{E7}$$

Refrences

1. Conwell, E., *High Field Transport in Semiconductors*, Academic, New York (1967).
2. Thornber, K. K., and R. P. Feynman, Phys. Rev. **B1**, (1970), 4099.
3. Thornber, K. K., Phys. Rev. **B3**, (1971), 1929.
4. Peeters, F. M., and J. T. Devreese, **B23**, (1981) 1936.
5. Seeger, K., ed. *Semiconductor Physics*, Springer Verlag, New York (1982).
6. Jacoboni, C., and L. Reggiani, Rev. Mod. Phys. **55**, (1983), 645.
7. Mahan, G. D., Phys.Rep. **145**, (1984), 251.
8. Lei, X.L., and C.S.Ting, Phys.Rev. **B32**, (1985), 1112.
9. Reggiani, L., Physica, **134**, (1985), 123.
10. Khan, F. S., J. H. Davies, and J. W. Wilkins, Phys. Rev. **B36**, (1987), 2578.
11. Lipavsky, P., F. S. Khan, A. Kalvova, and J. W. Wilkins, Phys. Rev. **B43**, (1991), 6650.
12. Lipavsky, P., F. S. Khan, and J. W. Wilkins, Phys. Rev. **B43**, (1991) 6665.
13. Su, Z. B., L. Y. Chen, and J. L. Birman, Phys. Rev. **B35**, (1987), 9744.
14. Su, Z. B., L. Y. Chen, and C.S. Ting, Phys. Rev. Lett., **60**, (1988), 2323.
15. Su, Z. B., L. Y. Chen, and C. S. Ting, Phys. Rev. **B38**, (1988), 3866.
16. Su, Z. B., and B. Sakita, Phys. Rev. **B40**, (1989), 9959.
17. Chen, L. Y., and Z. B. Su, Phys. Rev. **B40**, (1989), 9309.
18. Chen, L. Y., and Z. B. Su, Intern. J. Mod. Phys. **B4**, (1990), 293.
19. Feynman, R. P., and F. L. Vernon, Ann. Phys. (N.Y.) **24**, (1963), 118.
20. Chou, K. C., Z. B. Su, B. L. Hao and L. Yu, Phys. Rep. **118**, (1985), 1.
21. Su, Z. B., L. Y. Chen, X. T. Yu and K. C. Chou, Phys. Rev. **B37**, (1988), 9810.
22. Schwinger, J., J. Math. Phys. (New York) **2**, (1961), 407.

23. Kadanoff, L. P., and G. Baym, *Quantum Statistical Mechanics*, Benjiamin, New York (1962).

24. Keldysh, L. V., Zh. Eksp. Teor. Fiz., **47**, (1964), 1515.

25. Langreth, D., and J. W. Wilkins, Phys. Rev. **B6**, (1972), 3189.

26. Rammer, J., H.Smith, Rev.Mod.Phys. **58**, (1986), 323.

27. Lifshitz, E. ., and L. P. Pitaevskii, *Statistical Physics* (Pergamon, New York, 1980).

28. Haken, H., *Synergetics* (Springer-Verlag, New York, 1977).

International Journal of Modern Physics B, Vol. 6, No. 7 (1992) 1079–1098

Impurity Resistivity under Thermalized Condition

C. S. Ting
Department of Physics, University of Houston
Houston, Texas 77204 USA
and
L. Y. Chen
Department of Physics, Brown University
Providence, Rhode Island 02912 USA

The standard impurity resistivity based upon the force-balance equation is derived with use of the method of closed-time-path Green's functions. In this the effects from both the noncommutability of the center of mass fluctuations at different times and the exact noncannonical commutation relations between the coordinates and momentum of relative electrons are considered. In the presence of a short momentum-conserving inelastic scattering time due to electron-electron interaction, fast thermalization among charge carriers can be achieved. Under this condition, the (thermalized) impurity resistivity will have different form than the standard impurity resistivity and it is practically given by the lowest order electron-impurity term in the force balance equation. We also demonstrate that this conclusion is consistent with results based upon the Boltzmann equation in a relaxation time approximation.

I. Introduction

The development of the balance equation approach to high field transport by Lei and Ting[1] has opened the question of the relation of its low field linear limit for impurity scattering to corresponding results based on standard transport equations of the Boltzmann type[2] and the Kubo formula[3] in the absence of a heat bath and electron-electron interactions. In particular, Argyres and Sigel[4] and Huberman and Chester[5] have emphasized the fact that the latter results may be obtained by summing divergent perturbation series terms of all orders in van Hove's "$\lambda^2 t$" expansion[6] (λ measures impurity potential strength, t is real time), whereas the former balance equation approach yields results to lowest order in λ^2 which are different (except at zero temperature) from those in Refs.(4),(5) and (6). It is this point which we address in this review article, to clarify the physical difference between the two approaches, in identifying the latter approach[2,6] as corresponding to adiabatic transport, while the former one[1] describes thermalized or isothermal transport[7-10].

A brief perusal of the standard transport equation approach for impurity scattering is sufficient to confirm the impression that it is devoid of any coupling to a heat bath or thermalization mechanism, and is incapable of describing thermalized conduction. It predicts the linear adiabatic impurity resistivity as[2,3] (m, e, N are electron effective mass, charge and number density respectively)

$$\rho_i^A \simeq \frac{m}{Ne^2}\frac{\lambda^2}{<\tau>} \tag{1.1}$$

where τ is the relaxation time due to impurity scattering, and its average $<\tau>$ has been defined and fully discussed in Eqs.(15.38) and (15.39) of article 1 in this volume.

The linear resistivity predicted by the balance equation approach[1]

$$\rho_i^I \simeq \frac{m}{Ne^2} < \frac{\lambda^2}{\tau} >, \tag{1.2}$$

in contrast to Eq.(1.1), has been identified as a thermalized or isothermal resistivity[7] on the basis of a center of mass energy loss interpretation. The salient features about a heat bath in the context of thermalized conduction are that it drains dissipated energy essentially instantaneously as it is pumped into the the system by the external *dc* electric field[7] and it promotes a fast tendency toward thermalization by means of a rather strong bath-quantum/phonon mediated electron-electron interaction. The relaxation time for electron energy transfer to such an ideal heat bath must be relatively short, which bespeaks a relatively strong coupling to the bath. In another approach,[11] the authors demonstrate that the balance equations follow from the Zubarev *nonequilibrium statistical operator* method[12], with a density matrix closely related to a quasi-equilibrium drifted-temperature counterpart. The alternative approaches[7,12] cited here make clear the fact that the balance equation approach involves an additional constraint on the density matrix which is due to the presence of a heat bath and strong carrier-carrier collisions[1,12] originating from phonon mediated electron-electron interactions as well as Coulumb interactions among electrons. In linear response, one cannot address the role of the bath in draining dissipated Joule heat energy. Nevertheless, the role of bath quanta in providing a mechanism for a strong effective electron-electron interaction which promotes a tendency toward drifted thermalizaton can be a prominent feature of linear conduction. Such important physical features are omitted from the framework of the "$\lambda^2 t$" series analysis of adiabatic theory.[4,5,13]

The contents of this article are primarily based upon our previous publication in Ref.14. Here we shall study the linear impurity resistivity as it is influenced by an effective electron-electron interaction. This interaction may originate from the phonon-mediated or the direct electron-electron interaction which imparts to electron propagation in a finite momentum-conserving inelastic time τ_{in}. With this in view, we show that the leading divergencies of the high-order impurity resistivity contributions of adiabatic theory[4–6] are removed by the inverse energy relaxation time $1/\tau_{in}$. When the effective electron-electron interaction becomes strong or τ_{in} is shorter than the impurity scattering time, the impurity resistivity formula practically reduces to the thermalized form given by Eq.(1.2). Similar results have been obtained by Cai *et al*[15] and by Hu and O'Connell[16] using alternative interpretations. It is well known that the difference between the adiabatic resistivity (Eq.(1.1)) and the thermalized resistivity (Eq.(1.2))is vanishing at zero temperature and is very small at low temperature where the chemical potential μ of the electron system is much larger than temperature T and the inelastic scattering strength $1/\tau_{in} \sim (T/\mu)^2$ due to the Coulomb interaction among electrons is weak. This fact implies that the sum of all higher order divergent resistivity terms makes negligible contribution to the adiabatic resistivity formula at low temperature. However when the temperature becomes comparable to or larger than μ as in the case of a doped semiconductor, the effective electron-electron interaction is expected to be rather strong and all higher order divergent resistivity terms will be quenched by the inelastic scattering rate $1/\tau_{in}$. In this situation the correct resistivity formula to use should be given by Eq.(1.2).

In the following sections, we shall derive the impurity resistivity by using the method of closed-time-path Green's function[17]. A consistent treatment of the non-commutability[13] between the center of mass fluctuations (CMF) at different times and the exact noncannonical commutation relations between the variables of relative electrons will presented. In section II we show that the imposition of approximate cannonical commutation relations for relative coordinates results in complete cancellation of all the divergent terms generated by the non-commutability of CMF's. As a result the adiabatic impurity resistivity predicted in this way vanishes. This conclusion is consistent with that of Argyres[18] who has regarded the center of mass variables as commutative or classical. In section III, the macroscopic error of non-vanishing drift of relative electrons brought about by the imposition of approximate cannonical commutation relations on the relative electrons will be examined. After this relative electron drift is incorporated into the current, the standard adiabatic resistivity formula is seen to emerge. In section IV, we show that when the exact commutation relations between the relative coordinates and momenta are utilized, our result for adiabtic impurity resistivity again reduces to the standard form in Eq.(1.1). We would like to emphasize here that one a finite inelastic scattering time due to strong electron-electron interactions is introduced, the associated resistivity is devoid of the divergences of van Hove's "$\lambda^2 t$"-series expansion[6] and has the thermalized expression given by Eq.(1.2). In section V we show that the last conclusion can also be reached from the Boltzmann equation in a relaxation time approximation.[19] The final section contains a summary and discussion.

II. Quantum mechanical treatment of center of mass variables

In the paper of Fishman and Mahan,[13] the authors there focused attention on the quantum mechanical aspect of the heavy center of mass "particle" (with mass Nm, m–electron mass, N–electron number; N is a large number) for the purpose of determining adiabatic resistivity from the force balance equation[1] In this section, we shall show that their considerations were incomplete and that there exist an infinite number of divergent terms which they overlooked. Based on a consistent treatment to be presented below, it will become evident that all the divergent terms related to the non-commutability between the center of mass fluctuations at different time fully cancel among themselves and make no contribution to the resistivity formula. Therefore the center of mass fluctuations can be regarded as commutative and classical variables.

To begin our discussion, we briefly review the force balance equation method[1,13]. The Hamiltonian of an electron-impurity system in the presence of a constant electric field E is

$$H = H_e + \frac{\mathbf{P}^2}{2Nm} + Ne\mathbf{E}\cdot\mathbf{R} + H_i, \tag{2.1}$$

$$H_e = \sum_{i=1}^{N} \frac{p_i^2}{2m} + \sum_{i<j} v(r_i - r_j) \tag{2.2}$$

$$\begin{aligned} H_i &= \lambda \sum_{\mathbf{q}} \rho_{\mathbf{q}} e^{i\mathbf{q}\cdot\mathbf{R}} V(\mathbf{q}) \\ &= \lambda \sum_{\mathbf{q}} \rho_{\mathbf{q}} e^{i\mathbf{q}\cdot\delta\mathbf{R}} V(\mathbf{q}) e^{i\mathbf{q}\cdot\mathbf{R}_c}, \end{aligned} \tag{2.3}$$

where $\rho_{\mathbf{q}} = \sum_{\mathbf{k}} c^\dagger_{\mathbf{k}+\mathbf{q}} c_{\mathbf{k}}$ is the density operator of relative electrons. H_e is the relative electron Hamiltonian with electron-electron interaction $v(r_i - r_j)$, and p_i and r_i are relative momentum and coordinate of the i-th electron respectively. $c_{\mathbf{k}}(c^\dagger_{\mathbf{k}})$ is the destruction (creation) operator of a relative electron with momentum $\mathbf{k}$. $\mathbf{P}$ and $\mathbf{R}$ describe the momentum and position of the center of mass of N electrons, $\delta\mathbf{R} = \mathbf{R} - \mathbf{R_c}$ is the fluctuation of the center of mass from its classical trajectory $\mathbf{R_c}$. The electron-impurity potential of strength λ is

$$V(\mathbf{q}) = \sum_{\alpha} \phi(\mathbf{q}) e^{-i\mathbf{q}\cdot\mathbf{R}_\alpha}, \tag{2.4}$$

where $\mathbf{R}_\alpha$ is the α-th impurity site. In the following discussion, $\mathbf{R}$ and $\mathbf{P}$ will be treated quantum mechanically and exactly. However, approximate canonical commutation relations $[r_i, p_j] = i\delta_{ij}$ are assumed for the electrons' relative coordinates and momenta, and the exact noncanonical correction to these relations will be considered in the latter two sections.

Employing an interaction picture with H_i and $Ne\mathbf{E}\cdot\mathbf{R}$ being regarded as perturbative interactions, and separating the center of mass from the electrons' relative motion, the force balance equation can be written as[1,13]

$$Ne\mathbf{E} = Tr\rho(t)F_i(t), \tag{2.5}$$

where the density matrix $\rho(t)$ satisfies

$$i\partial_t\rho(t) = [H_i, \rho(t)] + [Ne\mathbf{E}\cdot\mathbf{R}, \rho(t)], \tag{2.6}$$

subject to the initial condition

$$\rho(t=0) = \rho_0 = e^{-\beta H_e}|P><P|, \tag{2.7}$$

in which $P|P> = Nmv_d|P>$ and v_d is the drift velocity of electrons. The frictional force operator due to electron impurity scattering is obtained as

$$\mathbf{F}_i(t) = \lambda \sum_{\mathbf{q}} i\mathbf{q}\rho_{\mathbf{q}} e^{i\mathbf{q}\cdot\delta\mathbf{R}} V(\mathbf{q}) e^{i\mathbf{q}\cdot\mathbf{R}_c}. \tag{2.8}$$

Expanding the right hand side of the force balance equation Eq.(2.5) in powers of λ and to linear order in $\mathbf{E}$, and then randomly averaging over impurity configurations, the force balance equation for impurity scattering has been expressed in the form[13]

$$NeE = F(\delta)v_d + \alpha(\delta)NeE, \tag{2.9}$$

where the long time (steady state) limit $\lim_{t\to\infty} f(t)$ is realized by the $\delta \to 0$ limit of the corresponding Laplacian component $\lim_{\delta\to 0} f(\delta) = \lim_{\delta\to 0} \delta \int_0^\infty dt f(t) exp(-\delta t)$. $F(\delta)$ and $\alpha(\delta)$ are (N_i is the impurity concentration)

$$\begin{aligned} F(\delta)v_d = &\sum_{n=1}^{\infty} \int_0^\infty dt_1 \delta e^{-\delta t_1} \int_0^{t_1} dt_2 \cdots \int_0^{t_{2n-2}} dt_{2n-1} \int_0^{t_{2n-1}} dt_{2n} \lambda^{2n} N_i^n \\ &\times \sum_{\mathbf{q}_1,\cdots\mathbf{q}_n} iq_{1x} |\phi(\mathbf{q}_1)|^2 e^{i\mathbf{q}_1\cdot v_d(t_1-t_2)} \ldots |\phi(\mathbf{q}_n)|^2 e^{i\mathbf{q}_n\cdot v_d(t_{2n-1}-t_{2n})} \\ &\times (i)^{2n} Tr(\rho_0[A(-\mathbf{q}_n, t_{2n}), [A(\mathbf{q}_n, t_{2n-1}), \cdots, [A(-\mathbf{q}_1, t_2), A(\mathbf{q}_1, t_1)] \cdots]]), \end{aligned} \tag{2.10}$$

and

$$\begin{aligned} \alpha(\delta) = &\sum_{n=1}^{\infty} \int_0^\infty dt_1 \delta e^{-\delta t_1} \int_0^{t_1} dt_2 \cdots \int_0^{t_{2n-2}} dt_{2n-1} \int_0^{t_{2n-1}} dt_{2n} \int_0^{t_{2n}} dt_{2n+1} \\ &\times \lambda^{2n} N_i^n \sum_{\mathbf{q}_1,\cdots\mathbf{q}_n} iq_{1x} |\phi(\mathbf{q}_1)|^2 e^{i\mathbf{q}_1\cdot v_d(t_1-t_2)} \ldots |\phi(\mathbf{q}_n)|^2 e^{i\mathbf{q}_n\cdot v_d(t_{2n-1}-t_{2n})} (i)^{2n+1} Tr(\rho_0 \\ &\times [R(t_{2n+1}), [A(-\mathbf{q}_n, t_{2n}), [A(\mathbf{q}_n, t_{2n-1}), \cdots, [A(-\mathbf{q}_1, t_2), A(\mathbf{q}_1, t_1)] \cdots]]]). \end{aligned} \tag{2.11}$$

Here $A(\mathbf{q},t)$ is defined in the interaction picture as

$$A(\mathbf{q},t) = \rho_{\mathbf{q}}(t)e^{i\mathbf{q}\cdot\delta\mathbf{R}(t)}. \tag{2.12}$$

It should be noted that the noncommutability between $R(t_{2n+1})$ and the $2n$ $\delta\mathbf{R}$'s involved through $A(\mathbf{q},t)$ in Eq.(2.11) yields the existence of the last term in Eq.(2.9). The authors of Ref.13 only take account of this part of the center of mass fluctuation effect and regard the 2n variables $\delta R(t_i)$ $(i = 1, 2, \cdots, 2n)$ in Eqs. (2.10) and (2.11) as mutually commutable. In fact, it is easy to verify that the noncommutabilities between each pair of the 2n δR's are of equal importance and their contributions must also be considered in a proper treatment.

Before a detailed calculation of the impurity resistivity is carried out, we need to establish a relation between $F(\delta)$ and $\alpha(\delta)$ in Eqs.(2.10) and (2.11) by utilizing the following commutation relation[13]

$$\left[R(t_{2n+1}), e^{iq\cdot\delta R(t_i)}\right] = -\frac{q_{ix}}{Nm}(t_i - t_{2n+1})e^{iq\cdot\delta R(t_i)} \tag{2.13}$$

Using the above relation to complete the commutations in Eq.(2.11) between $R(t_{2n+1})$ and all the factors $A(q,t_i)$ (see Eq.(2.12) with $i = 1, 2, \cdots, 2n$), Eq.(2.11) becomes

$$\begin{aligned}\alpha(\delta) = &\sum_{n=1}^{\infty}\int_0^\infty dt_1\delta e^{-\delta t_1}\int_0^{t_1}dt_2\cdots\int_0^{t_{2n-2}}dt_{2n-1}\int_0^{t_{2n-1}}dt_{2n}\int_0^{t_{2n}}dt_{2n+1}\\ &\times\lambda^{2n}N_i^n\sum_{\mathbf{q}_1,\cdots\mathbf{q}_n} iq_{1x}\frac{-1}{Nm}[q_{1x}(t_1-t_2)+q_{2x}(t_3-t_4)+\cdots+q_{nx}(t_{2n-1}-t_{2n})]\\ &\times|\phi(\mathbf{q}_1)|^2e^{i\mathbf{q}_1\cdot v_d(t_1-t_2)}\cdots|\phi(\mathbf{q}_n)|^2e^{i\mathbf{q}_n\cdot v_d(t_{2n-1}-t_{2n})}\\ &\times(i)^{2n}Tr(\rho_0[A(-\mathbf{q}_n,t_{2n}),[A(\mathbf{q}_n,t_{2n-1}),\cdots,[A(-\mathbf{q}_1,t_2),A(\mathbf{q}_1,t_1)]\cdots]]).\end{aligned} \tag{2.14}$$

Comparing the above equation with Eq.(2.10), it is straightforward to see that the structure of the two are almost parallel. Expanding Eqs.(2.10) and (2.14) respectively to the linear (and zeroth) order in v_d (drift velocity), and checking their time integrations, we can relate $\alpha(\delta)$ directly to $F(\delta)$ as follows:

$$\alpha(\delta) = \frac{F(\delta)}{Nm\delta}. \tag{2.15}$$

In order to calculate the frictional force $F(\delta)v_d$, we have to treat the retarded multi-correlation function of 2n A-operators in Eq.(2.10). Each A contains one electron density fluctuation operator ρ_q and one center of mass fluctuation operator δR. The Feynman diagram technique employing closed time path Green's functions[17] provides us with a convenient means for the management of such complicated correlation functions. In the Appendix, we present a brief introduction to the closed time path Green's functions and express the 2n-th retarded correlation function in a closed time path form. There we take n=2 as an example to demonstrate the detailed evaluation of the frictional force. We shall outline the rules for constructing the requisite Feynman diagrams. In this, we use arrowed lines with r, a, and c (Fig.1(a)) respectively to denote the retarded, advanced and correlation Green's functions:

$$\begin{aligned}iG_r(k,t-t') &= \theta(t-t')Tr\rho_0[c_k(t),c_k^\dagger(t')]_+\\ &= \theta(t-t')e^{-i\epsilon_k(t-t')-\eta|t-t'|},\\ iG_a(k,t-t') &= -\theta(t'-t)Tr\rho_0[c_k(t),c_k^\dagger(t')]_+\end{aligned} \tag{2.16}$$

$$= \theta(t'-t)e^{-i\epsilon_k(t-t')-\eta|t-t'|}, \tag{2.17}$$

$$iG_c(k,t-t') = Tr\rho_0[c_k(t), c_k^\dagger(t')]$$
$$= (1-2f(\epsilon_k))e^{-i\epsilon_k(t-t')-\eta|t-t'|}, \tag{2.18}$$

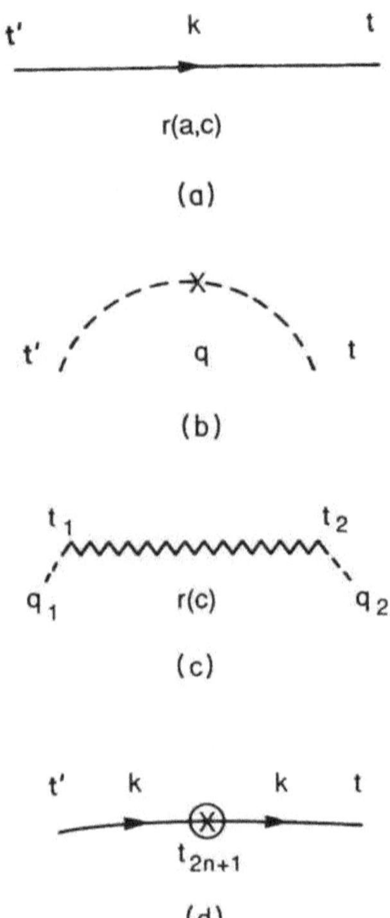

Fig. 1. Graph elements of the Feynman diagrams: (a) retarded (advanced, correlation) Green's function for electrons; (b) the factor from averaged impurity scattering; (c) retarded (correlation) Green's function for the center-of-mass fluctuations; (d) Green's functions influenced by the external electric field.

where the damping factor η for a single electron propagator is introduced to represent the effect of electron-electron interactions. $f(\epsilon)$ is the Fermi-Dirac distribution function with $\epsilon_k = k^2/2m$. We use a dashed line with a cross (Fig.1(b)) to represent the factor arising from averaged impurity scattering

$$\lambda^2 N_i|\phi(q)|^2 e^{iqv_d(t-t')}, \tag{2.19}$$

where the label q is the momentum transfer of an electron during the scattering process. The wavy lines with r and c symbols (Fig.1(c)) are employed to represent the retarded and correlation Green's functions for center of mass quantum fluctuation:

$$F_r(\mathbf{q}_1, \mathbf{q}_2, t_1 - t_2) = \theta(t_1 - t_2) Tr\rho_0[e^{i\mathbf{q}_1\cdot\delta\mathbf{R}(t_1)}, e^{i\mathbf{q}_2\cdot\delta\mathbf{R}(t_2)}]$$
$$= \theta(t_1 - t_2)(-i)\frac{i\mathbf{q}_1 \cdot i\mathbf{q}_2}{Nm}(t_1 - t_2) + 0((\frac{1}{N})^2), \tag{2.20}$$
$$F_c(\mathbf{q}_1, \mathbf{q}_2, t_1 - t_2) = Tr\rho_0[e^{i\mathbf{q}_1\cdot\delta\mathbf{R}(t_1)}, e^{i\mathbf{q}_2\cdot\delta\mathbf{R}(t_2)}]_+$$
$$= 2 + 0(\frac{1}{N}). \tag{2.21}$$

Here $[A, B] = AB - BA$ and $[A, B]_+ = AB + BA$.

In the present paper, only leading divergent terms like $\lambda^2(\lambda^2/\delta)^n$ $(n = 0, 1, 2, \cdots)$ arising in conjunction with van Hove's limiting prescription[9] will be considered. Considering the interaction structure of the three participating elements: electron (fermion) field, impurities and center of mass fluctuation (boson field), and utilizing the Feynman diagrammatic representation, it is straightforward to show that the frictional force $F(\delta)v_d$ in Eq.(2.10) can be expressed by the graphs in Fig.2(a), i.e.,

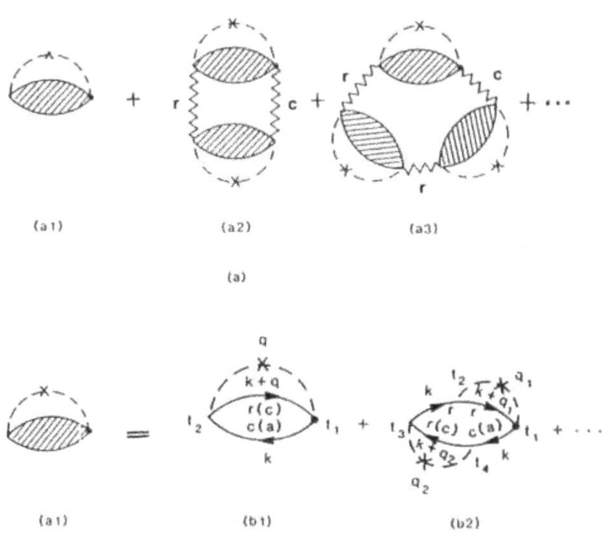

Fig.2. Feynman diagrams for the frictional force.

$$F(\delta)v_d = (a1) + (a2) + (a3) + \cdots, \tag{2.22}$$

where the black dot at time t_1 represents the factor iq_{1x} in Eq.(2.10). The shaded bubble with an impurity line appearing in each of (a)-series graphs corresponds to the summation of (b)-series graphs shown in Fig.2(b). For example, graph (a1) can be expressed as

$$(a1) = (b1) + (b2) + (b3) + \cdots + TSG. \tag{2.23}$$

Here TSG means time symmetrized graphs of $(b1), (b2), (b3), \ldots$, e.g., the TSG of (bn) includes all possible nonequivalent graphs obtained from (bn) by exchanging time points

$t_2, t_3, \ldots, t_{2n}$ with t_1 being fixed. Using the above rules for constructing Feynman diagrams, the evaluations of these graphs are quite straightforward. Summing up the (b)-series graphs (Fig.2(b)), we obtain the following result for graph (a1) in Fig.2(a):

$$(a1) = Nmv_d A(\delta + 2\eta), \tag{2.24}$$

where

$$\begin{aligned} A(\delta+2\eta) &= < \frac{\lambda^2}{\tau} > - < \frac{\lambda^2}{\tau}\frac{\lambda^2}{\tau(\delta+2\eta)} > + < \frac{\lambda^2}{\tau}(\frac{\lambda^2}{\tau(\delta+2\eta)})^2 > - \cdots \\ &= \left\langle \frac{\lambda^2/\tau}{1+\lambda^2/\tau(\delta+2\eta)} \right\rangle . \end{aligned} \tag{2.25}$$

Here τ is the relaxation time due to impurity scattering and its average $< \cdots >$ have been defined and fully discussed in the literature[13]. Using Eqs.(2.20) and (2.21) for wavy lines in other (a)-series graphs of Fig.2(a), the contributions from each of them can be similarly derived as

$$(a2) = Nmv_d[A(\delta+2\eta)]^2(-\frac{1}{\delta}), \tag{2.26}$$

$$(a3) = Nmv_d[A(\delta+2\eta)]^3(-\frac{1}{\delta})^2, \tag{2.27}$$

$$\cdots\cdots$$

It is very important to keep in mind that the higher order terms generated by the center of mass fluctuations are divergent like $(1/\delta)^n$ $(n = 1, 2, \cdots)$ and these divergencies can not be removed by the damping factor η. Substituting Eq.(2.24) and Eqs.(2.26),(2.27) back into Eq.(2.22), we obtain the final expression for the frictional force:

$$\begin{aligned} F(\delta)v_d &= Nmv_d A(\delta+2\eta)[1 - \frac{1}{\delta}A(\delta+2\eta) + (\frac{1}{\delta}A(\delta+2\eta))^2 - \cdots] \\ &= Nmv_d\frac{A(\delta+2\eta)}{1+A(\delta+2\eta)/\delta} . \end{aligned} \tag{2.28}$$

From the above result and Eq.(2.15), the expression for the coefficient $\alpha(\delta)$ is seen to be:

$$\alpha(\delta) = \frac{A(\delta+2\eta)/\delta}{1+A(\delta+2\eta)/\delta} . \tag{2.29}$$

With Eqs.(2.28) and (2.29), the electric field applied to maintain a steady transport state is related to the center of mass velocity v_d through the force balance equation (Eq.(2.9)):

$$NeE = \frac{F(\delta)v_d}{1-\alpha(\delta)} = Nmv_d A(\delta+2\eta). \tag{2.30}$$

If the the total momentum of the relative electrons is taken to vanish[1,13] and v_d is the conventional drift velocity, then the resistivity ρ_i deduced from Eq.(2.30) becomes

$$\rho_i = \frac{E}{Nev_d} = \frac{m}{Ne^2}\left\langle \frac{\lambda^2/\tau}{1+\lambda^2/\tau(\delta+2\eta)} \right\rangle . \tag{2.31}$$

In the above equation, the divergent terms generated from the center of mass fluctuations fully cancel among themselves and makes no net contribution to ρ_i. In the absence of electron-electron interactions $\eta = 0$, the adiabatic resistivity ρ_i vanishes as $\delta \to 0$. This is exactly the result obtained by Argyres[18]. However, when strong electron-electron interactions are present we have $2\eta = 1/\tau_{in} >> \lambda^2/\tau$, and in this case the thermalized resistivity of Refs.1 and 14 is immediately obtained.

III. Vanishing current and correct finite result for adiabatic resistivity

In the last section we demonstrated that the adiabatic resistivity derived from the force balance equation vanishes even though the center-of-mass fluctuation is treated quantum mechanically in the absence of electron-electron interactions. This unexpected result leads us to reconsider all the approximations made in the force balance formulation. The first is that the noncanonical commutation relations of the relative electron coordinates and momenta are approximated as canonical ones. The second is the assumption that the total momentum of relative electrons is zero so that electric current J is directly related to the center of mass velocity v_d as $J = Nev_d$. Unfortunately, these two assumptions are contradictory to each other. The latter assumption will be true only if the commutation relations of the relative electron variables are treated exactly. However, when one approximates the exact commutation relations, $[r_i, p_j] = i(\delta_{ij} - 1/N)$ as canonical ones: $[r_i, p_j] = i\delta_{ij}$, the space reflection symmetry for relative electrons is artificially broken (see the Hamiltonian in Eq.(2.1)). Consequently, the total momentum of relative electrons

$$\mathbf{C} = Tr\rho(t) \sum_p \mathbf{p} c_p^\dagger c_p \tag{3.1}$$

becomes nonvanishing in this approximation. Since this represents a macroscopic drift of the relative electrons, we incorporate it into the total electric current as

$$J = Nev_d + \frac{e}{m} C \tag{3.2}$$

in an effort to compensate for the use of incorrect canonical commutation relations for the relative electrons.

The calculation of the total relative momentum in Eq.(3.1) is parallel to that of the frictional force in Eq.(2.5). In this section we will treat center of mass variables quantum mechanically. Utilizing the perturbative solution of Eq.(2.6) for the density matrix $\rho(t)$, the leading order terms for C can be expressed as

$$C(\delta) = C_1(\delta) v_d + C_2(\delta), \tag{3.3}$$

in which

$$\begin{aligned} C_1(\delta) v_d = & \sum_{n=1}^{\infty} \int_0^{\infty} dt_1 \delta e^{-\delta t_1} \int_0^{t_1} dt_2 \cdots \int_0^{t_{2n-1}} dt_{2n} \int_0^{t_{2n}} dt_{2n+1} \lambda^{2n} N_i^n \\ & \times \sum_{\mathbf{q}_1, \cdots \mathbf{q}_n} \sum_{\mathbf{P}} ip_x |\phi(\mathbf{q}_1)|^2 e^{i\mathbf{q}_1 \cdot v_d (t_1 - t_2)} \cdots |\phi(\mathbf{q}_n)|^2 e^{i\mathbf{q}_n \cdot v_d (t_{2n-1} - t_{2n})} (i)^{2n} \\ & \times Tr(\rho_0 [A(-\mathbf{q}_n, t_{2n+1}), [A(\mathbf{q}_n, t_{2n}), \cdots, [A(-\mathbf{q}_1, t_3), [A(\mathbf{q}_1, t_2), c_p^\dagger c_p]] \cdots]]), \end{aligned} \tag{3.4}$$

$$\begin{aligned} C_2(\delta) = & NeE \sum_{n=1}^{\infty} \int_0^{\infty} dt_1 \delta e^{-\delta t_1} \int_0^{t_1} dt_2 \cdots \int_0^{t_{2n}} dt_{2n+1} \int_0^{t_{2n+1}} dt_{2n+2} \lambda^{2n} N_i^n \\ & \times \sum_{\mathbf{q}_1, \cdots \mathbf{q}_n} \sum_{\mathbf{P}} ip_x |\phi(\mathbf{q}_1)|^2 e^{i\mathbf{q}_1 \cdot v_d (t_1 - t_2)} \cdots |\phi(\mathbf{q}_n)|^2 e^{i\mathbf{q}_n \cdot v_d (t_{2n-1} - t_{2n})} (i)^{2n} \\ & \times Tr(\rho_0 [R(t_{2n+2}), [A(-\mathbf{q}_n, t_{2n+1}), [A(\mathbf{q}_n, t_{2n}), \cdots, [A(\mathbf{q}_1, t_2), c_p^\dagger c_p] \cdots]]]). \end{aligned} \tag{3.5}$$

Evidently, the structures of these two equations are parallel to those of Eqs.(2.10) and (2.11). In correspondence with the derivation of Eq.(2.15), we can use Eq.(2.13) to carry out the commutations between $R(t_{2n+2})$ and $2n\ \delta R$'s involved through A's in Eq.(3.5). We then find that $C_2(\delta)$ is related to $C_1(\delta)$ as

$$C_2(\delta) = \frac{C_1(\delta)}{Nm\delta} NeE. \tag{3.6}$$

The Feynman diagrams for $C_1(\delta)$ in Eq.(3.4) are demonstrated in Fig.3, i.e.,

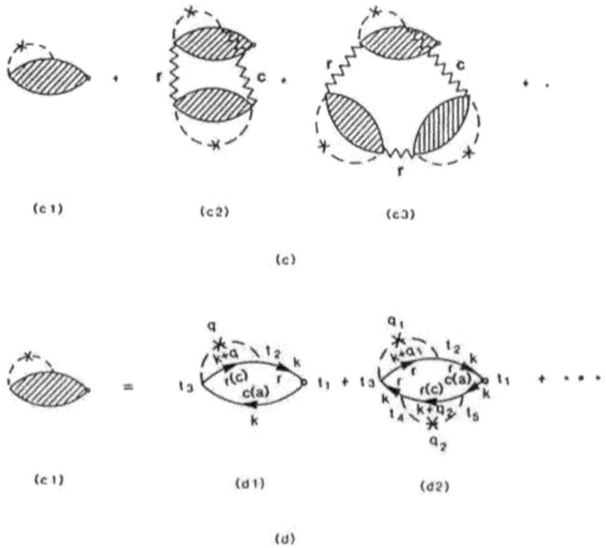

Fig.3. Feynman diagrams for total momentum of relative electrons.

$$C_1(\delta) = (\text{c1}) + (\text{c2}) + (\text{c3}) + \cdots. \tag{3.7}$$

In comparison with the Feynman graphs of Fig.2 for the frictional force $F(\delta)$ in Eq.(2.10), here in Fig.3 we have a small circle at the end of each bubble for the p_x factor in Eq.(3.4), while in Fig.2 we have a black dot for the q_{1x} in Eq.(2.10). Moreover, in Fig.3, we have one additional solid line for the one additional Green's function brought about by $c_p^\dagger c_p$ in Eq.(3.4) (comparing Fig.3(d) with Fig.1(b)). Except for the above mentioned differences, all other aspects of the graphs in Fig.3 are the same as those in Fig.2. Therefore the evaluation of them is fully analogous. Summing up the graph series in Fig.3(d), we get

$$\begin{aligned}(\text{c1}) &= (d1) + (d2) + (d3) + \cdots + TSG \\ &= -Nmv_d A(\delta + 2\eta)/(\delta + 2\eta).\end{aligned} \tag{3.8}$$

Furthermore, the evaluation of the other shaded bubble graphs with the wavy lines here follows exactly the same lines as with the results in the previous section,

$$(\text{c2}) = -Nmv - d\frac{A(\delta+2\eta)}{\delta+2\eta}\left(-\frac{A(\delta+2\eta)}{\delta}\right), \tag{3.9}$$

$$(\text{c3}) = -Nmv - d\frac{A(\delta+2\eta)}{\delta+2\eta}\left(-\frac{A(\delta+2\eta)}{\delta}\right)^2, \tag{3.10}$$

$\cdots\cdots.$

Substituting the results of Eqs.(3.8)-(3.10) into Eq.(3.7), we arrive at

$$C_1(\delta)v_d = -Nmv_d\frac{A(\delta+2\eta)}{1+A(\delta+2\eta)/\delta}. \tag{3.11}$$

Using this expression for $C_1(\delta)$ jointly with Eqs.(2.30),(3.6), we have

$$\begin{aligned} C_2(\delta) &= -NeE\frac{1}{\delta}\frac{A(\delta+2\eta)/(\delta+2\eta)}{1+A(\delta+2\eta)/\delta} \\ &= -Nmv_d\frac{A(\delta+2\eta)}{\delta}\frac{A(\delta+2\eta)}{1+A(\delta+2\eta)/\delta}. \end{aligned} \tag{3.12}$$

Employing Eqs.(3.2),(3.2) and (3.6), the electric current density is given by

$$J = Nev_d(1-\frac{A(\delta+2\eta)}{\delta+2\eta}) = Nev_d\left\langle\frac{1}{1+\lambda^2/\tau(\delta+2\eta)}\right\rangle. \tag{3.13}$$

The impurity resistivity is thus obtained from Eqs.(2.30) and (3.13) as

$$\rho_i = \frac{E}{J} = \frac{m}{Ne^2}\left\langle\frac{\lambda^2/\tau}{1+\lambda^2/\tau(\delta+2\eta)}\right\rangle\left\langle\frac{1}{1+\lambda^2/\tau(\delta+2\eta)}\right\rangle^{-1}. \tag{3.14}$$

In the adiabatic limit ($\eta \to 0$ and $\delta \to 0$), $\rho_i \simeq m\lambda^2/<\tau>Ne^2$, whereas, in the thermalized limit ($2\eta = 1/\tau_{in} >> \lambda^2/\tau$ and $\delta \to 0$), $\rho_i \simeq <\lambda^2/\tau> m/Ne^2$ We should point out that $2\eta = 1/\tau_{in}$ also renormalizes the value of $<1/\tau>$ in a measure expected to be of secondary importance in comparison with its effect in removing the leading divergencies of higher order resistivity contributions. Such a renormalization of $<1/\tau>$ should not involve any substantial change if the characteristic electron energy E (degenerate case Fermi energy or nondegenerate case thermal energy) is such that $E >> 1/\tau_{in} >> 1/\tau$. In this paper the vertex correction has not been considered, but we do not expect that such a correction would invalidate our conclusion. As a technical remark, we note that the density-density correlation function $\Pi(\mathbf{q},\omega)$ in this paper needs to be summed up over finite $\mathbf{q}$, and it is well known in many body theory that the dissipation part of the vertex correction strongly cancels the dissipation part of the self-energy only at $\mathbf{q} \to 0$. At finite $\mathbf{q}$, the cancellation is incomplete and this implies that the divergencies discussed in the present paper would still be removable in the dissipation of electrons. One essential feature of the above results is that the higher order divergent terms ($\eta \to 0$ and $\delta \to 0$) due to the noncommutability of the center of mass fluctuations again cancel among themselves and make no net contribution to the impurity resistivity. Thus the center-of-mass coordinates can be regarded as classical varibles[11-13]. This is, of course, in full accord with physical intuition.

It is also important to observe that, not withstanding the emergence of the correct adiabatic resistivity, the compensated current vanishes jointly with the resistive force (in the appropriate ratio). This anomalous feature requires further investigation.

IV. Consideration of the exact noncanonical relative electron commutation relations

The discussion in the previous sections shows that the error induced by using approximate canonical commutation relations for relative electrons must be compensated by accounting for their associated macroscopic drift in the calculation of adiabatic resistivity.

In this section, we shall derive the resistivity rigorously using the exact (noncanonical) relative electron commutation relations

$$[r_i, p_j] = i(\delta_{ij} - 1/N). \tag{4.1}$$

If $j (j = 1, 2, \ldots, N)$ is summed over in the above equation, one can easily observe that the total relative momentum properly vanishes, $\sum_j p_j \equiv 0$. The (1/N) term in the above equation is the deviation from the standard canonical commutation relations. Its impact, as will be shown below, can be represented by introducing an extra term in the Hamiltonian. In conjunction with this, the force balance equation is able to yield the correct adiabatic resistivity $\lambda^2 / < \tau >$ for a system in the absence of electron-electron interactions, with no anomalies.

Another lesson we learnt above is that the effect of the noncommutability of the CMF is not important, and it makes no net contribution to the final result for impurity resistivity. This conclusion is consistent with the conventional wisdom that the center-of-mass, with heavy mass Nm, can be treated in a semi-classical manner. In our derivation to follow, the center of mass coordinates shall be regarded as classical C-numbers accordingly.

Using the exact (noncanonical) commutation relations for the relative coordinates r_i and momentum p_j in Eq.(4.1), the following steady state kinetic equations for the center of mass and for the relative electrons may be obtained.

$$\dot{R} = P/Nm, \tag{4.2}$$

$$0 = \dot{P} = -NeE - \sum_{i\alpha} \nabla_i \phi(r_i + R - R_\alpha), \tag{4.3}$$

$$\dot{r}_i = p_i/m, \tag{4.4}$$

$$\dot{p}_i = -\nabla_i \sum_\alpha \phi(r_i + R - R_\alpha) - eE. \tag{4.5}$$

The derivation of Eq.(4.5) is partly based upon the expression given by Eq.(4.3). Considering Eqs.(4.2) to (4.5) one can easily observe that the total relative momentum is now assured to vanish when we keep the (1/N) term in the commutation relations (Eq.(4.1)). On the other hand it is rather inconvenient to study the many particle properties of relative electrons by employing the exact noncanonical commutation relations. Our scheme for incorporating the effects of the (1/N)-term in Eq.(4.1) is to replace the original Hamiltonian (Eq.(1.1)) with the following effective Hamiltonian

$$H_{eff} = H + eE \cdot \sum_i r_i = H + ieE \cdot \sum_k \frac{\partial c_k^\dagger}{\partial k} c_k. \tag{4.6}$$

This effective Hamiltonian describes two independent systems which consist of a heavy CM particle and N-electrons in an applied electric field. These two subsystems couple with each other through the electron-impurity interaction (Eq.(2.3)). Using the canonical commutation relations for both the heavy particle variables R, P and the electrons variables r_i and p_i $(i = 1, 2, \ldots, N)$:

$$[R, P] = i, \quad and \quad [r_i, p_j] = i\delta_{ij}, \tag{4.7}$$

we find that the same set of equations of motion (Eqs.(4.2) to (4.5)) are obeyed. Basing the study of impurity resistivity upon H_{eff} and Eq.(4.7), the total force acting on the many electrons $\sum_i \dot{p}_i$, determined from Eq.(4.5), always vanishes and thus the total "relative" momentum $< \sum_i p_i >= c_0$ is an **E** and t independent constant. Moreover, c_0

has to be equal to zero, since at **E**=0 the total momentum must vanish. This conclusion implies that the total current density $J = Nev_d$ comes entirely from the heavy particle motion, as we should expect in a fully proper theory. Although the net force acting on the total electrons is zero, each electron still experiences the force ($\dot{p}_i \neq 0$) due to **E** and the impurity scattering, even in the steady state. Having pointed out in previous sections that the noncommutability of CMF (the heavy particle fluctuations) does not contribute to the resistivity, we neglect its effects in the following discussion.

Employing the effective Hamiltonian in Eq.(4.6), the force balance equation may be written as

$$NeE = F(\delta)v_d + \beta(\delta)NeE \tag{4.8}$$

where the frictional force $F(\delta)v_d$ takes the same form as in Eq.(2.10) except that the heavy particle coordinate R will be treated classically. The coefficient $\beta(\delta)$ is brought about by the last term in H_{eff}. Employing the method described in section II, it is straightforward to show that

$$\begin{aligned} F(\delta)v_d = &\sum_{n=1}^{\infty}\int_0^{\infty} dt_1 \delta e^{-\delta t_1}\int_0^{t_1} dt_2 \cdots \int_0^{t_{2n-2}} dt_{2n-1}\int_0^{t_{2n-1}} dt_{2n}\lambda^{2n}N_i^n \\ &\times \sum_{\mathbf{q}_1,\cdots \mathbf{q}_n} iq_{1x}|\phi(\mathbf{q}_1)|^2 e^{i\mathbf{q}_1\cdot v_d(t_1-t_2)}\cdots|\phi(\mathbf{q}_n)|^2 e^{i\mathbf{q}_n\cdot v_d(t_{2n-1}-t_{2n})} \\ &\times (i)^{2n}Tr(\rho_0[\rho_{-\mathbf{q}_n}(t_{2n}),[\rho_{\mathbf{q}_n}(t_{2n-1}),\cdots,[\rho_{-\mathbf{q}_1}(t_2),\rho_{\mathbf{q}_1}(t_1)]\cdots]]), \end{aligned} \tag{4.9}$$

and

$$\begin{aligned} \alpha(\delta) = &\sum_{n=1}^{\infty}\int_0^{\infty} dt_1 \delta e^{-\delta t_1}\int_0^{t_1} dt_2 \cdots \int_0^{t_{2n-2}} dt_{2n-1}\int_0^{t_{2n-1}} dt_{2n}\int_0^{t_{2n}} dt_{2n+1} \\ &\times\lambda^{2n}N_i^n \sum_{\mathbf{q}_1,\cdots \mathbf{q}_n} iq_{1x}|\phi(\mathbf{q}_1)|^2 e^{i\mathbf{q}_1\cdot v_d(t_1-t_2)}\cdots|\phi(\mathbf{q}_n)|^2 e^{i\mathbf{q}_n\cdot v_d(t_{2n-1}-t_{2n})} \\ &\times (i)^{2n+1}Tr(\rho_0[-ieE\cdot\sum_k(\frac{\partial c_k^\dagger(t_{2n+1})}{\partial k}) \\ &\times c_k(t_{2n+1}),[\rho_{-\mathbf{q}_n}(t_{2n}),[\rho_{\mathbf{q}_n}(t_{2n-1}),\cdots,[\rho_{-\mathbf{q}_1}(t_2),\rho_{\mathbf{q}_1}(t_1)]\cdots]]]). \end{aligned} \tag{4.10}$$

Considering Feynman diagrams, the frictional force is simply represented by the (a)-series graphs of Fig.2(b) if the contribution from the heavy particle (center-of-mass) fluctuation is neglected, i.e.,

$$\begin{aligned} F(\delta)v_d &= (a1) + (a2) + (a3) + \cdots + TSG \\ &= Nmv_d\left\langle\frac{\lambda^2/\tau}{1+\lambda^2/\tau(\delta+2\eta)}\right\rangle. \end{aligned} \tag{4.11}$$

The diagrams for the coefficient $\beta(\delta)$ have one more structural feature as compared with those in Fig.2(b) for $F(\delta)$. It comes from outermost commutation in Eq.(4.10) and is represented by an arrowed line with a circled cross at time point t_{2n+1} (Fig.1(d)) which corresponds to the following factor:

$$(-ieE)\cdot\frac{\partial}{\partial k}(iG(k,t-t_{2n+1}))iG(k,t_{2n+1}-t'). \tag{4.12}$$

Then the leading order terms of $\beta(\delta)$ in Eq.(4.10) are represented by the graphs in Fig.4, i.e.,

$$\beta(\delta)NeE = (e1) + (e2) + (e3) + \cdots + TSG. \tag{4.13}$$

The evaluation of these graphs is completely parallel to that of the (b)-series graphs in Fig.2(b), and it yields the result

$$\beta(\delta)NeE = NeE\left\langle\frac{\lambda^2/\tau(\delta+2\eta)}{1+\lambda^2/\tau(\delta+2\eta)}\right\rangle. \tag{4.14}$$

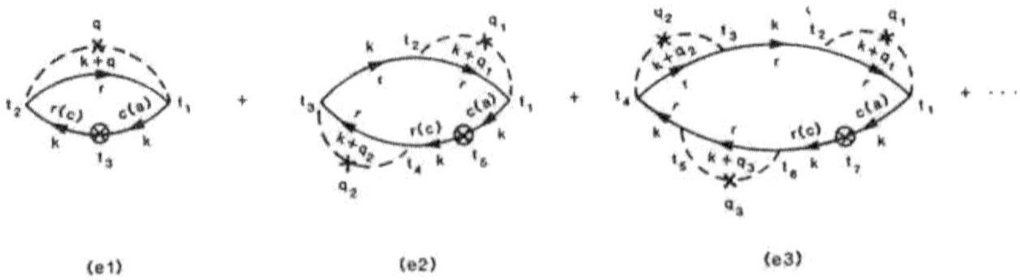

Fig.4. Feynman diagrams for the electric field coefficient in the force balance equation.

With Eqs.(4.11) and (4.14), the impurity resistivity can be shown to have identically the same form as Eq.(3.14). In the adiabatic limit $((\eta = 0))$, the resistivity becomes

$$\rho_i = \frac{F(\delta)}{(Ne)^2(1-\beta(\delta))}|_{\delta\to 0} = \frac{m}{Ne^2}\frac{\lambda^2}{<\tau>}. \tag{4.15}$$

Furthermore, we would like to emphasize that the force balance equation (Eq.(4.8)) also yields the thermalized resistivity expression when inelastic electron-electron scattering becomes strong: $\eta = 1/\tau_{in} >> \lambda^2/\tau$.

V. Boltzmann equation in relaxation time approximation

It is of interest to explore the role of the inelastic scattering interaction from an alternative point of view which leads us to expect that the divergencies of all the higher order resistivity contributions are indeed quenched by inelastic scattering. For this purpose we examine the impurity resistivity in the presence of a momentum-conserving inelastic scattering time τ_{in} using the Boltzmann equation in a relaxation time approximation. In this section, we adopt the method of Wingreen, Stanton, and Wilkins[19], in which the Boltzmann equation has been written in the form

$$e\mathbf{E}\cdot\frac{\partial}{\partial\mathbf{k}}f_0(\mathbf{k}) = \frac{f(\mathbf{k})-f_0(\mathbf{k})}{\tau_k} + \frac{f(\mathbf{k})-f_0(\mathbf{k}-<<\mathbf{k}>>)}{\tau_{in}}, \tag{5.1}$$

where $f_0(k)$ is the equilibrium distribution function in the absence of $\mathbf{E}$ and $f(k)$ is the actual distribuion function, τ_k is the elastic scattering time due to impurities or acoustic phonons. We take it that τ_{in} arises from phonon mediated electron-electron interactions as well as Coulomb interactions among electrons. The average carrier momentum $<<\mathbf{k}>>$ is defined as

$$<<\mathbf{k}>>= \frac{1}{N}\sum_k \mathbf{k}f(\mathbf{k}). \tag{5.2}$$

For linear transport we expect the drift momentum $<<k>>\sim E$ as $E\to 0$ and we have

$$f_0(\mathbf{k}-<<\mathbf{k}>>) = f_0(\mathbf{k}) - <<\mathbf{k}>>\cdot\frac{\partial}{\partial\mathbf{k}}f_0(\mathbf{k}). \tag{5.3}$$

From Eqs.(5.1) and (5.3), the distribution function $f(\mathbf{k})$ can be solved as

$$f(\mathbf{k}) = f_0(k) + [\frac{\tau_k \tau_{in}}{\tau_k + \tau_{in}} e\mathbf{E} - \frac{\tau_k}{\tau_k + \tau_{in}} << \mathbf{k} >>] \cdot \frac{\partial}{\partial \mathbf{k}} f_0(k). \tag{5.4}$$

From the above equation and Eq.(5.2), the electric current density takes the form

$$\mathbf{J} = -\frac{Ne << \mathbf{k} >>}{m} = \frac{\mathbf{E}}{\rho} \tag{5.5}$$

with the resistivity

$$\rho = \frac{m}{Ne^2} < \frac{1/\tau_k}{1 + \tau_{in}/\tau_k} > / < \frac{1}{1 + \tau_{in}/\tau_k} >, \tag{5.6}$$

where $< A(k) >$ is defined as

$$< A(k) > = \frac{1}{N} \sum_k A(k) \mathbf{k} \cdot [-\frac{\partial f_0(k)}{\partial \mathbf{k}}]. \tag{5.7}$$

It is readily verified that $< A(k) > = 1$ when $A(k) = 1$. The standard adiabatic resistivity (Eq.(1.1)) emerges by setting $1/\tau_{in} = \delta \to 0$ and $1/\tau_k = \lambda^2/\tau$ in Eq.(5.6). Alternatively, if we expand the left hand side of Eq.(5.6) in powers of $< (1/\tau_k)^n > /\delta^{n-1}$, the divergencies of van Hove's "$\lambda^2 t$" series expansion[4,5] are reproduced. However, these divergencies are removed in the presence of inelastic scattering. In the limit of $\tau_{in} << \tau_k$, the resistivity in Eq.(5.6) reduces to the expression (Eq.(1.2)) under thermalized condition[1].

VI. Summary and conclusions

The core of the balance equation method lies in the transformation to center-of-mass and relative electron coordinates. In this, the exact commutation relations of the relative electron coordinates and momenta are not canonical, differing by terms of order $0(1/N)$ that are required to assure that the total relative momentum shall vanish. In neglecting these terms and invoking a canonical approximation for the relative electron variables, a macroscopic drift of relative electrons is erroneously induced, and when this drift is compensated by incorporating it into the total current, we obtain the correct adiabatic resistivity. However, we have shown that this correction procedure involves the anomalous vanishing of the compensated current along with the resistive frictional force, such that their ratio yields the correct adiabatic resistivity. In section IV, we showed that this anomaly could be removed by a fully correct treatment of the exact noncanonical relative electron commutation relations. Thus, we have conclusively demonstrated the capacity of the balance equation method to yield the correct adiabatic resistivity. It is noteworthy that the center-of-mass variables are seen to behave in a purely classical manner in this demonstration, with no correction from their quantum mechanical commutation relations.[13]

Our consideration of the impurity resistivity in the thermalized limit devolved upon the "dressing" of the relative electron Green's functions by the strong electron-electron interaction (phonon mediated or direct) as manifested by an imaginary component of the electron self-energy given by $(2\tau_{in})^{-1}$ in all of the Feynman diagrams involved in the terms of van Hove's "$\lambda^2 t$" series, which are divergent in the adiabatic limit $\tau_{in} >> \tau$. Such dressing of the relative electron Green's function is represented in the present work through the relative electron linewidth $\eta = 1/2\tau_{in}$, and in the thermalized limit $\tau_{in} << \tau$

our analyses demonstrate that the divergencies of van Hove's "$\lambda^2 t$"-series are quenched, leading to the original thermalized balance equation result[1] for impurity resistivity $\rho_i^I = m < \lambda^2/\tau > /Ne^2$ (as opposed to the adiabatic result $\rho_i^A = m\lambda^2/Ne^2 < \tau >$). While it is true that our analysis neglects vertex corrections, there is no clear reason to expect difficulty from that quarter in the linear limit, and we have proven that the known divergent terms of van Hove's "$\lambda^2 t$"-series are rendered negligible in the thermalized limit.

In summary, we have shown that the balance equation method yields the correct commonly cited "Kubo formula" linear impurity resistivity in the adiabatic limit. While this adiabatic aspect of the Kubo formula is widely known, the Kubo formula prediction for linear impurity resistivity under thermalized conditions (strong electron-electron scattering) is not so well known. We expect that the balance equation prediction for linear thermalized impurity resistivity may also be obtained from the Kubo formula[3] by including the self-energy and vertex corrections due to strong electron-electron interactions.

Finally, we wish to emphasize that transport measurements are performed under the isothermal conditions. For the resistivity in a typical metal where the chemical potential of the carriers is usually much larger than temperature, there is no practical difference between Eqs.(1.1) and (1.2). However for appropriately doped semiconductors at relatively high temperature, it is usually expected[19] that $\tau_{in} << \tau$ and the correct resistivity formula to employ is that from the force balance equation [Eq.(1.2)],. We conclude that when the phonon mediated and the direct electron-electron interactions are explicitly considered, the higher order terms of the impurity resistivity do not diverge in the long time limit and such terms render negligible contributions to the *dc* resistivity. The considerations presented here confirm the basis for balance equation approach[1] and clarify its success in providing results which compare well quantitatively with experimental data[20,21] for wide range of electric field strengths and which also successfully reproduce results of Monter Carlo simulations for high field transport in doped silicon. The above conclusions also justify the success of the Bloch-Grüneisen formula[2] for the ideal resistivity, which is derived from the lowest order electron-phonon scattering process, in explaining the temperature dependences of the transport properties in a large class of metals.

Acknowledgements

We would like to thank Dr. Norman Horing for stimulating collaboration and helpful discussions and Dr. Zheng-Yu Weng for helpful discussions. This work has been supported by a Grant No. MDA 972-88-G-0002 from U.S. Defense Advanced Research Projects Agency and the State of Texas.

Appendix. Feynman diagram representation of Eq.(2.10) employing closed time path Green's functions

We extend the usual time path (from $-\infty$ to $+\infty$) into a closed one which runs from $-\infty$ to $+\infty$ and returns back to $-\infty$. The former branch (from $-\infty$ to $+\infty$) is regarded as a positive"+" branch, the latter one (from $+\infty$ to $-\infty$) is regarded as a negative "-" branch. Furthermore, the operators $\rho_q(t_+)$ and $\rho_q(t_-)$ are regarded as two different operators before the time ordering is completed. The generalized time ordering operator along the closed time path T_p behaves like the ordinary time ordering T on the positive branch and it behaves as the anti-time ordering $\tilde{T}$ on the negative branch; moreover, it orders all the operators on the negative branch to stand to the left of those on the positive branch. For example, if $A(t)$ and $B(t')$ are two arbitrary operators (Bosons), then

$$T_pA(t_+)B(t'_+) = TA(t)B(t'), \tag{A.1}$$
$$T_pA(t_-)B(t'_-) = \tilde{T}A(t)B(t'), \tag{A.2}$$
$$T_pA(t_+)B(t'_-) = B(t')A(t), \tag{A.3}$$
$$T_pA(t_-)B(t'_+) = A(t)B(t'). \tag{A.4}$$

Here the subscripts "+" and "-" mean the corresponding operators are on the "+" and "-" branches respectively. For convenience, we also introduce two symbols ξ_α and η_β with α and β being branch indices, and

$$\xi_+ = 1, \xi_- = 1, \tag{A.5}$$
$$\eta_+ = 1, \eta_- = -1. \tag{A.6}$$

Utilizing the definition of the generalized time ordering operator along the closed time path and taking notice of the time sequence $t_1 \geq t_2 \geq t_2 \geq \cdots \geq t_{2n}$ required by the time integrations in Eq.(2.10), the iterated commutation of 2n A operators can be expressed as

$$\begin{aligned}[A(-q_n, t_{2n}), [A(q_n, t_{2n-1}), \cdots, [A(-q_1, t_2), A(q_1, t_1)], \cdots]] \\ = -\frac{1}{2}\xi_{\alpha_1}\eta_{\alpha_2}\cdots\eta_{\alpha_{2n-1}}\eta_{\alpha_{2n}}T_p[A_{\alpha_{2n}}(-q_n, t_{2n}) \\ A_{\alpha_{2n-1}}(q_n, t_{2n-1}), \cdots A_{\alpha_2}(-q_1, t_2)A_{\alpha_1}(q_1, t_1)].\end{aligned} \tag{A.7}$$

Here $\alpha_i = +$ or $-$ $(i = 1, 2, \ldots, 2n)$ means operator $A(t_i)$ is on the α_i-th branch and all the repeated subscripts are summed over "+" and "−" branches. Eq.(A.7) may be verified by direct application of the definition of T_p. To exemplify this procedure, we take $A(t_1)$ and $B(t_2)$ with $t_1 > t_2$ as an example. Using Eqs.(A.5) and (A.6), we have

$$\begin{aligned}\frac{1}{2}\xi_{\alpha_1}\eta_{\alpha_2}T_pA_{\alpha_1}(t_1)B_{\alpha_2}(t_2) = \frac{1}{2}[T_pA_+(t_1)B_+(t_2) + T_pA_-(t_1)B_+(t_2) \\ -T_pA_+(t_1)B_-(t_2) - T_pA_-(t_1)B_-(t_2)].\end{aligned} \tag{A.8}$$

Now, making use of the definition of T_p in Eqs.(A.1)-(A.4), it becomes obvious that

$$\frac{1}{2}\xi_{\alpha_1}\eta_{\alpha_2}T_pA_{\alpha_1}(t_1)B_{\alpha_2}(t_2) = -[B(t_2), A(t_1)]. \tag{A.9}$$

Repeated application of Eqs.(A.1)-(A.6) readily confirms Eq.(A.7).

Employing the relations in Eq.(A.7), we are able to express the retarded correlation functions of Eq.(2.10) in terms of the generalized "contour" time ordered correlation functions as follows:

$$F(\delta)v_d = \sum_{n=1}^{\infty} F^{(n)}(\delta)v_d, \tag{A.10}$$

$$\begin{aligned}F^{(n)}(\delta)v_d = \int_0^{\infty} dt_1\delta e^{-\delta t_1}\int_0^{t_1} dt_2\cdots\int_0^{t_{2n-2}} dt_{2n-1}\int_0^{t_{2n-1}} dt_{2n}\lambda^{2n}N_i^n \\ \times \sum_{\mathbf{q}_1,\cdots\mathbf{q}_n} iq_{1x}|\phi(\mathbf{q}_1)|^2 e^{i\mathbf{q}_1\cdot v_d(t_1-t_2)}\cdots|\phi(\mathbf{q}_n)|^2 e^{i\mathbf{q}_n\cdot v_d(t_{2n-}-t_{2n})} \\ \times G^{(2n)}(q_1, q_2, \ldots, q_{2n}, t_1, t_2, \ldots, t_{2n}),\end{aligned} \tag{A.11}$$

$$G^{(2n)}(q_1, q_2, \ldots, q_{2n}, t_1, t_2, \ldots, t_{2n}) = -\frac{1}{2}\xi_{\alpha_1}\eta_{\alpha_2}\cdots\eta_{\alpha_{2n-1}}\eta_{\alpha_{2n}}$$
$$\times Tr[\rho_0 T_p(A_{\alpha_{2n}}(-q_n, t_{2n})A_{\alpha_{2n-1}}(q_n, t_{2n-1}), \cdots$$
$$\times A_{\alpha_2}(-q_1, t_2)A_{\alpha_1}(q_1, t_1))]. \tag{A.12}$$

For this 2n point closed time path Green's function in Eq.(A.12), we have the generalized Wick's theorem[8] to decouple it into a series of two point closed time path Green's functions. Noticing that each A (Eq.(2.12)) contains two single electron (fermion) operators (c_k and $c^\dagger_{k+q}$) and one center of mass (boson) operator ($e^{iq\cdot\delta R}$), each term of $G^{(2n)}$ contains $2n$ single electron Green's functions, $G_{\alpha_i,\alpha_j}(t_i, t_j)$, and n center of mass fluctuation Green's functions, $F_{\alpha_i,\alpha_j}(t_i, t_j)$. After the above mentioned decoupling process, we replace the two point closed time path Green's functions F_{α_i,α_j} and G_{α_i,α_j} with the familiar retarded, advanced and correlation Green's functions (Eqs.(16)-(21)) through the following relations:

$$G_{\alpha\beta} = \frac{1}{2}(\xi_\alpha\eta_\beta G_r + \eta_\alpha\xi_\beta G_a + \xi_\alpha\xi_\beta G_c), \tag{A.13}$$
$$F_{\alpha\beta} = \frac{1}{2}(\xi_\alpha\eta_\beta F_r + \eta_\alpha\xi_\beta F_a + \xi_\alpha\xi_\beta F_c). \tag{A.14}$$

With this, the representation of the frictional force $F(\delta)$ indicated in Fig.2 follows directly. In evaluating the graphs of Fig.2(a), the terms of higher order in (1/N) are dropped. Moreover, the time sequence order $t_1 \geq t_2 \geq \cdots \geq t_{2n}$ causes many terms, such as those containing $G_a(t_2 - t_3)$ to vanish automatically. The final expression for $F(\delta)$ emerges without complication.

In the following discussion, we take n=2 as an example to demonstrate the above-stated procedure and provide a detailed evaluation of the corresponding Feynman graphs for n=2,

$$G^{(4)}(q_1, q_2, t-1, \ldots, t_4) = -\frac{1}{2}\xi_{\alpha_1}\eta_{\alpha_2}\eta_{\alpha_3}\eta_{\alpha_4} Tr\rho_0 T_p \rho_{q_1\alpha_1}(t_1)e^{iq_1\cdot\delta R_{\alpha-1}(t_1)}$$
$$\times\rho_{-q_1\alpha_2}(t_2)e^{-iq_1\cdot\delta R_{\alpha_2}(t_2)}\rho_{q_2\alpha_3}(t_3)e^{iq_2\cdot\delta R_{\alpha_3}(t_3)}\rho_{-q_2\alpha_4}(t_4)e^{-iq_2\cdot\delta R_{\alpha_4}(t_4)}. \tag{A.15}$$

Decoupling the above four point correlation function into a series of two point functions:

$$iG_{\alpha\beta}(k, t-t') = Tr\rho_0 T_p c_{k\alpha}(t)c^\dagger_{k\beta}(t'), \tag{A.16}$$
$$F_{\alpha\beta}(q, q', t, t') = Tr\rho_0 T_p e^{iq\cdot\delta R_\alpha(t)}e^{iq'\cdot\delta R_\beta(t')}, \tag{A.17}$$

we may express the traced part on the right hand side of Eq.(A.14) as follows:

$$\sum_k [iG_{\alpha_1\alpha_2}(k+q_1, t_1-t_2)iG_{\alpha_2\alpha_3}(k, t_2-t_3)iG_{\alpha_3\alpha_4}(k+q_2, t_3-t_4)iG_{\alpha_4\alpha_1}(t_4-t_1)$$
$$+TSG]$$
$$\times[F_{\alpha_1\alpha_2}(q_1, -q_1, t_1-t_2)F_{\alpha_3\alpha_4}(q_2, -q_2, t_3-t_4) + TSG]$$
$$+\sum_{k_1}\sum_{k_2}[iG_{\alpha_1\alpha_2}(k_1+q_1, t_1-t_2)iG_{\alpha_2\alpha_1}(k_1, t_2-t_1)F_{\alpha_2\alpha_3}(-q_1, q_2, t_2-t_3)$$
$$\times iG_{\alpha_3\alpha_4}(k_2+q_2, t_3-t_4)iG_{\alpha_4\alpha_3}(k_2, t_4-t_3)F_{\alpha_4\alpha_1}(-q_2, q_1, t_4-t_1) + TSG]. \tag{A.18}$$

In Eq.(A.18), TSG (time symmetrized graphs) includes all possible nonequivalent terms obtained by exchanging the time points t_2, t_3, t_4 while fixing t_1. If we replace the closed time path Green's functions in Eq.(A.18) with the usual retarded, advanced and correlation functions through Eqs.(A.13) and (A.14), and then put all those terms back into Eq.(A.15), we arrive at a quite simple expression:

$$G^{(4)}(q_1, q_2, t_1 \ldots t_4) = \sum_k iG_r(k+q_1, t_1-t_2)iG_r(k, t_2-t_3)$$

$$\times\frac{1}{2}[iG_r(k+q_2, t_3-t_4)iG_c(k, t_4-t_1) + iG_c(k+q_2, t_3-t_4)iG_a(k, t_4-t_1)]$$

$$+\sum_k iG_r(k+q_1, t_1-t_2)\frac{1}{2}[iG_r(k, t_2-t_4)iG_c(k+q_2, t_4-t_3)$$

$$iG_c(k, t_2-t_4)iG_a(k+q_2, t_4-t_3)]iG_a(k, t_3-t_1)$$

$$+\sum_k iG_r(k, t_1-t_3)\frac{1}{2}[iG_r(k+q_2, t_3-t_4)iG_c(k, t_4-t_2)$$

$$iG_c(k+q_2, t_3-t_4)iG_a(k, t_4-t_2)]iG_a(k+q_1, t_2-t_1)$$

$$+\sum_k \frac{1}{2}[iG_r(k, t_1-t_4)iG_c(k+q_2, t_4-t_3) + iG_c(k, t_1-t_4)iG_a(k+q_2, t_4-t_3)]$$

$$\times iG_a(k, t_3-t_2)iG_a(k+q_1, t_2-t_1)$$

$$+\sum_k \frac{1}{2}[iG_r(k_1+q_1, t_1-t_2)iG_c(k_1, t_2-t_1) + iG_c(k_1+q_1, t_1-t_2)iG_a(k_1, t_2-t_1)]$$

$$\times F_r(-q_1, q_2, t_2-t_3)$$

$$\times\sum_k \frac{1}{2}[iG_r(k_2+q_2, t_3-t_4)iG_c(k_2, t_4-t_3)$$

$$+iG_c(k_2+q_2, t_3-t_4)iG_a(k_2, t_4-t_3)]. \quad \text{(A.19)}$$

In the process of deriving the above result, many terms such as those containing $G_a(t_3-t_4)$ vanish automatically because of the time sequence $t_1 > t_2 > t_3 > t_4$. Moreover, Eqs.(20) and (21) ($F_c = 2$) are applied here and all the terms which are of higher order in (1/N) are dropped.

With the use of Eq.(A.19) and remembering the Feynman diagram rules stated in section II, we observe that the n=2 term of the frictional force (Eq.(A.11)) corresponds to graph (b2) and its three TSG's and graph (a2) with the shaded bubble being replaced with (a1), i.e.,

$$F^{(2)}(\delta)v_d = [(b2) + TSG] + \textit{lowest order graph of } (a2). \quad \text{(A.20)}$$

Using the Green's functions defined in Eqs.(16)-(18) and (20) and in Eq.(A.19) for $G^{(4)}$ and carrying out the time and the momentum integrations, we obtain the following results for the Feynman graphs for $F^{(2)}(\delta)$:

$$[(b2) + TSG] = -\left\langle \frac{\lambda^2}{\tau}\frac{\lambda^2}{\tau(\delta+2\eta)}\right\rangle, \quad \text{(A.21)}$$

$$\textit{lowest order graph of } (a2) = -\left\langle\frac{\lambda^2}{\tau}\right\rangle\frac{1}{\delta}\left\langle\frac{\lambda^2}{\tau}\right\rangle. \quad \text{(A.22)}$$

References

1. X.L.Lei and C.S.Ting, Phys.Rev. **B32**, (1985) 1112.

2. J.M.Ziman, *Princ. Theory of Solids* Cambridge Univ. Press, London (1972).

3. R.Kubo, J.Phys.Soc.Japan **12**, (1957) 570.
4. P.N.Argyres and J.L.Sigel, Phys.Rev.Lett. **31**,1397(1973); also, Phys.Rev. **B9**, (1974) 3197.
5. M.Huberman and G.V.Chester, Adv.Phys. **24**, (1975) 489.
6. L.van Hove, Physica **21**,517(1955); **23**, (1957) 441.
7. N.J.M.Horing, X.L.Lei and H.L.Cui, Phys.Rev. **B33**, (1986) 6929.
8. R.B.Stinchcombe, in *Correlation Functions and Quasiparticle Interactions in Condensed Matter*, Ed. J.Woods Halley, NATO Advanced Studies Institute Series **35**, Plenum Press (1978); also, W.Bernard and H.B.Callen, Rev.Mod. Phys.**31**, (1959) 1017.
9. H.Falk, Phys.Rev. **165**, (1968) 602.
10. M.Suzuki, Physica **51**, (1971) 277; P.Mazur, Physica **43**, (1969) 533.
11. D.Y.Xing, P.Hu and C.S.Ting, Phys.Rev. **B35**, (1987) 6379.
12. D.N.Zubarev, *Nonequilibrium Statistical Thermodynamics*, Consultants Bureau, New York (1974); V.P.Kalashnikov, Physica (Utrecht) **48**, (1970) 93.
13. R.S.Fishman and G.D.Mahan, Phys.Rev.**B39**, (1988) 2990; R.S.Fishman, Phys. Rev. **B39**, (1988) 2994.
14. L. Y. Chen, C. S. Ting, and N. J. M. Horing, Phys. Rev. **B40**, (1989) 3756 and **B42**, (1990) 1129.
15. W. Cai, P. Hu, T. F. Zheng, B. Yudanin, and M. Lax, Phys. Rev. **B40**, (1989) 7671.
16. G. Y. Hu and R. F. O'Connel, Physica **A153**, (1989) 114; Phys. Rev. **B40**, (1989) 11701; **B39**, (1989) 12717.
17. K. C .Chou, Z. B. Su, B. L. Hao and L. Yu, Phys.Rep. **118**, (1985) 1; Z. B. Su, L. Y. Chen, X. T. Yu and K. C. Chou, Phys.Rev. **B37**, (1988) 9810.
18. P. N. Argyres, Phys. Rev. **B39**, (1989) 2982.
19. N. S. Wingreen, C. J. Stanton and J. W. Wilkins, Phys. Rev. Lett. **57**, (1986) 1084.
20. K. Hirakawa and H. Sakaki, J. Appl. Phys. **63**, (1988) 803.
21. M. Liu, D. Y. Xing, C. S. Ting and W. T. Xu, Phys. Rev. **B37**, (1988) 2997; and **B37**, (1988) 10283.

International Journal of Modern Physics B, Vol. 6, No. 7 (1992) 1099–1118

Nonequilibrium Green's Function Approach to Dynamic Properties of Resonant-Tunneling through Double-Barrier Structures

L. Y. Chen
Department of Physics, Brown University
Providence, Rhode Island 02912
and
C. S. Ting
Department of Physics, University of Houston
Houston, Texas 77204

In this paper, we review a recently developed approach to the dynamic properties of double barrier resonant tunneling systems, based on the nonequilibrium Green's function technique and the Feynman path integral theory. The transient behavior of tunneling current, immediately after the switch-on of a *dc* bias voltage, is characterized by the building up process of tunneling electrons in the quantum well. The novel negative differential conductance demonstrates itself as a function of frequency of the small *ac* signal imposed upon a *dc* bias. The imaginary part of admittance is shown to be related to the conductance via a Kronig-Kramers relation. The noise characteristics clearly demonstrate a strong time correlation derived from the Pauli exclusion principle and have evident dependence upon the structure parameters and the bias voltage. All those theoretical results are compared with experiments and numerical simulations.

I. Introduction

The novel transport characteristics of resonant-tunneling (RT) double-barrier (DB) structures (see Fig.1a) have been a focus of intensive experimental and theoretical interest for more than ten years. Since its conception by Tsu, Esaki and Chang[1–2] and the first realization of significant negative differential resistance by Sollner *et al*, [3] many aspects of this system have been studied, *e.g.*, *dc* characteristics,[1–12] time dependent behavior,[9,13–18] frequency response,[19–24] noise characteristics,[25,26] phonon and laser assisted tunneling,[27–35] effect of elastic scattering[36–38] and of inelastic scattering,[39–46] and charge accumulation effect.[47–49] The steady state properties of DB structures have been understood quite well and are mainly characterized by an I-V curve shown in Fig.1b where a regime of pronounced negative differential resistance (NDR) appears. In this paper, we review a recent theoretical investigation of the dynamic properties.

(i) Intrigued by Sollner *et al*'s experiment,[3] considerable theoretical efforts have been devoted to estimate the response speed limit.[9,13–18] The authors seem to reach the same conclusion that the response time is determined by the building up process of tunneling electrons in the quantum well and thus is inversely proportional to the tunneling broadening of the resonant level. In the nonequilibrium Green's function approach[18,24] to the transient behavior of tunneling current in response to a sudden switch on of a *dc* bias V (see section III), the same conclusion and a clear physical picture have be drawn. There the quantum and nonequilibrium properties of tunneling electrons are described in terms of the closed time path Green's functions[50–54] and the Feynman path integral theory.[55,56] The time evolution of the nonequilibrium Green's functions is derived to demonstrate time-dependent tunneling currents.

(ii) The frequency response to a small *ac* signal $u(t) \sim u_0 e^{-i\Omega t}$ superimposed upon a *dc* bias voltage V is a rather important issue. The frequency dependent conductance

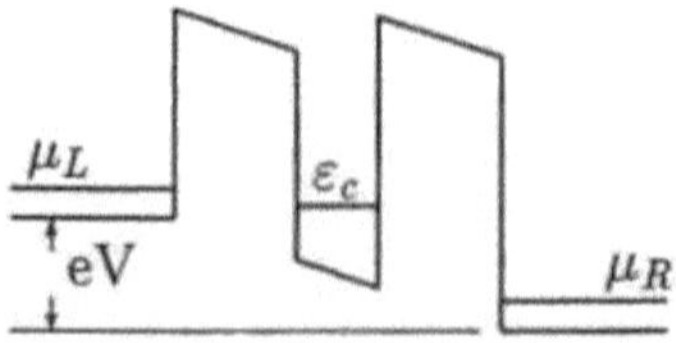

Fig.1a. A typical GaAs/AlGaAs double-barrier structure under bias V. From left to right are the emitter electrode of doped n^+GaAs, an $Al_xGa_{1-x}As$ barrier, an undoed GaAs layer as the quantum well, the second barrier of $Al_yGa_{1-y}As$, and the collector electrode of doped n^+GaAs. ε_c is the lowest resonant level, μ_L and μ_R are, respectively, the Fermi levels of two electrodes, and eV is the bias voltage.

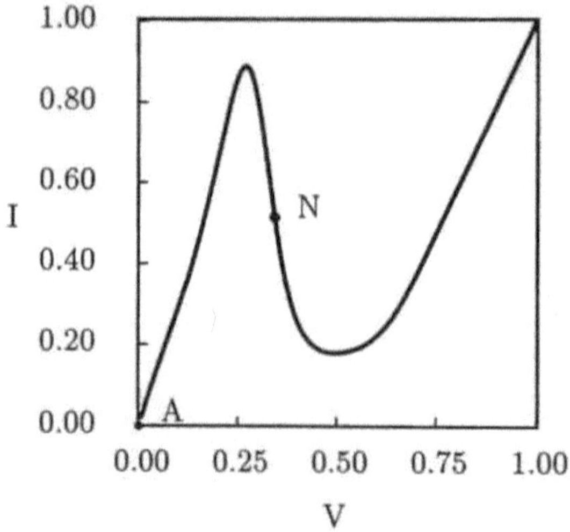

Fig.1b. A typical (qualitative) current-voltage characteristic of DBRTS. A—zero biased, N—biased so that the resonance level is equal to the conduction band bottom of the emitter electrode.

when $V = 0$ has been calculated by Jacoboni and Price.[22] The approach was based on a Kubo-like formula in terms of the othorgonalized scattering wave function basis. A characteristic frequency given by the tunneling broadening width emerges clearly out of their numerical results. In a numerical investigation based on the Wigner distribution function,[20] Frensley calculated both the dynamic conductance and the imaginary admittance in the negative differential resistance region and also the rectification at the resonance peak. The rectification function has also been calculated by Wingreen[21] based on the transfer Hamiltonian approach.[28] With help of the nonequilibrium Green's functions, Chen and Ting[23,24] performed an analytical study of the dynamic conductance and the imaginary admittance both at the zero bias point (point A in Fig.1b) and at the NDR (point N in Fig.1b). The results compare agreeably with that of Jacoboni and Price[22] when $V = 0$ and agree qualitatively with the numerical simulation[20]. This approach will be reviewed in section IV.

(iii) The study of noise characteristics in RT transport through DB structures was pioneered by Li *et al.*[25] In their experimental measurements of noise current power density, significant suppression of shot noise was observed. As strongly dependent upon the structure symmetry, the shot noise can be suppressed to the half full level. Chen and Ting applied the nonequilibrium approach to calculate analytically the noise frequency spectrum for the coherent tunneling case[26] and arrived at a rather simple suppression factor which clearly demonstrates a strong time-correlation between consecutive tunneling events derived from the Pauli exclusion.

In section II, the general description of the system will be presented together with a brief introduction to the nonequilibrium Green's functions. In sections III and IV, respectively, the transient response and the frequency response will be discussed. In section V, the noise characteristics will be investigated. Section VI contains a few concluding remarks. Before getting into detailed discussion, we emphasize that the present article only treats the coherent tunneling processes. However, the sequential tunneling in presence of strong phase-breaking scatterings can be studied in a generalized version of this approach.

II. General Description

In order to describe the essential ingredients of a double barrier resonant tunneling structure (DBRTS), we choose the following Hamiltonian for a one dimensional model system,

$$H_T = H + H'(t), \tag{2.1}$$

where

$$\begin{aligned} H &= \sum_k \varepsilon_k^L a_k^\dagger a_k + \varepsilon_c c^\dagger c + \sum_p \varepsilon_p^R b_p^\dagger b_p \\ &+ \sum_k \left(T_{Lk} c^\dagger a_k + T_{Lk}^* a_k^\dagger c\right) + \sum_p \left(T_{Rp} b_p^\dagger c + T_{Rp}^* c^\dagger b_p\right) \end{aligned} \tag{2.2}$$

is the Hamiltonian of a DBRTS under a *dc* bias voltage V and

$$H'(t) = c^\dagger c(-\alpha e u(t)) + \sum_p b_p^\dagger b_p(-eu(t)) \tag{2.3}$$

is the perturbation due to the *ac* signal $u(t)$ which is superimposed upon the bias V. a_k ($a_k^\dagger$), c ($c^\dagger$), and b_p ($b_p^\dagger$) are respectively the annihilation (creation) operators of electrons (fermions) in the left electrode, in the central quantum well, and in the right electrode. $\varepsilon_c = \varepsilon_0 - \alpha eV$ (α is structure dependent, $\alpha \simeq 0.5$ for a symmetrical structure) is the resonance level as affected by the *dc* bias. $\varepsilon_k^L = k^2/2m$ and $\varepsilon_p^R = p^2/2m - eV$ are the single particle energies of the left and right electrodes. The starting point for the energy level is chosen to be the conduction band bottom of the left electrode and αeV and eV are respectively the potential drops of the resonance level and the conduction band bottom of the right electrode caused by the bias V (Actually, the choice of the energy starting point is arbitrary. When it is chosen as the position of the resonance level, the difference thus brought is only a constant $\varepsilon_c(\sum_k a_k^\dagger a_k + c^\dagger c + \sum_p b_p^\dagger b_p)$ which does not yield any new physics). The fourth and the fifth terms of Eq.(2.2) describe the coupling between quantum well electrons and the two reservoirs. The tunneling matrices T_{Lk} and T_{Rp} depend on the barrier profile including effect of the bias V.

Since the electrode electrons respond to an applied field much faster than the quantum well electrons, they are generally treated as reservoirs. Thus the density matrix can be written as:

$$\rho = e^{-\beta \sum_k (\varepsilon_k^L - \mu_L) a_k^\dagger a_k} \rho_c e^{-\beta \sum_p (\varepsilon_p^R - \mu_R) b_p^\dagger b_p}, \tag{2.4}$$

where $1/\beta$ is temperature of the system. The left and right electrode subsystems are separately in their own equilibrium states with chemical potentials μ_L and μ_R respectively ($\mu_L - \mu_R = e(V + u(t))$). The central quantum well electrons are in a nonequilibrium state with the density matrix ρ_c to be determined by their coupling to the two reservoirs and to the applied electric field. In the practical calculation of the present problem, the path integral method[55,56] not only enables us to treat the tunneling coupling nonperturbatively, which is essential to the resonant phenomenon, it also allows us to work out analytically the statistical average with a nonequilibrium density matrix as shown in Eq.(2.4) in a tractable fashion. Moreover, the two density matrices for the two electrode subsystems (with different chemical potentials) can be incorporated into the effective action functional by replacing the free propagators of the lead subsystems with their thermo-counterparts.[54]

To describe the nonequilibrium state of electrons in the center quantum well, the retarded, advanced and distribution Green's functions are introduced as follows:

$$iG_r(t_1, t_2) = \theta(t_1 - t_2) < \{c(t_1), c^\dagger(t_2)\} >, \tag{2.5}$$

$$iG_a(t_1, t_2) = -\theta(t_2 - t_1) < \{c(t_1), c^\dagger(t_2)\} >, \tag{2.6}$$

$$iG^<(t_1, t_2) = - < c^\dagger(t_2) c(t_1) >, \tag{2.7}$$

where $< (\cdot) >$ represents the nonequilibrium statistical average $Tr(\cdot)\rho$. A more symmetrical form of the nonequilibrium closed time path Green's functions (CTPGF)[54] would be much easier to treat. They are defined along a closed time path which runs from $-\infty$ to $+\infty$ along the positive "+" branch and then returns from $+\infty$ back to $-\infty$ along the minus "–" branch:

$$iG_{\alpha\beta}(t_1, t_2) =< T_p c_\alpha(t_1) c_\beta^\dagger(t_2) >, \tag{2.8}$$

where $\alpha(\beta) = +$ or $-$ means $t_1(t_2)$ is located on the "+" or "–" branch respectively. T_p is the generalized chronological operator ordering physical operators along the closed time path:

$$iG_{++}(t_1, t_2) =< T c(t_1) c^\dagger(t_2) >, \tag{2.9}$$

$$iG_{+-}(t_1, t_2) = - < c^\dagger(t_2) c(t_1) >, \tag{2.10}$$

$$iG_{-+}(t_1, t_2) =< c(t_1) c^\dagger(t_2) >, \tag{2.11}$$

$$iG_{--}(t_1, t_2) =< \tilde{T} c(t_1) c^\dagger(t_2) >, \tag{2.12}$$

with T and $\tilde{T}$ being respectively the chronological and the anti-chronological ordering operators. The four components of $G_{\alpha\beta}$ are related to the physical G_r, G_a and $G^<$ as

$$G_r = \frac{1}{2}\xi_\alpha\eta_\beta G_{\alpha\beta}, \tag{2.13}$$

$$G_a = \frac{1}{2}\eta_\alpha\xi_\beta G_{\alpha\beta}, \tag{2.14}$$

$$G^< = G_{+-}, \tag{2.15}$$

$$0 = \frac{1}{2}\eta_\alpha\eta_\beta G_{\alpha\beta}. \tag{2.16}$$

which can be easily proven by using the definitions of them. Here $\xi_+ = \xi_- = \eta_+ = -\eta_- = 1$ and the repeated indices are summed over "+" and "−" branches.

For electrons in the left electrode, the Green's functions

$$A_{\alpha\beta}(k, t_1 - t_2) \equiv < T_p a_{k\alpha}(t_1) a^\dagger_{k\beta}(t_2) >$$

have the equilibrium form[54]

$$A_{r(a)}(k,\omega) = \frac{1}{\omega - \epsilon^L_k + (-)i\eta}, \quad A^<(k,\omega) = -f_L(\omega)\left(A_r(k,\omega) - A_a(k,\omega)\right), \tag{2.17}$$

while for right electrode electrons the Green's functions

$$B_{\alpha\beta}(p, t_1 - t_2) \equiv < T_p b_{p\alpha}(t_1) b^\dagger_{p\beta}(t_2) >$$

are given by[20]

$$B_{r(a)}(k,\omega) = \frac{1}{\omega - \epsilon^R_k + (-)i\eta}, \quad B^<(k,\omega) = -f_R(\omega)\left(B_r(k,\omega) - B_a(k,\omega)\right). \tag{2.18}$$

In Eqs.(2.17) and (2.18), $f_L(\omega)$ and $f_R(\omega)$ are respectively the Fermi-Dirac distribution functions with their chemical potentials as μ_L and μ_R. η is a positive infinitesimal.

In path integral formalism, the quantum statistical average over nonequilibrium ensemble for this DBRTS system can be expressed as[54–56]

$$< (\cdot) > = \int [da^\dagger_k][da_k] \int [db^\dagger_k][db_k] \int [dc^\dagger][dc](\cdot) e^{i\int_p (L(t) - H'(t))dt}, \tag{2.19}$$

where

$$L(t) = \sum_k a^\dagger_k(t)(i\partial_t)a_k(t) + \sum_k b^\dagger_k(t)(i\partial_t)b_k(t) + c^\dagger(t)(i\partial_t)c(t) - H(t) \tag{2.20}$$

is the Lagrangian of the DBRTS system without *ac* signal $u(t)$. $\int_p dt = \int_{-\infty}^{+\infty} dt_+ + \int_{+\infty}^{-\infty} dt_- = \eta_\alpha \int_{-\infty}^{+\infty} dt_\alpha$ is the time integration along the closed time path. Since the closed time path action $\int_p L(t)dt$ contains only bilinear terms of the a_k and $b_p$variables, these variables can be eliminated exactly in the following path integration formula by carrying out the Gaussian functional integrations of them.

$$\begin{aligned}
&\int [da^\dagger_k][da_k] \int [db^\dagger_k][db_k] \int [dc^\dagger][dc] O[c, c^\dagger; a_k, a^\dagger_k; b_p, b^\dagger_p] e^{i\int_p L(t)dt} \\
&= \int [dc^\dagger][dc] O\left[c, c^\dagger; i\frac{\delta}{\delta(T_{Lk}c^\dagger)}, -i\frac{\delta}{\delta(T^*_{Lk}c)}; i\frac{\delta}{\delta(T^*_{Rp}c^\dagger)}, -i\frac{\delta}{\delta(T_{Rp}c)}\right] e^{iI_{eff}[c,c^\dagger]} \\
&\equiv \left\langle O\left[c, c^\dagger; i\frac{\delta}{\delta(T_{Lk}c^\dagger)}, -i\frac{\delta}{\delta(T^*_{Lk}c)}; i\frac{\delta}{\delta(T^*_{Rp}c^\dagger)}, -i\frac{\delta}{\delta(T_{Rp}c)}\right]\right\rangle_{eff},
\end{aligned} \tag{2.21}$$

where $\delta/\delta(\cdot)$ is the functional derivative. $O[\cdots]$ is an arbitrary functional of c, a_k, and b_p Grassman variables. I_{eff} is the effective action of tunneling electrons:

$$I_{eff} = \int_p dt c^\dagger(t)(i\partial_t - \varepsilon_c)c(t) - \int_p dt \int_p dt' c^\dagger(t) \left(\sum_k |T_{Lk}|^2 A(k, t-t') + \sum_p |T_{Rp}|^2 B(p, t-t') \right) c(t') \quad (2.22)$$

including the quantum influence of reservoir (electrode) couplings. Therefore the effect of the density matrices of the two electrodes (see Eq.(2.4)) are incorporated into their propagators $A_{\alpha\beta}$ and $B_{\alpha\beta}$ ($A^<$ carries distribution f_L and $B^<$ carries f_R).

For later discussion, we present here four equations of functional integrations involving only one of the a_k and b_p variables:

$$\int [da_k^\dagger][da_k] \int [db_k^\dagger][db_k] \int [dc^\dagger][dc](\cdot)a_k(t)e^{i\int_p L(t)dt} = \int_p ds T_{Lk}^* A(k, t-s) \int [dc^\dagger][dc](\cdot)c(s)e^{iI_{eff}[c,c^\dagger]}, \quad (2.23)$$

$$\int [da_k^\dagger][da_k] \int [db_k^\dagger][db_k] \int [dc^\dagger][dc](\cdot)a_k^\dagger(t)e^{i\int_p L(t)dt} = \int_p ds T_{Lk} A(k, s-t) \int [dc^\dagger][dc](\cdot)c^\dagger(s)e^{iI_{eff}[c,c^\dagger]}, \quad (2.24)$$

$$\int [da_k^\dagger][da_k] \int [db_k^\dagger][db_k] \int [dc^\dagger][dc](\cdot)b_p(t)e^{i\int_p L(t)dt} = \int_p ds T_{Rp} B(p, t-s) \int [dc^\dagger][dc](\cdot)c(s)e^{iI_{eff}[c,c^\dagger]}, \quad (2.25)$$

$$\int [da_k^\dagger][da_k] \int [db_k^\dagger][db_k] \int [dc^\dagger][dc](\cdot)b_p^\dagger(t)e^{i\int_p L(t)dt} = \int_p ds T_{Rp}^* B(p, s-t) \int [dc^\dagger][dc](\cdot)c^\dagger(s)e^{iI_{eff}[c,c^\dagger]}. \quad (2.26)$$

III. Transient Response

In this section we examine the transient response process of a DBRTS system when a step bias voltage $V(t) = \theta(t)V$ is imposed ($u(t) = 0$). We will examine the transient response of the Green's function and then demonstrate how the tunneling current evolves onto its steady state limit. The time evolution of the system is mainly determined by the building up of a nonequilibrium distribution of tunneling electrons in the central quantum well and, therefore, the crucial time scale of the problem will be the inverse of the resonance level.

To derive the kinetic equations for $G_{\alpha\beta}$, we first take derivatives of Eq.(2.8) with respect to t_1 and t_2 respectively and utilizing the Hamiltonian in Eq.(2.2) and the Heisenberg equation to obtain a set of equations which involve the $c-$variable as well as a_k- and b_k-variables.

$$i\partial_{t_1} G_{\alpha\beta}(t_1, t_2) = \frac{1}{2}(\eta_\alpha + \eta_\beta)\delta(t_1 - t_2) - i\varepsilon_c < T_p c_\alpha(t_1)c_\beta^\dagger(t_2) > -i\sum_k T_{Lk} < T_p a_{k\alpha}(t_1)c_\beta^\dagger(t_2) > -i\sum_k T_{Lk}^* < T_p b_{p\alpha}(t_1)c_\beta^\dagger(t_2) >, \quad (3.1)$$

$$i\partial_{t_2} G_{\alpha\beta}(t_1, t_2) = -\frac{1}{2}(\eta_\alpha + \eta_\beta)\delta(t_1 - t_2) + i\varepsilon_c < T_p c_\alpha(t_1)c_\beta^\dagger(t_2) >$$

$$+i\sum_k T^*_{Lk} < T_p c_\alpha(t_1) a^\dagger_{k\beta}(t_2) > +i\sum_k T^*_{Lk} < T^*_p c_\alpha(t_1) b^\dagger_{p\beta}(t_2) > . \tag{3.2}$$

Expressing the nonequilibrium ensemble average in the form of Eq.(2.19), the statistical average becomes a path integration. Then carry out the Gaussian functional integrations over a_k- and b_k-variables by using Eqs.(2.21)-(2.26). As the result, we have

$$\begin{aligned} i\partial_{t_1} G_{\alpha\beta}(t_1,t_2) &= \frac{1}{2}(\eta_\alpha+\eta_\beta)\delta(t_1-t_2)+\varepsilon_c G_{\alpha\beta}(t_1,t_2) \\ &+\sum_k |T_{Lk}|^2 \int dt' A_{\alpha\nu}(k,t_1-t')\eta_\nu G_{\nu\beta}(t',t_2) \\ &+\sum_k |T_{Rk}|^2 \int dt' B_{\alpha\nu}(k,t_1-t')\eta_\nu G_{\nu\beta}(t',t_2), \end{aligned} \tag{3.3}$$

$$\begin{aligned} -i\partial_{t_2} G_{\alpha\beta}(t_1,t_2) &= \frac{1}{2}(\eta_\alpha+\eta_\beta)\delta(t_1-t_2)+\varepsilon_c G_{\alpha\beta}(t_1,t_2) \\ &+\sum_k |T_{Lk}|^2 \int dt' G_{\alpha\nu}(t_1,t')\eta_\nu A_{\nu\beta}(k,t'-t_2) \\ &+\sum_k |T_{Rk}|^2 \int dt' G_{\alpha\nu}(t_1,t')\eta_\nu B_{\nu\beta}(k,t'-t_2). \end{aligned} \tag{3.4}$$

It should be emphasized here that the Green's functions $A_{\alpha\beta}$, $B_{\alpha\beta}$ for the left and the right lead electrons are equilibrium functions and thus are time translationally invariant while the Green's functions $G_{\alpha\beta}$ are not. The relative time $t=(t_1-t_2)$-dependence of $G_{\alpha\beta}(t_1,t_2)$ describes the spectrum of the quantum well electrons, while its $T=(t_1+t_2)/2$-dependence corresponds to the transient evolution of their state. Therefore in the following discussion, we use (t,T) variables for $G(t_1,t_2)\equiv G(t,T)$. Fourier transforming the $t-$variable in $G(t,T)$ yields $G(\omega,T)$.

Making use of the relations between G_r, G_a, $G^<$ and$G_{\alpha\beta}$ as described in previous section (Eqs.(2.13)-(2.16)), we could get kinetic equations for G_r, G_a and $G^<$ Then changing the variables (t_1,t_2) to (t,T) and taking Fourier transform of the $t-$variable, we arrive at the following results:

$$(\omega-\varepsilon_c)G_{r(a)}(\omega)=1+\left(\sum_k |T_{Lk}|^2 A_{r(a)}(k,\omega)+\sum_k |T_{Rk}|^2 B_{r(a)}(k,\omega)\right)G_{r(a)}(\omega), \tag{3.5}$$

$$\begin{aligned} i\frac{\partial}{\partial T}G^<(\omega,T) &= \Big[\sum_k |T_{Lk}|^2 (A_r(k,\omega)-A_a(k,\omega)) \\ &+\sum_k |T_{Rk}|^2 (B_r(k,\omega)-B_a(k,\omega))\Big] G^<(\omega,T) \\ &-(G_r(\omega)-G_a(\omega))\left[\sum_k |T_{Lk}|^2 A^<(k,\omega)+\sum_k |T_{Rk}|^2 B^<(k,\omega)\right]. \end{aligned} \tag{3.6}$$

Eq.(3.5) yields the information of spectrum and dissipation:

$$G_{r(a)}(\omega)=\frac{1}{\omega-\varepsilon'_c+(-)i\gamma(\omega)}, \tag{3.7}$$

where ε'_c is the renormalized resonant level and

$$\gamma(\omega)=\gamma_L(\omega)+\gamma_R(\omega) \tag{3.8}$$

is its width and

$$\gamma_L(\omega) = \frac{i}{2}\sum_k |T_{Lk}|^2 (A_r(k,\omega) - A_a(k,\omega)), \tag{3.9}$$

$$\gamma_R(\omega) = \frac{i}{2}\sum_k |T_{Rk}|^2 (B_r(k,\omega) - B_a(k,\omega)). \tag{3.10}$$

The inverse of γ is the life time of the resonance state $\tau_R = 1/\gamma(\varepsilon_c')$. τ_R here is found to be dependent upon the barrier profile via the relative position of ε_c, μ_L and μ_R and via the tunneling coefficients in which the applied bias is included. The above result of the resonance level deduced from Eq.(3.5) does not have a dependence upon T since the resonance level only depends upon the barrier potential profile and the potential profile responds very fast to an applied bias voltage. This point has also been emphasized in section II when we introduce the nonequilibrium density matrix in Eq.(2.4).

From Eq.(3.6), we have the following time dependent solution which characterizes the transient tunneling process

$$G^<(\omega,T) = -\left[F(\omega) - (F(\omega) - F_0(\omega))e^{-2\gamma T}\right](G_r(\omega) - G_a(\omega)), \tag{3.11}$$

with

$$F(\omega) = \frac{f_L(\omega)\gamma_L(\omega) + f_R(\omega)\gamma_R(\omega)}{\gamma(\omega)} \tag{3.12}$$

being a Fermi-Dirac type distribution for the quantum well electrons. In Eq.(3.11) $F_0(\omega)$ is the initial distribution. When a *dc* bias is applied at $T = 0$, the two subsystems of lead electrons very soon reach their own new equilibrium states separately (with distribution f_L and f_R respectively). The evolution of quantum well electrons is determined here by the distribution Green's function $G^<$ given above. It takes a time $\tau_R = 1/\gamma$ for the transient part of Eq.(3.11) to die out after which the quantum well electrons arrive at a steady state with nonequilibrium distribution $F(\omega)$.

Just after a bias voltage V is applied at $t = 0$, the current flowing into the quantum well $I_L =< \hat{I}_L(t) >$

$$\hat{I}_L(t) \equiv -ie[H, \sum_k a_k^\dagger(t)a_k(t)] \tag{3.13}$$

is not equal to that flowing out of the well $I_R =< \hat{I}_R(t) >$

$$\hat{I}_R(t) \equiv ie[H, \sum_k b_k^\dagger(t)b_k(t)] \tag{3.14}$$

during the building up of electrons in the well. However, when a steady transport state is established after a certain long time, I_L and I_R should reach the same limit. Using Eq.(2.2) to make the commutators and Eqs.(2.19) and (2.21) to carry out the path integrals, we have the following time-dependent current (I_R as an example) expressed in terms of the nonequilibrium Green's functions discussed above:

$$\begin{aligned} I_R(t) = e\sum_k |T_{Rk}|^2 \int_0^\infty dt'[&B_r(k,t-t')G^<(t',t) - G^<(t,t')B_a(k,t'-t) \\ &+B^<(k,t-t')G_a(t'-t) - G_r(t-t')B^<(k,t'-t)]. \end{aligned} \tag{3.15}$$

After substitution of the Green's functions into Eq.(3.15), we arrive at the final result:

$$I_R(T) = I\left(1 + \frac{F_0 - F}{f_L - f_R}\frac{\gamma}{\gamma_L}e^{-2\gamma T}\right) \quad \text{for } T >> \tau_{\text{Lead}}. \tag{3.16}$$

Following the same procedure, we could derive an expression for $I_L(T)$,

$$I_L(T) = I\left(1 + \frac{F - F_0}{f_L - f_R}\frac{\gamma}{\gamma_R}e^{-2\gamma T}\right) \quad \text{for } T >> \tau_{\text{Lead}}. \tag{3.17}$$

In Eqs.(3.16) and (3.17), τ_{Lead} is the response time of the electrodes which is assumed to be much shorter than $1/\gamma$. I is steady state tunneling current

$$I = e\int \frac{d\omega}{2\pi}\frac{2[f_L(\omega) - f_R(\omega)]}{(\omega - \varepsilon_c')^2 + \gamma^2}2\gamma_L\gamma_R. \tag{3.18}$$

Obviously, it takes a time $\tau_R = 1/\gamma$ for $I_R(t)$ and $I_L(t)$ to reach their steady state limit I if $F_0 \neq F$, *i.e.*, the initial distribution of electrons in the quantum well is not the very nonequilibrium steady state distribution F. If initially the Fermi level of the two electrodes is below the resonance level before the bias is applied, then the initial distribution $F_0 = 0$ and $I_L(t)$ and $I_R(t)$ will have an overshoot and undershoot transient behavior respectively as shown in Fig.2.

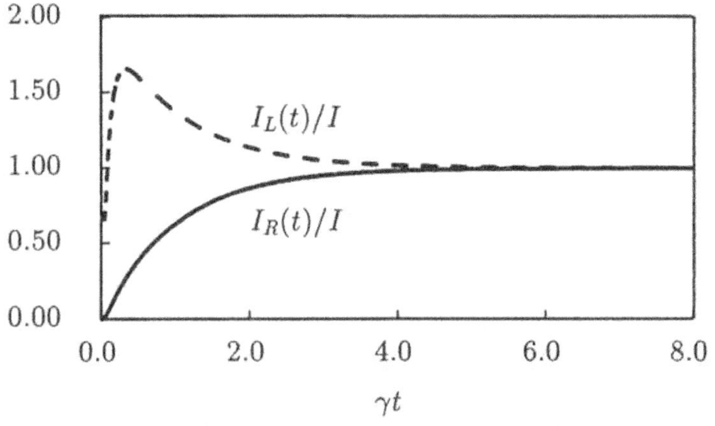

Fig.2. Transient behavior of tunneling current $I_L(t)$ and $I_R(t)$. $\gamma_L = \gamma_R$ after the bias is switched on. Initially before the bias is applied the Fermi level of the electrodes is below the resonance level so that $F_0 = 0$.

IV. Frequency Response

In the presence of a time-dependent field $u(t)$, the electric currents flowing into the quantum well and flowing out of the well $I_L(t)$ and $I_R(t)$ are not in balance. The terminal current is given by $I = (I_L + I_R)/2$, in accordance with the Ramo-Shockley theorem[57,58]

$$I(t) = -\frac{ie}{2}\left\langle \sum_k \left(T_{Lk}c^\dagger(t)a_k(t) - T_{Lk}^*a_k^\dagger(t)c(t)\right) + \sum_p \left(T_{Rp}b_p^\dagger(t)c(t) - T_{Rp}^*c^\dagger(t)b_p(t)\right)\right\rangle \tag{4.1}$$

where $< (\cdot) >$ is defined as in Eq.(2.19). In order to calculate the *ac* current $i(t)$ induced by the small signal $u(t)$, we expand the functional integral in Eq.(4.1) as defined by

Eq.(2.19) to the linear order in H' (thus linear in $u(t)$). Then it is straightforward to obtain the following linear response result for $i(t)$,

$$i(t) = \frac{e^2}{4\pi}\int dt'\Pi_r(t-t')u(t'), \tag{4.2}$$

where Π_r is a retarded correlation function and its closed time path form is

$$\Pi_{\alpha\beta}(t-t') = 2\pi\left\langle\left(\sum_k\left[T_{Lk}c^\dagger(t)a_k(t) - T^*_{Lk}a^\dagger_k(t)c(t)\right] + \sum_p\left[T_{Rp}b^\dagger_p(t)c(t)\right.\right.\right.$$
$$\left.\left.\left. - T^*_{Rp}c^\dagger(t)b_p(t)\right]\right)_\alpha\left(\alpha c^\dagger(t')c(t') + \sum_p b^\dagger_p(t')b_p(t')\right)_\beta\right\rangle. \tag{4.3}$$

In Eq.(4.3), $<(\cdot)>$ represents the path integral in Eq.(2.19) with $H'=0$. Carrying out the path integrations in Eq.(4.3) will enable us to find the desired admittance $Y(\Omega)$: $i(\Omega) = Y(\Omega)u(\Omega)$, as a function of frequency Ω, *i.e.*,

$$Y(\Omega) = \frac{e^2}{4\pi}\Pi_r(\Omega). \tag{4.4}$$

Since

$$\Pi_r(t-t') = \frac{1}{2}\xi_\alpha\eta_\beta\Pi_{\alpha\beta}(t-t') \tag{4.5}$$

is a retarded function, its Fourier transform $\Pi_r(\Omega)$ and thus $Y(\Omega)$ will be analytical in the upper half plane of complex variable Ω. Consequently, the conductance $\sigma(\Omega) = ReY(\Omega)$ and the imaginary admittance $ImY(\Omega)$ can be related through the Kronig-Kramers relations.

In order to calculate the correlation function Π in Eq.(4.3), we utilize Eq.(2.21) to eliminate the a_k and b_p variables and then carry out the path integrations of c variable. As the result, we have

$$\begin{aligned}
&\Pi_{\alpha\beta}(t-t')\\
&= 2\pi\alpha\eta_\nu\int ds\Big[A_{\alpha\nu}(t-s)G_{\nu\beta}(s-t')G_{\beta\alpha}(t'-t)\\
&-G_{\alpha\beta}(t-t')G_{\beta\nu}(t'-s)A_{\nu\alpha}(s-t)\\
&+G_{\alpha\beta}(t-t')G_{\beta\nu}(t'-s)B_{\nu\alpha}(s-t) - B_{\alpha\nu}(t-s)G_{\nu\beta}(s-t')G_{\beta\alpha}(t'-t)\\
&+G_{\alpha\nu}(t-s)\sum_p B_{\nu\beta}(p,s-t')|T_{Rp}|^2B_{\beta\alpha}(p,t'-t)\\
&-\sum_p B_{\alpha\beta}(p,t-t')|T_{Rp}|^2B_{\beta\nu}(p,t'-s)G_{\nu\alpha}(s-t)\Big]\\
&+2\pi\eta_\mu\eta_\nu\eta_\rho\int ds_1\int ds_2\int ds_3\\
&\Big[A_{\alpha\mu}(t-s_1)G_{\mu\nu}(s_1-s_2)\sum_p B_{\nu\beta}(p,s_2-t')|T_{Rp}|^2B_{\beta\rho}(p,t'-s_3)G_{\rho\alpha}(s_3-t)\\
&-G_{\alpha\mu}(t-s_1)\sum_p B_{\mu\beta}(p,s_1-t')|T_{Rp}|^2B_{\beta\nu}(p,t'-s_2)G_{\nu\rho}(s_2-s_3)A_{\rho\alpha}(s_3-t)\\
&+G_{\alpha\mu}(t-s_1)\sum_p B_{\mu\beta}(p,s_1-t')|T_{Rp}|^2B_{\beta\nu}(p,t'-s_2)G_{\nu\rho}(s_2-s_3)B_{\rho\alpha}(s_3-t)\\
&-B_{\alpha\mu}(t-s_1)G_{\mu\nu}(s_1-s_2)\sum_p B_{\nu\beta}(p,s_2-t')|T_{Rp}|^2B_{\beta\rho}(p,t'-s_3)G_{\rho\alpha}(s_3-t)\Big].
\end{aligned} \tag{4.6}$$

In Eq.(4.6),

$$A_{\alpha\beta}(t) \equiv \sum_k |T_{Lk}|^2 A_{\alpha\beta}(k,t), \quad B_{\alpha\beta}(t) \equiv \sum_p |T_{Rp}|^2 B_{\alpha\beta}(p,t). \tag{4.7}$$

Next we use the relations in Eqs.(2.13)-(2.16) to express the retarded correlation Π_r in terms of the retarded, advanced and distribution Green's functions. In such substitution procedure, the following four relations are repeatedly utilized. The first one is

$$\frac{1}{2}\xi_\alpha \eta_\beta C_{\alpha\beta}(t-t')D_{\beta\alpha}(t'-t) = C^<(t-t')D_a(t'-t) + C_r(t-t')D^<(t'-t), \tag{4.8}$$

where C and D could be combinations of Green's functions A, B and G. For the combination defined as

$$C_{\alpha\beta}(t-t') \equiv \int ds_1 \int ds_2 \cdots \int ds_n C^{(1)}_{\alpha\mu}(t-s_1)\eta_\mu C^{(2)}_{\mu\nu}(s_1-s_2)\eta_\nu \cdots \eta_\rho C^{(n)}_{\rho\beta}(s_n-t'),$$

we have the three more identities:[53]

$$\begin{aligned} C^<(t-t') = \int ds_1 \cdots \int ds_n \Big[& C^{(1)<}(t-s_1)C^{(2)}_a(s_1-s_2)\cdots C^{(n)}_a(s_n-t') \\ & + C^{(1)}_r(t-s_1)C^{(2)<}(s_1-s_2)\cdots C^{(n)}_a(s_n-t') \\ & + \cdots\cdots \\ & + C^{(1)}_r(t-s_1)C^{(2)}_r(s_1-s_2)\cdots C^{(n)<}(s_n-t')\Big], \end{aligned} \tag{4.9}$$

$$C_r(t-t') = \int ds_1 \cdots \int ds_n \left[C^{(1)}_r(t-s_1)C^{(2)}_r(s_1-s_2)\cdots C^{(n)}_r(s_n-t')\right], \tag{4.10}$$

$$C_a(t'-t) = \int ds_1 \cdots \int ds_n \left[C^{(1)}_a(t'-s_1)C^{(2)}_a(s_1-s_2)\cdots C^{(n)}_a(s_n-t)\right]. \tag{4.11}$$

To prove these four identities (Eqs.(4.8)-(4.11)) is quite straightforward by using the basic relations in Eqs.(2.13)-(2.16). With the help of the above mentioned relations, it is easy to show that each term of $\Pi_r(t-t')$ (see Eq.(4.5) and Eq.(4.6)) has the common factor $\theta(t-t')$ coming either from $G_r(t-t')$ or from $G_a(t'-t)$, which explicitly confirms that $\Pi_r(t-t')$ is retarded and thus $\Pi_r(\Omega)$ is analytical in the upper half plane of complex Ω.

The Fourier transform $\Pi_r(\Omega)$ is found to be

$$\begin{aligned} \Pi_r(\Omega) = \alpha \int_{-\infty}^{\infty} d\omega \, \{ & 2i[\gamma_L(+)f_L(+) - \gamma_R(+)f_R(+)]G_a(+)G_a \\ & -2i[\gamma_L f_L - \gamma_R f_R]G_r(+)G_r \\ & +[A_r(+) - A_a - B_r(+) + B_a][G^<(+)G_a + G_r(+)G^<]\} \\ + \int_{-\infty}^{\infty} d\omega \, \{ & G^<(+)\Gamma_a(\omega,\Omega) - G^<\Gamma_r(\omega,\Omega) \\ & -2i[\gamma_R(+)f_R(+) - \gamma_R f_R]\frac{1}{\Omega}[G_r(+) - G_a] \\ & -2i[\gamma_R(+)f_R(+) - \gamma_R f_R]G_r(+)G_a\frac{1}{\Omega}[A_r(+) - A_a - B_r(+) + B_a] \\ & +2i[\gamma_L(+)f_L(+) - \gamma_R(+)f_R(+)]G_a(+)G_a\Gamma_a(\omega,\Omega) \\ & -2i[\gamma_L f_L - \gamma_R f_R]G_r(+)G_r\Gamma_r(\omega,\Omega) \\ & +\Gamma_a(\omega,\Omega)G^<(+)G_a[A_r(+) - A_a - B_r(+) + B_a] \\ & +\Gamma_r(\omega,\Omega)G_r(+)G^<[A_r(+) - A_a - B_r(+) + B_a]\}, \end{aligned} \tag{4.12}$$

where $\gamma_{L(R)}(+)$ and $\gamma_{L(R)}$ *etc.* stand for respectively $\gamma_{L(R)}(\omega+\Omega)$ and $\gamma_{L(R)}(\omega)$ *etc.*. Γ_r and Γ_a are introduced as

$$\Gamma_r(\omega,\Omega) \equiv \sum_p B_r(p,\omega+\Omega)|T_{Rp}|^2 B_r(p,\omega), \tag{4.13}$$

$$\Gamma_a(\omega,\Omega) \equiv \sum_p B_a(p,\omega+\Omega)|T_{Rp}|^2 B_a(p,\omega). \tag{4.14}$$

In deriving Eq.(4.12), we have also used

$$A^<(\omega) = 2if_L(\omega)\gamma_L(\omega), \quad B^<(\omega) = 2if_R(\omega)\gamma_R(\omega), \tag{4.15}$$

$$\sum_p B^<(p,\omega+\Omega)|T_{Rp}|^2 B_a(p,\omega) = -\frac{2i}{\Omega} f_R(\omega+\Omega)\gamma_R(\omega+\Omega), \tag{4.16}$$

$$\sum_p B_r(p,\omega+\Omega)|T_{Rp}|^2 B^<(p,\omega) = \frac{2i}{\Omega} f_R(\omega+\Omega)\gamma_R(\omega+\Omega). \tag{4.17}$$

Eq.(4.12) is our central result for the frequency response characteristics. Using the nonequilibrium steady state Green's functions derived in Eqs.(3.7) and (3.11) and Eqs.(3.17 and (3.18) to carry out the frequency integrations, we can obtain the final expressions of conductance $\sigma(\Omega)$ and imaginary admittance $ImY(\Omega)$ for various cases.

When the temperature $1/\beta = 0$, the *dc* bias $V = 0$ (point A in Fig.1) and for a system whose resonance level is equal to the Fermi level of the two electrodes, *i.e.*, $\varepsilon_c = \mu_L = \mu_R$, the distribution of tunneling electrons $F(\omega) = f_L(\omega) = f_R(\omega) = \theta(\varepsilon_c - \omega)$. Approximately, $\gamma_L(\omega)$ and $\gamma_R(\omega)$ are assumed to be constants γ_L and γ_R respectively. Further carrying out the ω-integrations in Eq.(4.12), the conductance and the imaginary part of the admittance can be shown to have the following compact forms

$$\frac{\sigma(\Omega)}{|\sigma(0)|} = \frac{(16\gamma_L\gamma_R+\Omega^2)\gamma^3}{4\gamma_L\gamma_R(\Omega^2+4\gamma^2)\Omega} tan^{-1}(\frac{\Omega}{\gamma}) + \frac{(\gamma_L-\gamma_R)^2\gamma^2}{4\gamma_L\gamma_R(\Omega^2+4\gamma^2)} ln(1+\frac{\Omega^2}{\gamma^2}), \tag{4.18}$$

$$\frac{ImY(\Omega)}{|\sigma(0)|} = \frac{(16\gamma_L\gamma_R+\Omega^2)\gamma^3}{8\gamma_L\gamma_R(\Omega^2+4\gamma^2)\Omega} ln(1+\frac{\Omega^2}{\gamma^2}) - \frac{(\gamma_L-\gamma_R)^2\gamma^2}{2\gamma_L\gamma_R(\Omega^2+4\gamma^2)} tan^{-1}(\frac{\Omega}{\gamma}). \tag{4.19}$$

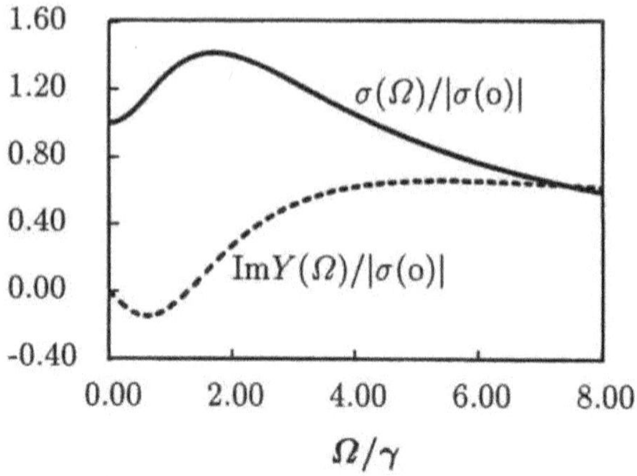

Fig.3. Frequency dependent conductance and imaginary part of admittance of DBRTS with zero bias. The system is symmetrical so that $\gamma_L = \gamma_R = 0.5\gamma$.

The numerical results for Eqs.(4.18) and (4.19) are shown in Fig.3. In response to an applied *ac* voltage, the electrons tunnel into the center well from one electrode and tunnel out of it into the other electrode. This response decreases monotonically as shown in Fig.3 with the increasing frequency when $\Omega > \Omega_0$. The characteristic frequency for this behavior is given by the resonance level broadening $\Omega_0 = 2\gamma$. When $\Omega >> \Omega_0$, tunneling electrons will not be able to follow the applied field and the tunneling current becomes vanishing. For an asymmetrical system such as that shown in Fig.3 ($\gamma_L = 9\gamma_R$), the low frequency conductance shows a capacitance behavior since the terminal current is mainly controlled by one (left) barrier, which is in good agreement with a Kubo formula study by Jacoboni and Price.[22]

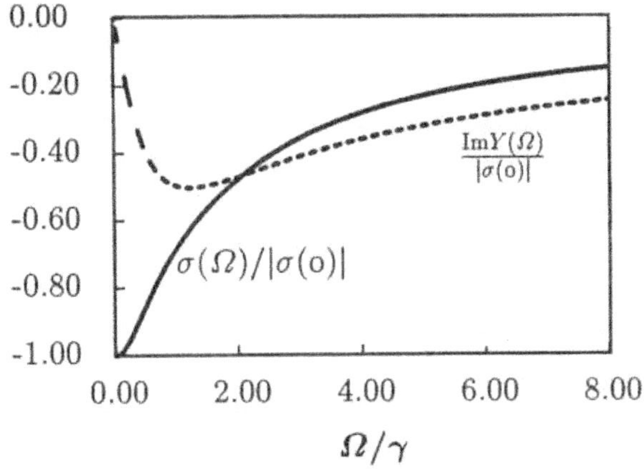

Fig.4. Frequency dependent conductance and imaginary part of admittance of DBRTS biased at NDR ($V = V_n$ so that $\varepsilon_c = \varepsilon^L_{k=0} = 0$). The system is symmetrical and it becomes very asymmetrical when biased: $\gamma_L = 0.001\gamma$ and $\gamma_R = 0.999\gamma$.

Also at zero temperature, when the DBRTS is biased at V_N (point N in Fig.1) so that the resonance level is equal to the conduction band bottom, *i.e.*, $\varepsilon_c = \varepsilon^L_{k=0} = 0$, $\gamma_L(\omega)$ and $\gamma_R(\omega)$ can be approximated as $\gamma_L\theta(\omega)$ and γ_R respectively. Since $\mu_L >> 0$ and $\mu_R << 0$, the distribution functions, $f_L \simeq 1$, $f_R \simeq 0$, and $F(\omega) \simeq \gamma_L(\omega)/\gamma(\omega)$. With these approximations, the negative conductance $\sigma(\Omega)$ and imaginary part of the admittance $ImY(\Omega)$ are derived from Eq.(4.12) to have the following expressions:

$$\frac{\sigma(\Omega)}{|\sigma(0)|} = -\frac{\gamma_R + \xi(\gamma+\gamma_R)}{4\gamma_R}\frac{\gamma}{\Omega}tan^{-1}(\frac{\Omega}{\gamma}) + \frac{\xi\gamma}{8\gamma_R}ln\left(1+\frac{\Omega^2}{\gamma^2}\right)$$
$$-\frac{\gamma + \xi(\gamma+\gamma_R)}{4\gamma_R}\frac{\gamma}{\Omega}tan^{-1}(\frac{\Omega}{\gamma_R}) + \frac{\xi\gamma}{8\gamma_R}ln\left(1+\frac{\Omega^2}{\gamma_R^2}\right), \quad (4.20)$$

$$\frac{ImY(\Omega)}{|\sigma(0)|} = -\frac{\gamma_R + \xi(\gamma+\gamma_R)}{4\gamma_R}\frac{\gamma}{2\Omega}ln\left(1+\frac{\Omega^2}{\gamma^2}\right) + \frac{\xi\gamma}{4\gamma_R}tan^{-1}(\frac{\Omega}{\gamma})$$
$$-\frac{\gamma + \xi(\gamma+\gamma_R)}{4\gamma_R}\frac{\gamma}{2\Omega}ln\left(1+\frac{\Omega^2}{\gamma_R^2}\right) + \frac{\xi\gamma}{4\gamma_R}tan^{-1}(\frac{\Omega}{\gamma_R}), \quad (4.21)$$

with

$$\xi = \frac{2\gamma_R^2 - \gamma_L(\gamma + \gamma_R)}{\Omega^2 + (\gamma + \gamma_R)^2}. \tag{4.22}$$

For an asymmetrical DBRTS, if the right barrier is higher and/or thicker than the left one, $\gamma_L >> \gamma_R$ at V=0. However, it may become symmetrical $\gamma_L = \gamma_R$ when a proper *dc* bias is applied. In Fig.4 we plot the conductance and imaginary part of the admittance as functions of frequency Ω for such an asymmetrical structure. The differential conductance (negative) $\sigma(\Omega)$ rolls off with a characteristic frequency given by the width of resonance level $\Omega_0 = \gamma + \gamma_R$ and keeps to be significantly negative until $\Omega \sim$ several Ω_0. This behavior agrees with a previous numerical investigation[20] and is consistent with the main feature of experiments.[2,19] The imaginary part of the admittance is shown as the dashed curve in Fig.4.

V. Noise Characteristics

To demonstrate the noise characteristics of a DBRTS in steady state we put $H' = 0$. Since the number of electrons in the central quantum well fluctuates, the unaveraged currents $\hat{I}_L(t)$ (Eq.(3.13)) and $\hat{I}_R(t)$ (Eq.(3.14)) are generally different from each other even in a steady state. The terminal current is given by $\hat{I} = (\hat{I}_L + \hat{I}_R)/2$. The noise power density of the tunneling current fluctuation then can be written as the Fourier transform $S_i(\Omega)$ of the following correlation function

$$S_i(t-t') = \langle\{(I(t) - <I>), (I(t') - <I>)\}\rangle \tag{5.1}$$

where $<(\cdot)>$ is defined as in Eq.(2.19) with $H' = 0$. Through Eq.(2.19), the correlation function of the current fluctuation S_i is expressed in terms of path integrals. Utilizing Eq.(2.21) to eliminate the a_k- and b_p-variables and carrying out the path integrations of the $c-$variable, the fluctuating current correlation can be formulated as:

$$S_i(t-t') = \frac{1}{2}\xi_\alpha \xi_\beta S_{\alpha\beta}(t-t') \tag{5.2}$$

with

$$\begin{aligned}
\frac{S_{\alpha\beta}(t-t')}{e^2/4} = &\,[A_{\alpha\beta}(t-t') + B_{\alpha\beta}(t-t')]G_{\beta\alpha}(t'-t) \\
&- G_{\alpha\beta}(t-t')[A_{\beta\alpha}(t'-t) + B_{\beta\alpha}(t'-t)] \\
&+ \eta_\mu \eta_\nu \int ds_1 \int ds_2 \\
&\{[A_{\alpha\mu}(t-s_1) - B_{\alpha\mu}(t-s_1)]\, G_{\mu\beta}(s_1-t') \\
&\times [A_{\beta\nu}(t'-s_2) - B_{\beta\nu}(t'-s_2)]\, G_{\nu\alpha}(s_2-t) \\
&- G_{\alpha\beta}(t-t')\,[A_{\beta\mu}(t'-s_1) - B_{\beta\mu}(t'-s_1)] \\
&\times G_{\mu\nu}(s_1-s_2)\,[A_{\nu\alpha}(s_2-t) - B_{\nu\alpha}(s_2-t)] \\
&- [A_{\alpha\mu}(t-s_1) - B_{\alpha\mu}(t-s_1)]\, G_{\mu\nu}(s_1-s_2) \\
&\times [A_{\nu\beta}(s_2-t) - B_{\nu\beta}(s_2-t)]\, G_{\beta\alpha}(t'-t) \\
&+ G_{\alpha\mu}(t-s_1)\,[A_{\mu\beta}(s_1-t') - B_{\mu\beta}(s_1-t')] \\
&\times G_{\beta\nu}(t'-s_2)\,[A_{\nu\alpha}(s_2-t) - B_{\nu\alpha}(s_2-t)]\}
\end{aligned} \tag{5.3}$$

in which $A_{\alpha\beta}(t)$ and $B_{\alpha\beta}(t)$ are introduced as in Eq.(4.7). The subscripts α, β in Eq.(5.2) and μ, ν in Eq.(5.3) are summed over "+" and "−" branches.

In order to express the right hand side of Eq.(5.3) in terms of the known retarded, advanced and distribution Green's functions, we need to use the identities in Eqs.(4.9)-(4.11) and another identity, similar to Eq.(4.8),

$$\begin{aligned}\frac{1}{2}\xi_\alpha\xi_\beta C_{\alpha\beta}(t-t')D_{\beta\alpha}(t'-t) = 2C^<(t-t')D^<(t'-t)\\ +C^<(t-t')\left(D_r(t'-t)-D_a(t'-t)\right)\\ +\left(C_r(t-t')-C_a(t-t')\right)D^<(t'-t)\end{aligned} \tag{5.4}$$

which can be easily proven with the basic relations in Eq.(2.13)-(2.16). After those substitutions and the Fourier transform of the time variables, the power spectrum of noise current becomes

$$\begin{aligned}\frac{S_i(\Omega)}{e^2/2} = \int\frac{d\omega}{2\pi}\{&-2\left[F(+)+F-2FF(+)\right]\gamma(\omega)Im\left[G_r(+)+G_r\right]\\ &-16\left[f_L(+)-F(+)\right]\left[f_L-F\right]\gamma_L^2(\omega)Im\left[G_r(+)\right]Im\left[G_r\right]\\ &-4\left[f_L(+)+f_L-2f_Lf_L(+)\right]\gamma_L^2(\omega)Re\left[G_r(+)G_a\right]\\ &-16\left[f_R(+)-F(+)\right]\left[f_R-F\right]\gamma_R^2(\omega)Im\left[G_r(+)\right]Im\left[G_r\right]\\ &-4\left[f_R(+)+f_R-2f_Rf_R(+)\right]\gamma_R^2(\omega)Re\left[G_r(+)G_a\right]\\ &+4\gamma_L(\omega)\gamma_R(\omega)Re\left[G_r(+)G_a\right]\\ &-2\left[(1-2f_L(+))(1-2f_R)+(1-2f_L)(1-2f_R(+))\right]\\ &\qquad\times\gamma_L(\omega)\gamma_R(\omega)Re\left[G_r(+)G_r\right]\\ &+4\left[(1-2F)(2f_L(+)+2f_R(+)-1-2F(+))+(1-2F(+))\right.\\ &\quad\left.\times\,(2f_L+2f_R-1-2F)\right]\gamma_L(\omega)\gamma_R(\omega)Im\left[G_r(+)\right]Im\left[G_r\right]\},\end{aligned} \tag{5.5}$$

where $G_{r(a)}$, $f_{L(R)}$, and F represent respectively $G_{r(a)}(\omega)$, $f_{L(R)}(\omega)$ and $F(\omega)$. $G_{r(a)}(+)$, $f_{L(R)}(+)$ and $F(+)$ stand for $G_{r(a)}(\omega+\Omega)$, $f_{L(R)}(\omega+\Omega)$ and $F(\omega+\Omega)$ respectively. In Eq.(5.5), $G_{r(a)}(\omega)$ is the retarded (advanced) Green's function given in Eq.(3.7) and $F(\omega)$ is the nonequilibrium distribution introduced in Eq.(3.12). When deriving Eq.(5.5), the relations in Eq.(4.15) are frequently used to replace A and B functions with proper combinations of γ_L and γ_R. Eq.(5.5) is the general expression for the power spectrum of noise current in DBRTS. To obtain the final numerical result, one need only to evaluate a number of frequency integrations numerically or analytically whenever possible such as the two cases we shall discuss below.

In the linear response region, the applied voltage $eV << \gamma$, the distribution function $f_L(\omega) = f_R(\omega) = F(\omega) = \theta(\varepsilon_c - \omega)$, and the $\gamma_L(\omega)$ and $\gamma_R(\omega)$ can be approximated as two constants γ_L and γ_R. Then the noise current power density $S_i(\Omega)$ is related to the linear conductance $\sigma(\Omega)$ (see section IV) by

$$S_i(\Omega) = \frac{e^{\beta\Omega}+1}{e^{\beta\Omega}-1}2\Omega\sigma(\Omega), \tag{5.6}$$

which produces the well known Einstein relation if the frequency is much smaller than that given by temperature, *i.e.*, $\Omega << 1/\beta$.

When the temperature $1/\beta$ is zero and the system is biased in the resonant tunneling region ($eV >> \gamma$), the system of tunneling electrons becomes very nonequilibrium. The distribution functions $f_L \simeq 1$, $f_R \simeq 0$ and $F \simeq \gamma_L/\gamma$. The $\gamma_L(\omega)$ and $\gamma_R(\omega)$ are approximately constant. After working out the integrations in Eq.(5.5), the noise current power density becomes

$$S_i(\Omega) = e < I > \left(1 + \left(1 - \frac{4\gamma_L\gamma_R}{\gamma^2}\right)\frac{4\gamma^2}{\Omega^2 + 4\gamma^2}\right), \tag{5.7}$$

where $< I >$ is the steady state tunneling current induced by the bias. Eq.(5.7) clearly demonstrates the shot noise suppression as a quantum many body effect. In a coherent tunneling process, the tunneling event of one electron through one channel is correlated to those of other electrons on the same channel because every electron stays in the central quantum well for a finite time (life time of resonant state) when getting across the double barrier region. Therefore when one electron proceeds to tunnel through the other electrons (fermions) of the same quantum number can not follow up within a correlation time given by the life time of resonance state. This correlation suppresses the shot noise and yields a non-white spectrum (Lorentzian) with the right characteristics frequency $\Omega_0 = 2\gamma$. Moreover, Eq.(5.7) tells that the high frequency components of noise current are suppressed more than the low frequency ones, which is also attributed to the Pauli exclusion derived time correlation. Obviously, the Pauli exclusion tends to force the tunneling events distributing evenly in a time path. Since higher frequency of a noise process implies smaller time distance between consecutive tunneling events and more inhomogeneous distribution in its time path, the high frequency components are more suppressed and asymptopically approach the half full level.

Usually, the frequency of interest $\Omega << \Omega_0$ (Ω_0 is around $10^{11} sec^{-1}$ for typical structures) and the low frequency limit of Eq.(5.7) can be experimentally measured. The dependence of the low frequency characteristics upon the structure parameters and the applied bias voltage is mainly through the symmetry factor $4\gamma_L\gamma_R/\gamma^2$:

$$\frac{S_i(0)}{e < I >} = 2 - \frac{4\gamma_L\gamma_R}{\gamma^2}. \tag{5.8}$$

When a system of proper parameters is biased at such a point that $\gamma_1 = \gamma_2$, the shot noise is suppressed most. The noise power density is only at the half-full level. In this situation, the average time for a tunneling electron to spend in the QW $\tau_0 = 2/(\gamma_1+\gamma_2)$ reaches its maximum over the average overall transport time $\tau = (\gamma_1+\gamma_2)/\gamma_1\gamma_2$. So the Pauli exclusion correlation has the strongest effect. When the system is very asymmetrical, either $\gamma_1 >> \gamma_2$ or $\gamma_1 << \gamma_2$, the ratio τ_0/τ is very small. Then the transport is similar to that through a single-barrier junction and nearly full shot noise is produced. γ_L and γ_R depend respectively on the left and the right barrier width and height including the effect of the bias voltage. Thus a symmetrical structure ($\gamma_L = \gamma_R$ at zero bias) could become very asymmetrical, *e.g.*, $\gamma_L << \gamma_R$ under a certain reverse bias (by convention, reverse bias means the left barrier is used as emitter and right barrier as collector). Accordingly, the noise characteristics (Eq.(5.8)) will change with a varying bias voltage. Actually, our theoretical result for the dependence of low frequency noise characteristics upon the structure parameters and the bias voltage, given by Eq.(5.8), can be used directly to interpret the experimental measurements[25] Take sample 2 of Ref.25 as an example. The left barrier is an 85Å layer of $Al_{0.5}Ga_{0.5}As$, the right barrier 85Å of $Al_{0.4}Ga_{0.6}As$. The left barrier is higher than and has the same width as the right one and thus $\gamma_L << \gamma_R$. Under a forward bias, the left barrier will be relatively lowered and thus γ_L will be growing closer to γ_R. Therefore Eq.(5.8) tells that $\kappa = S_i(0)/e\langle I\rangle$ is firstly close to 2 for small bias and becomes suppressed with the increasing forward bias (see the dashed curve in Fig.5). Under an increasing reverse bias, the right barrier is pulled further lower than the left barrier and thus γ_R becomes even larger than γ_L. Therefore κ gets closer to 2 which corresponds to full shot noise (see the solid curve in Fig.5).

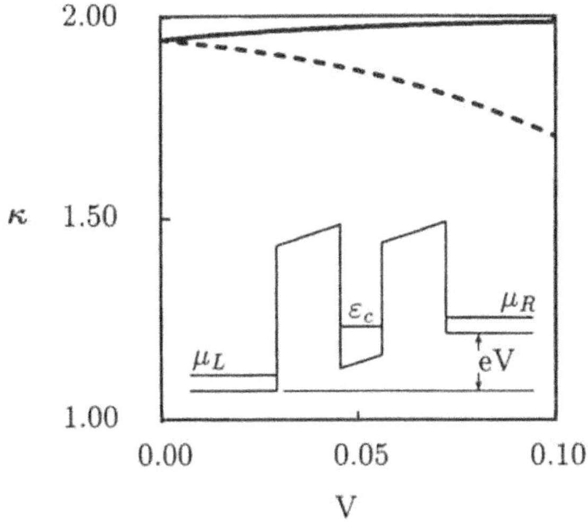

Fig.5. The shot noise suppression factor $\kappa = S_i/eI$ as a function of bias voltage V. The left barrier is 85Å thick and 0.65eV high. The right one is also 85Å thick but 0.5eV high. The quantum well is 56Å wide. The solid curve corresponds to reverse bias and the dashed curve corresponds to forward bias. Convention: the left barrier is the emitter under reverse bias and is the collector under forward bias. Inset: the DB structure under forward bias V.

If not using the Ramo-Shockley theorem as in Eq.(5.1), we can also calculate the self-correlation and inter-correlation functions of the emitter current I_L and the collector current I_R.

$$S_{LL}(t-t') = \langle\{(I_L(t) - \langle I\rangle), (I_L(t) - \langle I\rangle)\}\rangle, \tag{5.9}$$

$$S_{RR}(t-t') = \langle\{(I_R(t) - \langle I\rangle), (I_R(t) - \langle I\rangle)\}\rangle \tag{5.10}$$

carry the information of the emitter and the collector noise currents, respectively, while

$$S_{LR}(t-t') = \langle\{(I_L(t) - \langle I\rangle), (I_R(t) - \langle I\rangle)\}\rangle, \tag{5.11}$$

$$S_{RL}(t-t') = \langle\{(I_R(t) - \langle I\rangle), (I_L(t) - \langle I\rangle)\}\rangle \tag{5.12}$$

describe the correlation between the two noise currents. In a completely parallel way to the calculation in this section, we can find their frequency spectra:

$$S_{LL} = S_{RR} = e\langle I\rangle\left(2 - \frac{4\gamma_L\gamma_R}{\gamma^2}\frac{4\gamma^2}{\Omega^2+4\gamma^2}\right), \tag{5.13}$$

$$S_{LR} = S_{RL} = e\langle I\rangle\left(2 - \frac{4\gamma_L\gamma_R}{\gamma^2}\right)\frac{4\gamma^2}{\Omega^2+4\gamma^2} \tag{5.14}$$

which confirm the above discussion of the time correlation originated from the Pauli exclusion principle and $S_i = (S_{LL} + S_{LR} + S_{RL} + S_{RR})/4$.

VI. Concluding Remarks

To conclude, we have presented a nonequilibrium Green's function treatment for the time dependent properties of double barrier resonant tunneling systems. With the help of Feynman path integral theory, the tunneling couplings are treated exactly (nonperturbatively), which is essential to the resonant phenomena. The other essential of the problem, the nonequilibrium statistics is also well accounted in this approach. The distribution of tunneling electrons in the central quantum well, driven far from equilibrium by the applied field, is determined by the couplings to the two reservoirs (electrodes) with different chemical potentials. Especially, in response to an oscillating voltage, the tunneling electrons oscillate into and out of the well and so does their distribution. The duration time for the tunneling electrons to follow the varying applied voltage yields a characteristic frequency given by the resonance level broadening at which the differential conductance begins to roll off. The noise characteristics uniquely demonstrate the time correlation between consecutive tunneling electrons derived from the Pauli exclusion principle. The shot noise suppression is directly related to this quantum many body effect. Finally the present time-dependent quantum statistical approach can be generalized to treat three dimensional systems[59–60] including both the elastic and inelastic scattering effects and to study quantum dot systems incorporating charging and Coulomb repulsion effect.[61–63]

Acknowledgements

The authors thank Dr. Gregory Levine, Professor Zhao-Bin Su, Dr. Zheng-Yu Weng, and Dr. Jian Yang for helpful discussions and are grateful to Dr. Peter Price for a careful reading of the manuscript and valueable comments. This work has been supported by a Grant No. MDA 972-88-G-0002 from U.S. Defense Advanced Research Projects Agency and the State of Texas.

References

1. R. Tsu and L. Esaki, Appl. Phys. Lett. **22**, 562 (1973).
2. L. L. Chang, L. Esaki, and R. Tsu, Appl. Phys. Lett. **24**, 593 (1974).
3. T.C.L.G. Sollner, W. D. Goodhue, P.E. Tannenwald, C. D. Parker and D. D. Peck, Appl. Phys. Lett. **43**, 588(1983).
4. B. Ricco and M. Ya. Azbel, Phys. Rev. **B29**, 1970(1984).
5. E. R. Brown, T.C.L.G. Sollner, C.D. Parker, W.D. Goodhue, and C. L. Chen, Appl. Phys. Lett. **55**, 1777 (1989).
6. F. Capasso, K. Mohammed, and A. Y. Cho, IEEE J. Quantum Electronics, **QE-22**, 1853 (1986).
7. For a recent review, see, F. Capasso, F. Beltram, S. Sen, and A. Y. Cho, in *Physics of Quantum Electron Devices*, ed. F. Capasso, Springer-Verlag, Heidelberg (1990).
8. P. J. Price, Phys. Rev. **B38**, 1994 (1988).
9. S. Luryi, Appl. Phys. Lett. **47**, 490 (1985).
10. A. R. Bonnefoi, T.C. McGill, and R. D. Burnham, Phys. Rev. **B37**, 8754 (1988).

11. G. Kim and G. B. Arnold, Phys. Rev. **B38**, 3252 (1988).
12. W. R. Frensley, Phys. Rev. **B36**, 1570 (1987); Rev. Mod. Phys. **62**, 745 (1990).
13. G. Garcia-Calderon and A. Rubio, Phys. Rev. **B36**, 4462 (1987).
14. J. D. Bruno, T.B.Bahder and C.A.Morrison, Phys. Rev. **B37**, 7098 (1988).
15. H. Guo, K. Diff, G. Neofotistos, and J. D. Gunton, Appl. Phys. Lett. **53**, 131 (1988).
16. W. Trzeciakowski, D. Sahu, and T. F. George, Phys. Rev. **B40**, 6058 (1989).
17. B. Gu and L. Gu, Phys. Rev. **B40**, 6126 (1989).
18. L. Y. Chen and C. S. Ting, Phys. Rev. **B41**, 8533 (1990).
19. J. M. Gering, D. A. Crim, D. G. Morgan, P. D. Coleman, W. Kopp, and H. Morkoc, J. Appl. Phys. **61**, 271 (1987).
20. W. R. Frensley, Superlattice and Microstructure **4**, 497 (1988).
21. N. S. Wingreen, Appl. Phys. Lett. **56**, 253 (1990).
22. C. Jacoboni and P. J. Price, Solid State Comm. **75**, 193 (1990).
23. L. Y. Chen, and C. S. Ting, Phys. Rev. Lett. **64**, 3159 (1990).
24. L. Y. Chen, and C. S. Ting, Phys. Rev. **B43**, 2097 (1991).
25. Y. P. Li, A. Zaslavsky, D. C. Tsui, M. Santos, and M. Shayegan, Phys. Rev. **B41**, 8388 (1990).
26. L. Y. Chen, and C. S. Ting, Phys. Rev. **B43**, 4534(1991).
27. V. J. Goldman, D. C. Tsui, and J. Cunningham, Phys. Rev. **B36**, 7635 (1987).
28. N. S. Wingreen, K.W. Jacobsen and J.W. Wilkins, Phys. Rev. Lett. **61**, 1396 (1988).
29. D. Sokolovski, Phys. Rev. **B37**, 4201 (1988).
30. M. Jonson, Phys. Rev. **B39**, 5924 (1989).
31. W. Cai, T. F. Zheng, P. Hu, and M. Lax, Phys. Rev. Lett. **63**, 418 (1989).
32. P. Johansson, Phys. Rev. **B41**, 9892 (1990).
33. A. P. Jauho, Phys. Rev. **B41**, 12327 (1990).
34. W. Cai, T. F. Zheng, P. Hu, M. Lax, K. Shum, and R. R. Alfano, Phys. Rev. Lett. **65**, 104 (1990).
35. G. S. Boebinger, A. F. J. Levi, S. Schmitt-Rink, A. Passner, L. N. Pfeiffer, and K. W. West, Phys. Rev. Lett. **65**, 775 (1990).
36. H. A. Fertig, and S. Das Sarma, Phys. Rev. **B40**, 7410 (1989).
37. J. Leo, and A. H. MacDonald, Phys. Rev. Lett. **64**, 817 (1990).
38. P. A. Schulz, Phys. Rev. **B43**, 4548 (1991).

39. A. D. Stone and P. A. Lee, Phys. Rev. Lett. **54**, 1196 (1985).

40. T. Weil, and B. Vinter, Appl. Phys. Lett. **50**, 1281 (1987).

41. M. Jonson, and A. Grincwajg, Appl. Phys. Lett. **51**, 1729 (1987).

42. V. J. Goldman, D. C. Tsui, and J. E. Cunningham, Solid State Elec. **31**, 731 (1988).

43. M. Büttiker, IBM J. Res. Dev. **32**, 63 (1988).

44. W. R. Frensley, Solid State Elec. **31**, 739 (1988).

45. S. Luryi, Superlatt. Microstr. **5**, 375 (1989).

46. T. K. Woodward, D. S. Chemla, I. Bar-Joseph, H. U. Baranger, D. L. Sivco, and A. Y. Cho, Phys. Rev. **B44**, 1353 (1991).

47. V. J. Goldman, D. C. Tsui, and J. E. Cunningham, Phys. Rev. Lett. **58**, 1256 (1987).

48. C. Presilla, G. Jona-Lasinio, and F. Capasso, Phys. Rev. **B43**, 5200 (1991).

49. A. Nakano, P. Vashishta, and R. K. Kalia, Phys. Rev. **B43**, 10928 (1991).

50. J. Schwinger, J. Math. Phys. (New York) **2**, 407 (1961).

51. L. V. Keldysh, Zh. Eksp. Teor. Fiz., **47**,1515 (1964).

52. L.P.Kadanoff and G. Baym, *Quantum Statistical Mechanics* (Benjiamin, New York 1962).

53. D. Langreth and J. W. Wilkins, Phys. Rev. **B6**, 3189 (1972).

54. For recent review and the path integral formulation, see K.C. Chou, Z. B. Su, B. L. Hao and L. Yu, Phys. Rep. **118**,1 (1985) and Z. B. Su, L. Y. Chen, X. T. Yu and K. C. Chou, Phys. Rev. **B37**, 9810 (1988).

55. A.R. Hibbs and R.P. Feynman, *Quantum Mechanics and Path Integrals* (McGraw-Hill, New York, 1965).

56. R. P. Feynman and F. L. Vernon, Ann. Phys. (N.Y.) **24**, 118 (1963).

57. S. Ramo, Proceedings of the IRE, **27**, 584 (1939).

58. W. Shockley, J. Appl. Phys. **9**, 639 (1938).

59. L. Y. Chen and C. S. Ting, preprint (1991).

60. L. Y. Chen and C. S. Ting, preprint (1991).

61. A. N. Korotkov, D. V. Averin, and K. K. Likharev, Physica, **B165&166**, 927 (1990).

62. A. Groshev, T. Ivanov, and V. Valtchinov, Phys. Rev. Lett. **66**, 1082 (1991).

63. L. Y. Chen and C. S. Ting, Phys. Rev. **B44**, 5916 (1991).

www.ingramcontent.com/pod-product-compliance
Ingram Content Group UK Ltd.
Pitfield, Milton Keynes, MK11 3LW, UK
UKHW061932100926
8556IPUK00007B/884